6th International Conference on Optical, Optoelectronic and Photonic Materials and Applications

(ICOOPMA 2014)

Journal of Physics: Conference Series Volume 619

Leeds, United Kingdom
27 July - 1 August 2014

ISBN: 978-1-5108-0800-3
ISSN: 1742-6588

6th International Conference on Optical, Optoelectronic and Photonic Materials and Applications

(ICOOPMA 2014)

Journal of Physics: Conference Series Volume 619

Leeds, United Kingdom
27 July - 1 August 2014

ISBN: 978-1-5108-0800-3
ISSN: 1742-6588

IOP Institute of Physics

Printed from e-media with permission by:

Curran Associates, Inc.
57 Morehouse Lane
Red Hook, NY 12571

Printed by Curran Associates, Inc. (2015)

For permission requests, please contact the Institute of Physics
at the address below.

Institute of Physics
Dirac House, Temple Back
Bristol BS1 6BE UK

Phone: 44 1 17 929 7481
Fax: 44 1 17 920 0979

techtracking@iop.org

Additional copies of this publication are available from:

Curran Associates, Inc.
57 Morehouse Lane
Red Hook, NY 12571 USA
Phone: 845-758-0400
Fax: 845-758-2634
Email: curran@proceedings.com
Web: www.proceedings.com

Table of contents

Volume 619

6th International Conference on Optical, Optoelectronic and Photonic Materials and Applications (ICOOPMA) 2014
27 July to 1 August 2014, Leeds, UK

Accepted papers received: 29 April 2015
Published online: 17 June 2015

Preface

Papers

Session I: Inorganic Semiconductor Materials, Devices and Applications

Session II: Carbon and Polymeric Materials, Devices and Applications

Session III: Inorganic Glass and Crystal Materials, Devices and Applications

Session IV: Metamaterials and Plasmonic Devices

Journal of Physics: Conference Series **619** (2015) 011001 doi:10.1088/1742-6596/619/1/011001

Proceedings of the International Conference on Optical, Optoelectronic and Photonic Materials and Applications (ICOOPMA) 2014

Stephen J. Sweeney[1] and Animesh Jha[2]

[1] Advanced Technology Institute and Department of Physics, University of Surrey, Guildford, GU2 7XH, UK

[2] Institute for Materials Research, University of Leeds, Leeds, LS2 9JT, UK

Preface

Developments in the connected fields of optics, optoelectronics and photonics have had a profound effect on the emergence of modern technologies and their influence on our lives. In all of these fields, understanding and improving the basic underlying materials is of crucial importance for the development of systems and applications. The International Conference on Optical, Optoelectronic and Photonic Materials and Applications (ICOOPMA) has successfully married these fields and become a regular feature in the conference calendar. The 6th conference in the series, held at the University of Leeds from 27th July – 1st August, 2014, continued the ICOOPMA tradition and attracted 220 international delegates with a diverse range of disciplines and interests. The 59 papers in this Proceedings provide an excellent overview of the topics presented. The conference consisted of four thematic areas in the fields of *inorganic semiconductors, carbon and polymeric materials, inorganic glasses and crystalline materials, and metamaterials and plasmonics* where each theme area included research on basic materials through to device applications.

The conference began with a Workshop organised by Professor Dan Hewak (University of Southampton) with speakers covering *Organic Optoelectronic Complexes* (Dr Richard Curry, University of Surrey), *Metamaterials* (Dr Vassili Fedotov, University of Southampton), *Graphene* (Dr Monica Craciun, University of Exeter) and *Amorphous Semiconductors* (Dr Jiri Orava, University of Cambridge and Tohoku University). This provided an excellent overview of a representative range of the important topics that were discussed further at the conference. The conference included a banquet which successfully combined excellent food with a relaxing opportunity for the conference delegates to socialise and network. The pinnacle of the evening was the after dinner speech given by our distinguished guest, Professor Sir David Payne (University of Southampton), who gave a highly informative and humorous history of the development of photonics and the important role played by optical fibres, of which he was one of the key pioneers. Emerging scientific talent was also given a platform, with a particular feature of the conference being the Young Scientists Forum, organised by Dr Senthil Ganapathy (University of Southampton), consisting of a series of themed sessions providing up-and-coming researchers an opportunity to present and discuss their work.

Each technical theme of the conference was introduced with Plenary lectures delivered by leading international figures in their respective fields. Professor Jim

Harris (Stanford University) discussed the development of epitaxially grown dilute-nitride semiconductors and the important role that they play in high efficiency solar cells. This theme was continued by Professor Wolfgang Stolz (Philipps University, Marburg) who described the use of novel III-V semiconductors for the development of lasers on silicon. Dr Jerry Meyer (Naval Research Laboratories) and Dr. Kumar Patel (Pranalytica) discussed the important field of mid-infrared devices highlighting the significant developments that have been made in inter-band cascade and quantum cascade based devices, respectively and which have wide-ranging applications in sensing. Professor Neil Greenham (University of Cambridge) discussed solution-based semiconductors based on conjugated polymers and nanocrystals offering low cost routes to the wide-scale deployment of photonics. Professor Ortwin Hess (Imperial College) gave an overview of the physics and fascinating technological potential of negative refractive index materials. Professor Stephen Elliott (University of Cambridge) discussed the optoelectronic properties of phase change materials for memories. These areas were expanded upon in over 70 invited talks from international speakers highlighting recent developments in each of these areas. Further detailed discussions were presented in many parallel Oral and Poster sessions.

This Proceedings provides an exciting and wide-ranging discussion of the topics presented at ICOOPMA 2014. We are very grateful to the many people who helped with the organisation of the conference, with particular thanks to Alison Whiteley and her team who enabled the smooth running of the conference, Dr Senthil Ganapathy for his many and wide ranging contributions to the conference and finally Professor Safa Kasap for his wisdom, advice and strong support. The conference could not have happened without the commitment of the Local Organising Committee, who helped in many ways to assemble and run the conference, and for the International Programme, Advisory and Steering committees who guided the technical direction of the conference and assisted with the conference programme and proceedings. We are also extremely pleased to have received generous sponsorship from a large number of organisations and companies. Finally, we are also very grateful to Sarah Toms and the Editorial staff at the Institute of Physics, and the many contributors and reviewers for helping us to put-together this proceedings.

Stephen Sweeney, Guildford, UK
Animesh Jha, Leeds, UK

UNIVERSITY OF SURREY UNIVERSITY OF LEEDS

 IOP Publishing
doi:10.1088/1742-6596/619/1/011001

Sponsored by

APPLIED MATERIALS®

BRUKER

SOCIETY OF GLASS TECHNOLOGY

GTS
SPECIALIST KNOWLEDGE IN GLASS

WILEY
WILEY-VCH

M SQUARED

MINERVA
improved medical diagnostics
MId- to NEaR infrared spectroscopy
for improVed medical diAgnostics
A European Commission
Framework Seven (FP7) Project (317803)
www.minerva-project.eu

UNIVERSITY OF LEEDS

Journal of Physics: Conference Series **619** (2015) 011001 doi:10.1088/1742-6596/619/1/011001

Supported by

ASH SCIENTIFIC

AMERICAN ELEMENTS

HAMAMATSU
PHOTON IS OUR BUSINESS

IOP Institute of Physics
Quantum Electronics and Photonics Group

LOT
QuantumDesign

THORLABS

UNIVERSITY OF SURREY

Journal of Physics: Conference Series **619** (2015) 011002 doi:10.1088/1742-6596/619/1/011002

Peer review statement

All papers published in this volume of *Journal of Physics: Conference Series* have been peer reviewed through processes administered by the proceedings Editors. Reviews were conducted by expert referees to the professional and scientific standards expected of a proceedings journal published by IOP Publishing.

Thin film germanium on silicon created via ion implantation and oxide trapping

R Anthony[1] and A P Knights[1]

[1] Department of Engineering Physics, McMaster University, 1280 Main Street West, Hamilton, Ontario, Canada L8S 4L7

Email: anthonre@mcmaster.ca

Abstract. We present a novel process for integrating germanium with silicon-on-insulator (SOI) wafers. Germanium is implanted into SOI which is then oxidized, trapping the germanium between the two oxide layers (the grown oxide and the buried oxide). With careful control of the implantation and oxidation conditions this process creates a thin layer (current experiments indicate up to 20-30nm) of almost pure germanium. The layer can be used potentially for fabrication of integrated photo-detectors sensitive to infrared wavelengths, or may serve as a seed for further germanium growth. Results are presented from electron microscopy and Rutherford back-scattering analysis, as well as preliminary modelling using an analytical description of the process.

1. Introduction

Silicon photonics is seen as the best platform to solve current connectivity issues associated with copper interconnects [1]. Silicon is transparent at optical communication wavelengths, and is ubiquitous throughout the semiconductor industry. Silicon photonic devices require on-chip photo-detection which is usually achieved by integrating germanium onto the chip. In principle, germanium deposition is a scalable and low cost solution providing a high absorption coefficient for infrared wavelengths. All germanium photodetectors are grown currently via Molecular Beam Epitaxy [2], reduced pressure or ultra-high vacuum Chemical Vapour Deposition [3], or low-energy Plasma-Enhanced Chemical Vapour Deposition [4]. However, due to a 4% lattice mismatch between silicon and germanium, these processes require precise control to avoid excessive defects at the silicon/germanium interface and unacceptable dark current in subsequently fabricated photo-detectors [5]. Another method for introducing germanium into silicon is ion implantation. Previous work done on the oxidation of deposited silicon-germanium and pure germanium implanted into silicon has demonstrated that silicon is oxidized preferentially over germanium [6, 7]. Further, if the rate of oxidation exceeds the rate of germanium diffusion, the oxidation process leads to a "pile-up" of germanium at the surface. In this work we describe a novel method for germanium integration with silicon where implanted germanium is trapped between the growing and buried oxide layers on an SOI substrate. The process is shown schematically in Figure 1.We show experimentally that the germanium concentration is highly dependent on the oxidation process used, and that germanium concentrations as high as 87% can be achieved.

2. Experimental

Samples of single crystal 2.5 μm, or 220nm, silicon thin film thick SOI were implanted using Ge^+ ions with fluences of $3x10^{16}$ cm^{-2}, $7x10^{16}$ cm^{-2}, or 10^{17} cm^{-2}. All samples were implanted at 33 KeV.

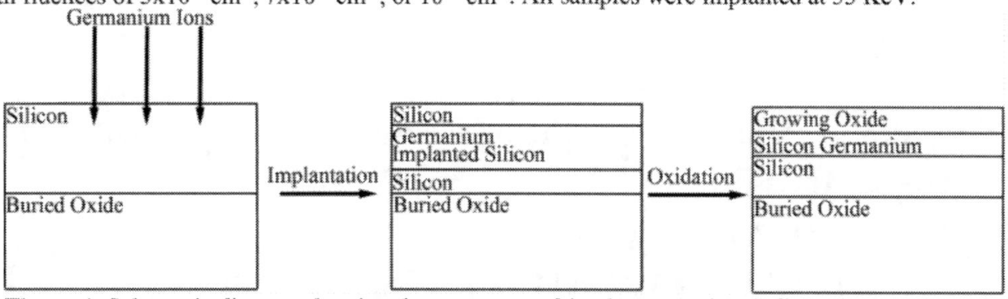

Figure 1. Schematic diagram showing the process resulting in germanium "pile-up".

Subsequent to implantation, thermal oxidation was performed in a tube furnace. The length of oxidation was determined by consideration of the Deal-Grove Model [8]. This allowed us to produce samples with approximately equal oxide thickness despite the variation in oxidation temperature. For the 2.5 μm SOI samples were initially oxidized at 900°C for ninety minutes, 1000°C for 24 minutes or 1100°C for 9 minutes. The samples were then etched using buffered HF to remove the grown oxide. The etched samples were oxidized a second time at 900°C, 1000°C or 1100°C for 90 minutes (with each second oxidation temperature matching that from the first). The 220 nm samples were also oxidized twice and due to their initial and relative thin silicon thickness, the majority of the silicon was removed through these oxidations processes.This provides for a significantly increased germanium concentration compared to the thicker samples.

Confirmation of germanium "pile-up" was obtained via Transmission Electron Microscopy (TEM). Cross-sectional samples were prepared by mechanical polishing and argon ion milling. The germanium concentration of the samples was measured using Rutherford backscattering (RBS) with 1.5 MeV He^+ atoms. RBS was performed after each oxidation step.

3. Results and Discussion

Figure 2a shows an example of a TEM image of a 2.5 μm sample implanted with a fluency of $7x10^{16}$ after an initial oxidation of 90 minutes at 900°C, while Figure 2b shows an example TEM image of a 220 nm sample implanted with a fluency of $7x10^{16}$ after a 45 minute oxidation at 1000°C. Figure 2a shows the "pile-up" of germanium whereas Figure 2b shows the trapping of germanium between the two oxide layers.

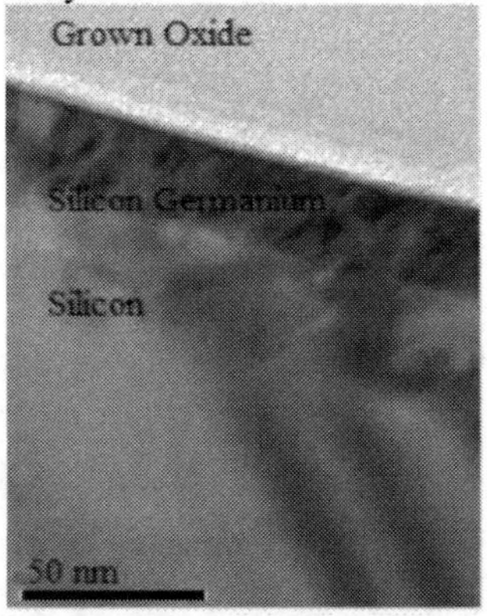

Figure 2a. TEM image of 2.5 μm sample

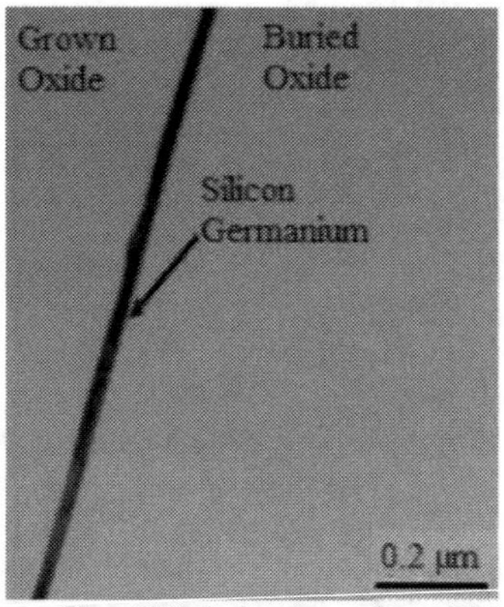

Figure 2b. TEM image of a 220 nm sample.

Figure 3 presents the experimental RBS data for a 2.5 μm sample after implantation with $1x10^{17}cm^{-2}$ Ge^+ ions and subsequently oxidized at 900 °C. The data was fit using a layer stack model similar to that shown in the right panel of Figure 1. Analysis of the RBS data after the initial oxidation for samples oxidized at 900°C showed that they had the highest concentration of germanium in the SiGe layer, ranging from 72% to 80%. The samples oxidized at 1000 °C had germanium concentrations ranging from 45% to 53% and the samples oxidized at 1100 °C had the lowest germanium concentrations, ranging from 29% to 35%. These results demonstrate the sensitivity of the process to the balance between oxidation rate and Ge diffusion.

Figure 4 presents the RBS data of the same sample as in Figure 3, but after a second oxidation. As can be seen in the figure there is a much smaller silicon-germanium peak indicating a loss of germanium and the formation of a film of $Si_xGe_yO_z$. The $Si_xGe_yO_z$ is due to a high initial flux of oxygen which is normally reduced by the growing oxide layer [9], whereas the loss of germanium in the silicon germanium region is due to evaporation. These features are not apparent after only one oxidation due to the presence of a thin layer of silicon above the implanted germanium layer which acts as an initial oxide barrier. This barrier decreases the flux of oxygen to the oxidizing Ge-interface and prevents evaporation.

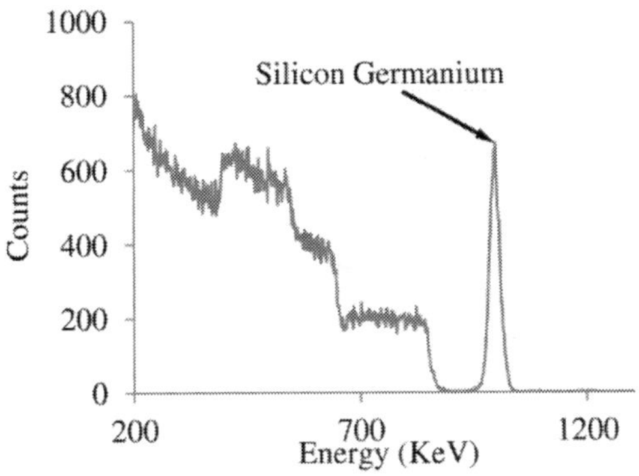

Figure 3. RBS spectra of the initial oxidation of a 2.5 μm sample.

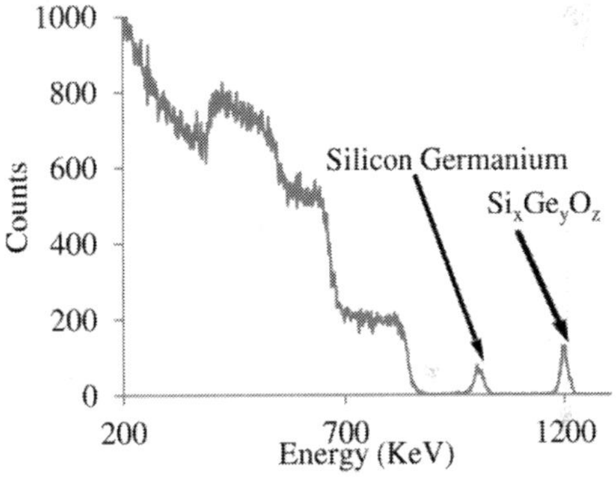

Figure 4. RBS spectra of second oxidation step of a 2.5 μm sample

Analysis of the RBS data of the 220 nm samples indicated that the majority of the silicon was removed for the sample implanted with a fluency of $7x10^{16}$ cm^{-2} after the second oxidation at 1000°C. The peak concentration of the germanium in this sample was found to be 87%.

RBS data also indicated that the grown oxide on all samples in this study was thicker than that predicted from the Deal-Grove Model. This is due to a combination of three effects: the weakness of the silicon-germanium bond compared to the silicon-silicon bond, the lack of a stable germanium oxide as the germanium atom in the GeO that is formed is replaced by a silicon, and the suppression of interstitial injection [6, 9].

4. Preliminary Analytical Model

An analytical model is under development in order to offer a physical description for the thin film formation. The model uses finite differences to solve the diffusion equation with the boundary conditions being set so that there is zero flux of germanium out of the film. The initial condition is that of the solution to the diffusion equation in an infinite medium with a an imitial implantation fluency Q and an effective diffusion coefficient for Ge of $D(T)$, where T is the temperature of oxidation, and C is the concentration of germanium at a point x in the film at time t. The model is described by equation (1):

$$C(x,t) = \frac{Q}{2(\pi D(T)t)^{1/2}} e^{\left(-\frac{x^2}{4Dt}\right)} \quad (1)$$

During the simulation the x-position of one boundary is moved to simulate the growing of the oxide layer. The movement of the boundary is based on a modified Deal Grove Model (to account for increased oxidation rates). This modification is used to describe the effect of the germanium at the oxide interface specifically the difference in the binding energies of silicon-silicon and silicon-germanium [3].

Results of the model of peak germanium concentration and film thicknesses for a 1.5 hour oxidation at 900°C are given in Table I. The model assumes D is the only temperature dependent parameter, in Table I fixed for 900°C at 4.5×10^{-18} cm^{2}s^{-1}.

Table I. Model results for a 2.5 um sample 1.5 hour oxidation at 900°C

Dose (cm^{-2})	Peak Germanium Concentration (%)	Film Thickness (nm)	D(cm^{2}s^{-1})	Experimental Peak Concentration (%)
3×10^{16}	86	15	4.5×10^{-18}	80
7×10^{16}	90	33	4.5×10^{-18}	74
1×10^{17}	93	50	4.5×10^{-18}	77

As can be seen from the table the peak concentrations calculated from the model are greater than that of the experimentally determined concentrations. This is due to the effective rate of diffusion which according to this model is dependent only on the temperature. In reality the effective rate of diffusion is also concentration dependent meaning higher concentrations will leads a larger effective rate of diffusion. Further plans to improve the model will require incorporating this effect.

5. Conclusion and future work

The trapping of germanium between a growing and buried oxide layer has been demonstrated. The germanium has been found to be crystalline in nature and has concentrations greater for 220 nm SOI than for 2.5 μm SOI samples. Future work will involve developing a more rigorous model that will predict oxide growth rate as a function of germanium concentration at the interface as well indicating processing conditions which will provide for the maximum germanium concentration.

References

[1] Soref R. 2006 *IEEE J. of Selected Topic in Quant. Electron.* **12** 1678-87
[2] Jutzi M, Berroth M, Wohl G, Oehme M and Kasper, E 2005 *IEEE Photon. Technol. Lett.* **17** 1510-1512
[3] Loh T, Nguyen H, Tung C, Trigg A, Lo G, Balasubrmanian N, Kwong D and Tripathy S 2007 *Appl. Phys. Lett.* **90** 092108
[4] Osmond J, Isella G, Chrastina D, Kaufmann R, Acciarri M, and Kanel H V 2009 *Appl. Phys. Lett.* **94** 201106
[5] Wang J and Lee S 2011 *Sensors* **11** 696-718
[6] Fathy D, Holland O W and White C W 1987 **51** 1337-39
[7] Holland O W, White C W and Fathy D 1987 **51** 520-22
[8] Plummer J D, Deal M D and Griffin P B 2000 *Silicon VLSI Technology Fundamentals, Practice and Modelling* (Englewood Cliffs, NJ: Prentice-Hall)
[9] Rabie M A, Haddara Y M and Carette J 2005 A kinetic model for the oxidation of silicon germanium alloys *J. Appl. Phys.* **98** 07904

Depth-dependent crystallinity of nano-crystalline silicon induced by step-wise variation of hydrogen dilution during hot-wire CVD

C J Arendse[1], B A van Heerden[1], T F G Muller[1], F R Cummings[1], C J Oliphant[2], G F Malgas[1] and D E Motaung[3]
[1] Department of Physics, University of the Western Cape, Private Bag X17, Bellville 7535, South Africa
[2] National Metrology Institute of South Africa, Private Bag X34, Lynnwood Ridge, Pretoria 0004, South Africa
[3] DST/CSIR National Centre for Nano-Structured Materials, P. O. Box 395, Pretoria 0001, South Africa

E-mail: cjarendse@uwc.ac.za

Abstract. To induce an amorphous surface in a nano-crystalline silicon (nc-Si:H) thin film, the hydrogen dilution was reduced step-wise at fixed time intervals from 90 – 50% during the hot-wire chemical vapour deposition process. This contribution reports on the structural properties of the resultant nc-Si:H thin film as a function of the deposition time. Raman spectroscopy, confirmed by high resolution transmission spectroscopy, indicates crystalline uniformity in the growth direction, accompanied by the progression of an amorphous surface layer as the deposition time is increased. The silicon- and oxygen bonding configurations were probed using infrared spectroscopy and electron energy loss spectroscopy. The growth mechanism is ascribed to the improved etching rate by atomic hydrogen in nano-crystalline silicon towards the film/substrate interface region. The optical properties were calculated by applying the effective medium approximation theory, where the existence of bulk and interfacial layers, as inferred from cross-sectional microscopy, were taken into account.

1. Introduction

The use of hydrogenated nano-crystalline silicon (nc-Si:H) thin films as the absorber intrinsic layer in silicon thin film solar cells has been extensively investigated, mainly due its improved photo-induced stability [1-2]. The hot-wire chemical vapour deposition (HWCVD) technique, based on the catalytic decomposition of the SiH_4 and H_2 gas by a heated transition metal filament, has been established as a viable deposition technique for nc-Si:H [3]. The structural and opto-electronic properties of the thin films are dependent on the deposition parameters, of which the hydrogen (H) dilution and substrate temperature are the most crucial. It has been reported that the best nc-Si:H based solar cells, with optimal opto-electronic properties, are deposited near to the amorphous/nano-crystalline silicon transition regime [4]. It is well known that nc-Si:H has a distinct structural inhomogeneity in the growth direction, which initiates with an amorphous incubation layer that transitions into an increased crystallite size and fraction as the growth progresses. This non-uniformity in the crystallinity with thickness results in the formation of cracks within the thin films and a reduced short-circuit current in solar cells, which eventually leads to deterioration in the solar cell performance [5]. The hydrogen profiling technique, where the hydrogen dilution is continuously decreased as a function of deposition time, has been proposed to maintain a fixed crystallite size and fraction as a function of depth [6-7]. In this contribution, we propose a HWCVD hydrogen profiling sequence where the H dilution ratio is changed from 90 to 50% at fixed time intervals to enable an investigation into the evolution of the structural properties as a function of deposition time and the eventual optical properties.

2. Experimental

The nc-Si:H thin films were deposited by HWCVD, using mixtures of SiH_4 and H_2 gas, decomposed over a tantalum filament heated to 1600 °C. A detailed description of the experimental set-up is given elsewhere [8].With the exception of the H dilution ratio R (defined as ratio of the flow rate of H_2 to the total gas flow rate), the substrate temperature, total gas flow rate and deposition pressure were fixed at 240 °C, 30 sccm and 60 μbar, respectively. The film series were deposited sequentially at R = 90%, 83%, 80% and 50% for 900 s, 720s, 480 s and 300 s, respectively.

Fourier transform infrared (FTIR) absorption spectra were collected in transmission geometry from 400 – 4000 cm^{-1} with a spectral resolution of 1 cm^{-1}, using a Perkin-Elmer Spectrum 100 FTIR spectrophotometer. The structural properties were investigated using a Horiba Jobin-Yvon T64000 micro-Raman spectrometer in backscattering geometry at room temperature. The Raman spectra were collected in the region 100 – 1000 cm^{-1} with a spectral resolution of 0.4 cm^{-1}, using excitation wavelengths of 532 and 647 nm. A cross-section of the final thin film was prepared by a FEI Helios NanoLab 650 Dual Beam focussed-ion beam SEM (FIBSEM), which were subsequently analysed in a FEI Tecnai F20 FEGTEM in bright field mode at an accelerating voltage of 200 kV. Electron energy loss spectroscopy (EELS) spectra were collected using a Gatan Image Filter (GIF2001) at an energy dispersion of 5eV/channel and resolution of 1.3 eV. Optical transmission spectra were measured using a CARY 1E UV/VIS spectrometer in the range 200 – 900 nm with a spectral resolution of 1 nm. Raman and FTIR spectra were collected on the films deposited after each time interval (900 s, 1620 s, 2100 s and 2400 s), whereas TEM, EELS and UV/VIS spectroscopy were performed only on the final thin film (2400 s).

3. Results and Discussion

The Raman spectra of the sample series at different thicknesses, using an excitation wavelength of 532 nm, are given in figure 1. The film thickness was extracted by interferometry from the UV/VIS transmission spectra. The initial thin film deposited at R = 90% reveal a distinct crystalline signature, as indicated by the narrow c-Si transverse-optic (TO) peak at 515 cm^{-1}. The crystalline fraction of this thin film amounts to 38%. A further growth in the thickness, accompanied by a reduction in R, results in a large reduction in the intensity of the c-Si TO peak, which eventually disappears. As the Raman signal produced by the excitation wavelength of 532 nm is dominated by the surface of the nc-Si:H thin films, it is evident that the crystalline fraction at the surface decreases with a growth in the film thickness, eventually leading to a breakdown in nc-Si:H growth. The width of the amorphous silicon (a-Si) TO peak at 480 cm^{-1} for the final thin film (933 nm-thick) amounts to 53 cm^{-1}, which denotes a highly ordered, dense a-Si surface layer [9]. To probe deeper into the thin films, an excitation wavelength of 647 nm was used, for which the Raman spectra are given in figure 2. Analysis of the Raman spectra reveals that the crystalline fraction remains relatively constant at ~ 30% as the film thickness increases, indicative of crystalline uniformity in the growth direction.

FTIR spectroscopy is the technique of choice to probe into the Si, H and O bonding configuration in nc-Si:H. Figure 3 shows the absorption spectra of the stretching vibrations for the sample series at different thicknesses. The mode centred around 2000 cm^{-1} is assigned to the Si-H monohydrides in the compact isolated phase of nc-Si:H, whereas the shoulder peak at 2100 cm^{-1} is attributed to the Si-H_2

bonds in the clustered phase, associated with the presence of microvoids. In the initial growth step, H is bonded preferentially in the clustered phase, which signifies a void-rich layer. As the growth progresses, the H content bonded in the compact, isolated phase increases and is accompanied with a reduction of that in the clustered phase. Therefore, as the growth progresses the micro-structure and compactness of the nc-Si:H thin film improves, thereby resulting in a more dense and stable thin film. This observation is confirmed by the reduction in oxidation, as inferred from the disappearance of Si-O-Si vibration at around 1070 cm^{-1} (not shown).

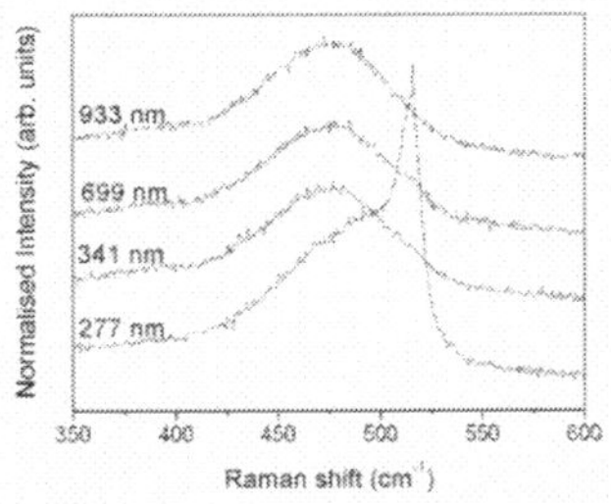

Figure 1. Raman spectra using 532 nm excitation.

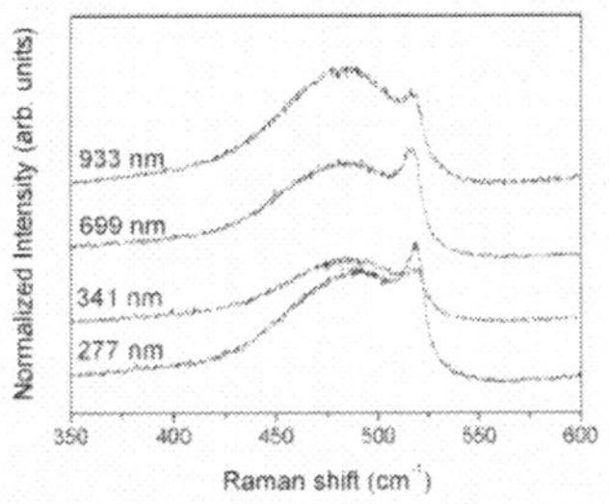

Figure 2. Raman spectra using 647 nm excitation.

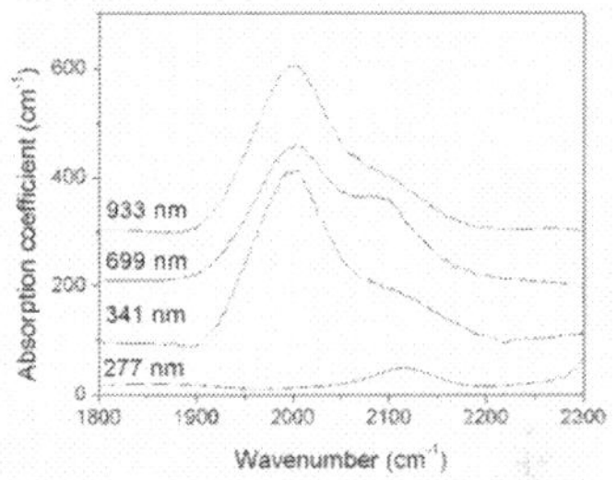

Figure 3. FTIR spectra of sample series.

Raman and infrared spectroscopy therefore suggests that the reduction in the hydrogen dilution as the thin film growth progresses, favours the formation of a compact, homogeneous nano-crystalline structure with a dense, highly ordered amorphous silicon surface. The reduction in the hydrogen dilution during the deposition infers a reduction in the amount of atomic hydrogen available for the etching of strained Si-Si surface bonds, thereby promoting the growth of an amorphous silicon surface layer at low *R*-values. Furthermore, the continued expansion of the nc-Si:H region at high *R*-values is attributed to the preferred etching by atomic hydrogen in the original crystalline regions.

Figure 4. TEM micrograph of the final nc-Si:H thin film. Inset shows the overlapping oxygen and silicon EELS maps (Si-green and O-red).

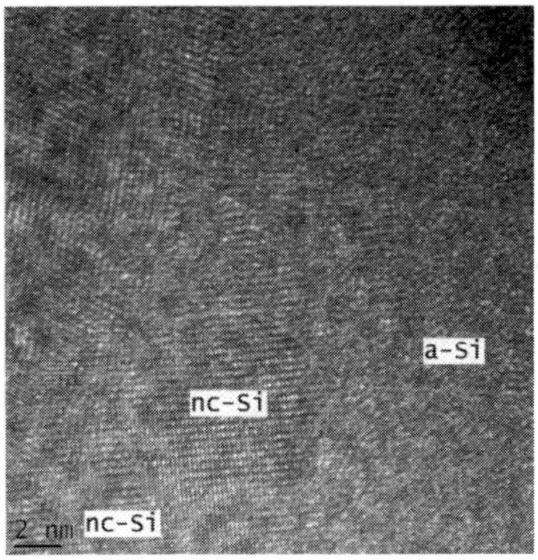

Figure 5. HRTEM micrograph showing nc-Si and a-Si regions.

Figure 4 shows the bright field TEM micrograph of the final nc-Si:H thin film, with a thickness of ~ 900 nm, which agrees with the interferometry calculations. A 10 nm-thick amorphous incubation layer is observed at the film/substrate interface, which is characteristic of nc-Si:H grown at high H dilution ratios. Thereafter, a well-defined crystalline region extends 500 nm in the growth direction with a uniform crystalline distribution, which corroborates the Raman analysis. Finally, the thin film is capped with a thick, compact amorphous region. The corresponding overlapping oxygen and silicon EELS maps (inset in Figure 4) shows no detectable oxygen present in the thin film, which confirms the FTIR results. Closer inspection into the crystalline structure by HRTEM reveals distinct nano-crystallites embedded in an a-Si matrix with diameters ranging from 2 to 10 nm (see figure 5).

The optical properties of the final thin film were calculated by applying the effective medium approximation theory, where the existence of all the phases in the bulk and interfacial layers, as inferred from cross-sectional TEM, were taken into account [10]. The refractive index at zero energy (n_o) amounts to ~ 3.30, indicative of a dense nc-Si:H thin film. Regardless of its relatively high crystalline fraction, the absorption coefficient (at 2 eV) amounts to 2.1×10^4 cm^{-1}, which is comparable to that of device-quality hydrogenated amorphous silicon.

4. Conclusion

A step-wise decrease in the hydrogen dilution was used to demonstrate the deposition of a structurally improved nc-Si:H/a-Si bilayer structure by hot-wire CVD. Raman spectroscopy, confirmed by TEM, shows the initial formation of a thin a-Si incubation layer at the substrate interface followed by a uniform crystalline structure in the growth direction with a crystalline fraction of 30%. The structure is capped with a dense, highly ordered amorphous silicon layer. The superior density of the thin film is corroborated by the absence of oxidation and the enhanced refractive index.

Acknowledgements
The authors acknowledge the financial support of the National Research Foundation. The authors would like to thank the Zernike Institute at Groningen University for the Raman spectroscopy measurements.

References

[1] Meier J, Flückiger R, Keppner H and Shah A 1994 *Appl. Phys. Lett.* **65** 860
[2] Yan B J, Yue G Z, Banerjee A, Yang J and Guha S 2004 *Mater. Res. Soc. Symp. Proc.* **808** A9.41
[3] Oliphant C J, Arendse C J, Knoesen D, Muller T F G, Prins S and Malgas G F 2011 *Thin Solid Films* **519** 4437
[4] van Veen M 2003 *Tandem solar cells deposited using hot-wire chemical vapour deposition* (Utrecht University, The Netherlands)
[5] Yan B J, Yue G Z, Yang J, Guha S, Williamson D L, Han D X and Jiang C S 2004 *Mater. Res. Soc. Symp. Proc.* **808** A8.5
[6] Yan B, Yue G, Yang J, Guha S, Williamson D L, Han D and Jiang C S 2004 *App. Phys. Lett.* **85** 1955
[7] Li H, Franken R H, Stolk R L, van der Werf C H M, Rath J K and Schropp R E I 2008 *J. Non-Cryst. Solids* **354** 2087
[8] Arendse C J, Knoesen D and Britton D T 2006 *Thin Solid Films* **501** 92
[9] Beeman D, Tsu R and Thorpe M F 1985 *Phys. Rev. B* **32** 874
[10] Muller T F G, Knoesen D, Arendse C J, Halindintwali S, Malgas G F, Houweling Z S and Schropp R E I 2010 *Phys. Status Solidi C* **7** 750

Effects of interface roughness scattering on device performance of indirectly pumped terahertz quantum cascade lasers

Seyed Ghasem Razavipour[1, a], Emmanuel Dupont[2], Zbig R. Wasilewski[1], Dayan Ban[1, b]

[1]Department of Electrical and Computer Engineering, University of Waterloo, Waterloo Institute for Nanotechnology, 200 University Ave. W, Waterloo, Ontario, N2L 3G1, Canada
[2]National Research Council, 1200 Montreal Rd., Ottawa, Ontario, K1A 0R6, Canada

E-mail: [a]sgrazavi@uwaterloo.ca, [b]dban@uwaterloo.ca

Abstract. The impacts of interface roughness (IR) scattering on device performance of indirectly-pumped (IDP) terahertz quantum cascade lasers are studied. Three different active region designs with almost the same lasing frequency at threshold and comparable oscillator strength are experimentally investigated and the measurement data are analyzed and compared with numerical simulation. The simulation results show that all structures suffer from the detrimental effect of intersubband roughness scattering in terms of threshold current density, and probably operating temperature. The intrasubband IR scattering time could also to be a limiting factor in the IDP structures due to the employed high energetic barrier.

1. Introduction

After more than one decade of the invention of first terahertz (THz) quantum cascade laser (QCL) [1], this device has not satisfied the basic requirement of most of industrial applications for room temperature operation. Even though different design schemes were theoretically studied and some of them experimentally implemented, none of them could operate without an inconvenient cryogenic cooling system. In 2012, Fathololoumi *et al.* [2] presented a structure based on resonant-phonon (RP) scheme and improved the operating temperature of THz QCLs to $\sim$ 200 K. The oscillator strength study on three-well RP-QCLs showed that any further improvement in this design scheme needs detailed understanding of carrier transport in THz QCLs [3]. The tall barrier study in the hope of decreasing the leakage currents in the RP structures could only lead to some minimal device performance improvements [4].

A drawback of the RP structure is the fundamental limit of less than 50% population inversion [5, 6, 7, 8]. To overcome this limitation, the indirectly-pumped (IDP) scheme, originally implemented by Yamanishi in mid-infrared (MIR) QCLs [9], emerged to be a good candidate for high operating temperature applications. Kumar *et al.* [6], Fujita *et al.* [10], and Dupont *et al.* [8], proposed different IDP schemes based on different material systems, while the best performance was achieved by Kumar *et al.* with a structure that lased up to 163 K at 1.8 THz ($k_B T/\hbar\omega = 1.9$). Contrary to the first two structures that employed a resonant tunneling followed by a phonon relaxation transition to depopulate carriers from the lower lasing state, Dupont *et al.* used a direct phonon relaxation transition, also called phonon-photon-phonon (3P) scheme, in order to avoid the broadening of gain induced by the resonant tunneling. Shortly after that, two different structures based on the 3P scheme were implemented with different figures of merit but the same material system, however the improvement in operating temperature was not as good as expected [11, 12]. In this paper, we investigate how interface roughness

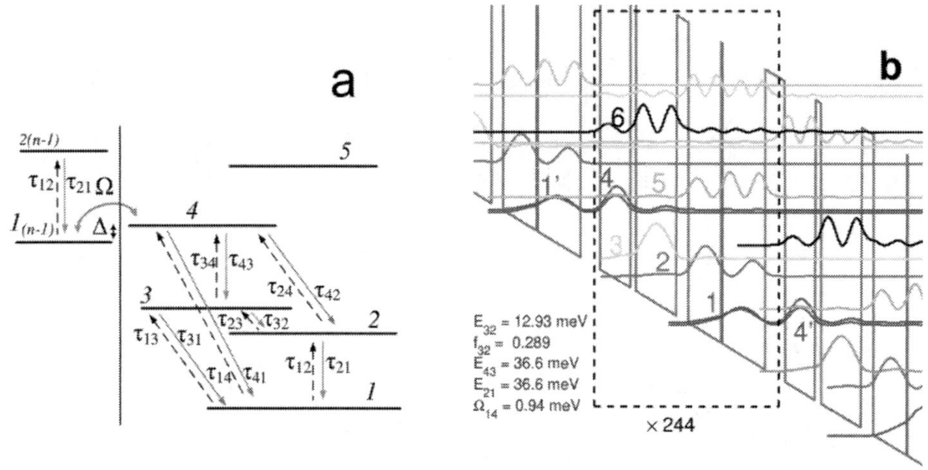

Figure 1. a) Schematic diagram of a 3P-QCL active region based on a phonon-photon-phonon configuration. The solid lines show the forward scatterings, while the dashed lines show the back scatterings. Δ and Ω are the detuning and the coupling between states *1* and *4*, respectively. The green arrows represent the correct injection and extraction, while the red arrows represent the wrong injection and extraction in each module. b) Conduction band diagram and the moduli squared wavefunctions of the THz 3P-QCL, V895, at 21 kV/cm. The quantum wells and barriers of V895, starting with the injector barrier, are **44**/66.85/**12.55**/88.95/**25.55**/73.4/**3.45**/94 Å in thickness, where the bold font indicates the $Al_{0.25}Ga_{0.75}As$ barriers. The center of the injector barrier was delta-doped with Si to 3.67×10^{10} cm^{-2}.

scattering affects the device performance of 3P based THz QCLs. In the 3P structures, since both injection of carriers to the upper lasing state (ULS) and their depopulation from the lower lasing state (LLS) are mediated by phonon relaxation transition, the voltage drop of each module is higher than that of conventional RP structures. This enforces to increase the energetic barrier height (from 15% aluminum composition in conventional RP structures to 25% in the 3P structures) in order to reduce electron leakage to the continuum band. On the other hand, a higher energetic barrier introduces more IR scattering, which could significantly degrade the device performance. To quantitatively investigate the IR effects on the different parameters of each structure, we selected three of our 3P-QCLs that have almost the same lasing frequency, number of modules in the active region, and doping concentration per module. The similarity in the lasing frequency at threshold and the doping concentration roughly ensure a same cavity loss among the selected three structures.

2. Effect of IR scattering on effective lifetime

The three QCL structures under study are named after their wafer numbers, respectively. V845 [11] and V895 are four-well structures with three-dimensional doping concentration of 9×10^{15}cm^{-3} while V962 [12] is a five-well structure with three-dimensional doping concentration of 7.6×10^{15}cm^{-3}. Since the MBE growth and the fabrication process of all presented structures were done in a similar manner, it is reasonable to assume that they have the same growth and waveguide quality. The detail of quantum design and physical processes of V845 and V962 were discussed in [11] and [12], respectively. We analyze the design structure and carrier transport of the 3P structure by presenting the schematic design scheme and conduction band diagram of V895 in Figure 1. Levels *1*, *2*, *3*, and *4* are the ground state (also called the extraction state), the lower lasing state, the upper lasing state, and the upper phonon state (also called the injection state) of the structure, respectively. Electrons, tunneling from level *1'* to the level *4*, will be injected to the level *3* (ULS) via a phonon scattering relaxation mechanism. The lasing transition occurs between the levels *3* and *2*. Due to sub-picosecond scattering time of *4* $\rightarrow$ *3* and *2* $\rightarrow$ *1* transitions, the current density of the structure is largely regulated by tunneling time (*1'* $\rightarrow$ *4*) and the scattering time between the lasing state (*3* $\rightarrow$ *2*). At low

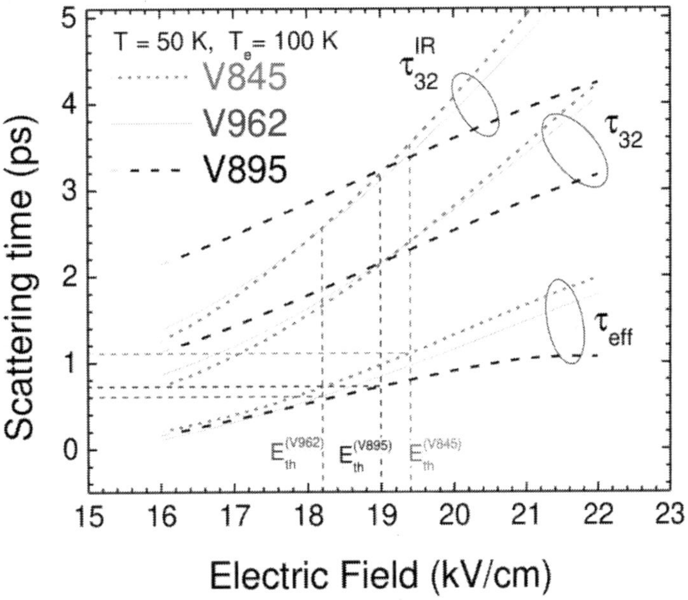

Figure 2. Different scattering times between the lasing states (level *3* and *2*) of devices V845 (dashed red lines), V895 (dashed black lines), and V962 (solid blue line) at lattice temperature of 50 K. The electronic temperature in all subbands was assumed at 100 K.

temperatures, the scattering time between the levels *3* and *2* before the threshold electric field is mainly attributed to the interface roughness scattering time [13]. The energy spacing between those states is less than 15 meV for all three structures which results in a phonon scattering time of more than 10 ps at 50 K. In addition, the impurity scattering time between the lasing states at low temperature is also more than 20 ps for all 3P-QCLs. The effective lifetime (τ_{eff}), interface roughness scattering time (τ_{IR}), and total scattering time between the lasing states (τ_{32}) were calculated based on the model presented in [8]. The expression of effective lifetime, which links population inversion with current density, when including the effect of *2* $\rightarrow$ *3* backscattering, reads as:

$$\tau_{\text{eff}} = \frac{g_{43}(g_2 - g_{32}) - g_{42}(g_3 - g_{23})}{g_4(g_3 g_{21} + g_{23} g_{31})}, \tag{1}$$

Table 1. Design parameter (volume doping concentration: N_{3D}), measured results (threshold current density: J_{th}, threshold electric field: E_{th}, maximum lasing temperature: T_{max}, and lasing frequency: ν) , and calculated parameters (effective lifetime: τ_{eff}, oscillator strength: f_{23}, IR scattering lifetime: τ_{32}^{IR}, and lifetime between ULS and LLS: τ_{32}) at threshold electric field and lattice temperature of 50 K in the 3P-QCLs.

Device	N_{3D} (cm^{-3})	J_{th}(kA/cm^2)	E_{th}(kV/cm)	T_{max}(K)	ν(THz)	τ_{eff}(ps)	f_{23}	τ_{32}^{IR}(ps)	τ_{32}(ps)
V845	9×10^{15}	0.89	19.4	152	2.4	1.1	0.33	3.53	2.52
V895	9×10^{15}	1.33	19	134	2.5	0.72	0.35	3.23	2.28
V962	7.6×10^{15}	1.47	18.2	151	2.4	0.61	0.39	2.54	1.87

where g_{ij} is the scattering rate from state i to state j and g_i is the total intersubband scattering from state i. IR scattering was identically applied on both types of interfaces: well-on-barrier and barrier-on-well with the parameters $\Delta = 0.28$ nm for the mean height of roughness and $\Lambda =$ 6.5 nm for the correlation length. These times are plotted as a function of electric field in Figure 2. It shows that all three structures have an IR scattering time of less than 4 ps at threshold, which strongly affects the total scattering time between the lasing states and consequently, the effective carrier lifetime. From Figure 2, the effective lifetime of these devices at their threshold electric field at 50 K are extracted; their values vary considerably, i.e, 1.1 ps in V845, 0.72 ps in V895 and 0.61 ps in V962. Extensive variations are also observed in threshold current density at 50 K: 0.89 kA/cm^2 in V845, 1.33 kA/cm^2 in V895 and 1.47 kA/cm^2 in V962. It clearly shows that a higher threshold current density is associated with a shorter effective lifetime. This can be explained as following. Considering the fact that the gain of the structure will be pinned to the cavity loss at threshold point, the product of threshold current density, the effective lifetime, inverse of superperiod, and the oscillator strength of these three structures should remain roughly constant if we assume that the product between cavity loss and gain bandwidth are comparable. As stated earlier, the cavity losses of all structures can be assumed to be the same due to the comparable lasing frequency, doping concentration, and waveguide quality. The detail of design parameters, the experimental results, and some calculated parameters including different scattering times and oscillator strength are presented in Table 1.

In order to further verify that experimental $J_{th,exp}$ scales with the inverse of IR-limited effective lifetime and also, to illustrate this hypothesis graphically, we defined a function LB (in $THz.cm^{-1}$) versus electric field (E):

$$LB(E) = 43.72 \times f_{23}(E) \times \tau_{eff}(E) \times J_{th,exp}/L_p, \tag{2}$$

where 43.72 is a constant value ($10^{-12}e/2\pi m\epsilon_0 nc$) that depends on the effective mass, the permittivity of the material and the speed of light, τ_{eff} is given in ps, $J_{th,exp}$ in A/cm^2, L_p is the superperiod in Angstrom. Figure 3 represents the calculated LB as a function of electric field for the three structures. The three LB values at threshold electric field, marked by square dots, are reasonably comparable among the three structures, confirming a constant cavity loss-bandwidth product of these devices. The small variation of the LB values of the marked dots in Figure 3 (ranging from ~36 to ~39 $THz.cm^{-1}$) may come from an uncertainty in gain bandwidth. This indicates that IR scattering is the governing factor for the threshold current at low temperature [13]. The derived loss-bandwidth products are slightly smaller than those derived in [11] with another method. Further improvement of the threshold current density, the dynamic range, and the output power of the structures can be achieved by designing a structure with a longer effective lifetime. Minimizing the IR scattering time between the lasing states should be a promising approach to realize a high performance 3P-QCL [14].

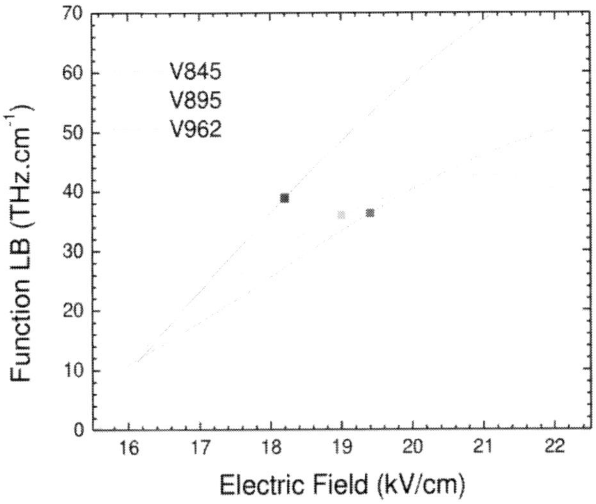

Figure 3. Function LB versus electric field at 50 K. The LB value at threshold electric field, marked by square dot, gives an estimation of the cavity loss-gain bandwidth product at 50 K. The three structures have similar loss-bandwidth product of 36-39 THz.cm^{-1}.

3. Intrasubband IR scattering in 3P-QCLs

In conventional RP structures, the gain bandwidth of the design is mainly determined by the tunneling coupling between either the injection state and the upper lasing state or that between the extraction state and the lower lasing state [15, 16]. In the 3P-QCLs, there is no state to be in resonance with either the upper or the lower lasing state, then the gain spectrum is not perturbated by tunneling effect. To compare how the pure dephasing time between the lasing states ($\tau^* = 2\hbar/\Gamma^{ul}_{\text{intra}}$ [17]) changes from a RP-QCL to a 3P-QCL, we calculate the intrasubband IR scattering time between the lasing states in both RP and 3P-QCLs.

The intrasubband ion-impurity scattering time and the IR scattering time were calculated at designed electric field and different temperatures. Our calculation shows that the intrasubband ion-impurity scattering time of the RP-QCL presented in [2] is 0.75 ps at 50 K while the IR scattering time is 1.45 ps at 50 K. By using the dephasing time equation ($\tau^* = 2\hbar/\Gamma^{ul}_{\text{intra}}$), the value of 1 ps is obtained which is comparable with the value reported in [2]. We calculated the intrasubband IR scattering time between the lasing states of the 3P-QCLs and the values of 0.16 ps, 0.24 ps, and 0.43 ps were obtained in V845, V962, and V895, respectively. The much shorter intrasubband IR scattering time between the lasing states in the 3P-QCLs is attributed to a higher energetic barrier and a lower oscillator strength compared to the RP-QCL presented in [2]. Our calculation shows that not only does IR scattering degrade the performance of the structure by lowering the effective lifetime and hence the population inversion of the structure, but it also may increase the gain bandwidth of the structure. However, these predictions on the effect of IR on gain bandwidth need to be confirmed experimentally, for instance by time-domain spectroscopy. Of course, impurity scattering is also at play in the bandwidth [18], therefore the position of doping should be carefully optimized [19].

4. Conclusion

The interface roughness scattering mechanism in THz QCLs based on 3P designs was investigated. The results indicate that the IR scattering might be the limiting factor on the device performance, in particular for threshold current density at low temperature. Our modeling shows that not only can IR scattering increase the threshold current density of the structure by reducing the effective lifetime, but it also would increase the gain bandwidth of the design due to a shorter intrasubband IR scattering time between the lasing states. Any further improvement in THz 3P-QCLs may require precisely minimizing the IR scattering impacts on both effective lifetime and gain broadening of the quantum design.

References

[1] R. Kohler, A. Tredicucci, F. Beltram, H. E. Beere, E. H. Linfield, A. G. Davies, D. A. Ritchie, R. C. Iotti, and F. Rossi. Terahertz semiconductor-heterostructure laser. *Nature*, 417(6885):156–159, 2002.
[2] S. Fathololoumi, E. Dupont, C.W.I. Chan, Z.R. Wasilewski, S.R. Laframboise, D. Ban, A. Mátyás, C. Jirauschek, Q. Hu, and H. C. Liu. Terahertz quantum cascade lasers operating up to ∼ 200 K with optimized oscillator strength and improved injection tunneling. *Opt. Express*, 20(4):3866–3876, 2012.
[3] S. Fathololoumi, E. Dupont, Z. R. Wasilewski, C. W. I. Chan, S. G. Razavipour, S. R. Laframboise, Shengxi Huang, Q. Hu, D. Ban, and H. C. Liu. Effect of oscillator strength and intermediate resonance on the performance of resonant phonon-based terahertz quantum cascade lasers. *J. Appl. Phys.*, 113(11):113109, 2013.
[4] C. W. I. Chan, Q. Hu, and J. L. Reno. Tall-barrier terahertz quantum cascade lasers. *Appl. Phys. Lett.*, 103, 2013.
[5] H. Yasuda, T. Kubis, P. Vogl, N. Sekine, I. Hosako, and K. Hirakawa. Nonequilibrium Green's function calculation for four-level scheme terahertz quantum cascade lasers. *Appl. Phys. Lett.*, 94(15):151109, 2009.
[6] S. Kumar, C. W. I. Chan, Q. Hu, and J. L. Reno. A 1.8-THz quantum cascade laser operating significantly above the temperature of 1.9 $\hbar\omega/k_B$. *Nat. Phys.*, 7:166–77, 2010.
[7] T. Kubis, S. R. Mehrotra, and G. Klimeck. Design concepts of terahertz quantum cascade lasers: Proposal for terahertz laser efficiency improvements. *Appl. Phys. Lett.*, 97(26):261106, 2010.
[8] E. Dupont, S. Fathololoumi, Z. R. Wasilewski, G. Aers, S. R. Laframboise, M. Lindskog, S. G. Razavipour, A. Wacker, D. Ban, and H. C. Liu. A phonon scattering assisted injection and extraction based terahertz quantum cascade laser. *J. Appl. Phys.*, 111(7):073111, 2012.
[9] M. Yamanishi, K. Fujita, T. Edamura, and H. Kan. Indirect pump scheme for quantum cascade lasers: dynamics of electron-transport and very high T_0-values. *Opt. Express*, 16(25):20748–20758.
[10] K. Fujita, M. Yamanishi, S. Furuta, K. Tanaka, T. Edamura, T. Kubis, and G. Klimeck. Indirectly pumped 3.7 THz InGaAs/InAlAs quantum-cascade lasers grown by metal-organic vapor-phase epitaxy. *Opt. Express*, 20(18):20647–20658.
[11] S. G. Razavipour, E. Dupont, S. Fathololoumi, C. W. I. Chan, M. Lindskog, Z. R. Wasilewski, G. Aers, S. R. Laframboise, A. Wacker, Q. Hu, D. Ban, and H. C. Liu. An indirectly pumped terahertz quantum cascade laser with low injection coupling strength operating above 150 K. *J. of Appl. Phys*, 113, 2013.
[12] S. G. Razavipour, E. Dupont, C. W. I. Chan, C. Xu, Z. R. Wasilewski, S. R. Laframboise, Q. Hu, and D. Ban. A high carrier injection terahertz quantum cascade laser based on indirectly pumped scheme. *Appl. Phys. Lett.*, 104(4):041111, 2014.
[13] Y. Chiu, Y. Dikmelik, P. Q. Liu, N. L. Aung, J. B. Khurgin, and C. F. Gmachl. Importance of interface roughness induced intersubband scattering in mid-infrared quantum cascade lasers. *Appl. Phys. Lett.*, 101(17):171117, 2012.
[14] Y. Chiu, Y. Dikmelik, Q. Zhang, J.B. Khurgin, and C.F. Gmachl. Engineering the intersubband lifetime with interface roughness in quantum cascade lasers. In *Conference on Lasers and Electro-Optics (CLEO), 2012*, pages 1–2, May 2012.
[15] S. Kumar, Q. Hu, and J. L. Reno. 186 K operation of terahertz quantum-cascade lasers based on a diagonal design. *Appl. Phys. Lett.*, 94(13):131105, 2009.
[16] E. Dupont, S. Fathololoumi, and H. C. Liu. Simplified density-matrix model applied to three-well terahertz quantum cascade lasers. *Phys. Rev. B*, 81(20):205311, 2010.
[17] Romain Terazzi. Ph.D. dissertation, ETH Zurich, 2012.
[18] R. Nelander and A. Wacker. Temperature dependence and screening models in quantum cascade structures. *J. of Appl. Phys.*, 106(6):063115, 2009.
[19] F. Carosella R. Ferreira C. Ndebeka-Bandou, A. Wacker and G. Bastard. Dopant engineering of inter-subband linewidth and lineshape in multiwell heterostructures. *Appl. Phys. Express*, 6(9):094101, 2013.

Synthesis and characterization of CdTe/CdSe core/shell nanoparticles with high emission from 570 to 610 nm

O. G. Rojas Valencia[1], J. L. Casas Espinola[2,*], M. Corea Tellez[1]

[1]ESIQIE– Instituto Politécnico Nacional, México D. F. 07738, México,
[2]ESFM – Instituto Politécnico Nacional, México D. F. 07738, México

Email: jlcasas@esfm.ipn.mx

Abstract. The paper presents the advance process of synthesis of CdTe/CdSe core/shell Quantum Dots (QDs) with emission wavelength from 570 to 610 nm. This synthesis was performed using modified Hot-Injection method. The CdTe/CdSe core-shell QDs have been characterized by the transmission electron microscopy (TEM) with the aim to control the QDs shape and the average size. The optical absorption and photoluminescence (PL) spectra are investigated in four samples obtained at different reaction times (5, 10, 15 and 20 s) of core growth after the CdSe passivation of the CdTe cores. Obtained results have shown that synthesized QDs have a good crystallinity, spherical shape and bright emission at 610 nm. The reaction time enlargement permits increasing the average size of CdTe/CdSe core-shell QDs, together with "red" shifting the PL wavelength and increasing the PL intensity. The advantages of synthesized CdTe/CdSe core/shell QDs are discussed.

1. Introduction

To date, QDs have various nanostructured configurations, typically as core/shell heterostructure QDs, where two different semiconductors are incorporated into a single colloidal QD. The CdTe/CdSe core/shell QDs, have been synthesized via colloidal chemical routine [1-5] and studied intensively owing to the quantum confinement effects, size-dependent photoemission, narrow emission spectrum and high-quantum yield photoluminescence (up to 75%) [6-7]. These QDs were applied as well for biomedical imaging, in photovoltaic devices, for the bioconjugation to antibodies [8-9] and in the early detection of a set of important diseases including cancer. Band gap engineering in CdTe/CdSe core/shell QDs can open a new window for the various QD material choices. In these core/shell heterostructured nanoparticles one of carriers is mostly confined in the core of QDs, while other one remains in the shell, giving a significant increase in the exciton lifetime [10]. For CdTe/ CdSe QDs the hole is mostly confined to the CdTe core, while the electron is located mostly in the CdSe shell.

2. Experimental details

2.1. Materials

All the following material were purchased from Aldrich®: Cadmium acetate hydrate ($(C_2H_3O_2)_2Cd$ xH_2O, 99.99%), tellurium (Te, 99.99%), selenium (Se, 99.99%), tri-n-octylphosphine ($C_{24}H_{51}P$, 97.00%) (TOP), tri-n-octylphosphine oxide ($C_{24}H_{51}PO$, 90.00%) (TOPO), 1-octadecene ($C_{18}H_{36}$, 95%)

(ODE) and oleic acid ($C_{18}H_{34}O_2$, 90%). All materials were used directly without any further purification.

2.2. Synthesis of CdTe core

The CdTe-core was prepared via a modified Hot-Injection method published earlier in [11]. Cadmium precursor were prepared by dissolving 40 mg of Cadmium acetate hydrate, 0.8 g of TOPO, 8 ml of TOP, 0.6 ml of oleic acid and 10 ml of octadecene in a flask. The mixture was heated at 350 °C during 90 minutes and shaken vigorously. The tellurium precursor was prepared by mixing 26 mg of tellurium powder, 3 ml of TOP at temperature of 200 °C during 2 hours and shaken vigorously. The CdTe-core solution was produced by injecting 1 ml room-temperature tellurium precursor into the cadmium precursor at 350 °C. The nucleation of CdTe immediately started and the aliquots amount of reaction solution were extracted every 10 seconds, namely 10, 20, 30 and 40 s in order to get variation in the size core. The reactions of aliquots amount were quenched by introducing them into 50 ml of liquid nitrogen. All steps in the reaction were carried out under N_2 atmosphere

2.3. Synthesis of CdTe/CdSe core/shell

After the CdTe-core solution was produced, the CdSe-shell was grown on the CdTe-core surface using the selenium precursor which was prepared mixing 26 mg of selenium powder, 3 ml of TOP at temperature of 200 °C during 2 hours and shaken vigorously. To produce CdTe/CdSe core/shell, 1 ml of the CdTe-core solution and 1 ml of selenium precursor were injected at the same time into hot solution (350 °C) of cadmium precursor. The formation of CdSe-shell immediately started and one aliquots amount of reaction solution was extracted after 20s. The reaction of the aliquot amount was quenched by introducing it into 50 ml of liquid nitrogen. The reaction was carried out under N_2 atmosphere. For further purification, 1 ml of $CHCl_3$ and 5 ml of CH_3OH were added to 2 ml of CdTe/CdSe core/shell solution, this mixture was shaken during15 min. The $CHCl_3$ and CH_3OH were used as solvent to separate CdTe/CdSe core/shell nanocrystals from byproducts and unreacted precursors.

2.4. Characterization

The photoluminescence signal was dispersed by a spex 500M monochromator with focal length 0.5 m and detected by Hamamatsu GaAs photomultiplier tube (PMT) model R943-02 at room temperature. A He-Cd laser 325 nm with 50 mW power was used for PL excitation. To control the shape and size of nanocrystals a JOEL JEM 6300 transmission electron microscope was used. The optical absorptions were measured using a Perkin Elmer UV-Vis/ lambda 24/deuterium lamp spectrometer.

3. Experimental results and discussion

The first step in the nanocrystal growth is the nucleation in a solution at constant temperature and pressure. This process is driven by the difference in the free energy between the crystalline phase and the solution phase. The synthesis of monodisperse colloids via homogeneous nucleation requires a temporal separation of nucleation and growth of the seeds. Initially the concentration of monomers constantly increases by addition from exterior. In stage I no nucleation occurs even in supersaturated solution ($S>1$), due to the high energy barrier for spontaneous homogeneous nucleation. The latter is overcome in stage II for a yet higher degree of supersaturation ($S>S_c$), where nucleation and formation of stable nuclei take place. As the rate of monomer supply, the monomer concentration and hence the supersaturation decrease below S_c, the level at which the nucleation rate becomes zero. In the following stage III, the particle growth continues under further monomer consumption as long as the system is in the supersaturated regime. When the formation of the CdTe core was achieved, the CdSe-shell was grown on the CdTe-core surface. The mismatch (7.1%) of the lattice constant between the core (CdTe) and shell (CdSe) facilitates the epitaxial growth of the CdSe shell around the CdTe core. Using the hot-injection synthesis method, it is possible to obtain samples with very low standard deviation from the mean size as shown in TEM images of the as-prepared CdTe/CdSe nanocrystal (figure 1) from aliquots (namely 10, 20, 30 and 40 s) and its analysis presented in the histogram of particle size distribution in figure 2 where QDs average size increase from 3 to 6 nm.

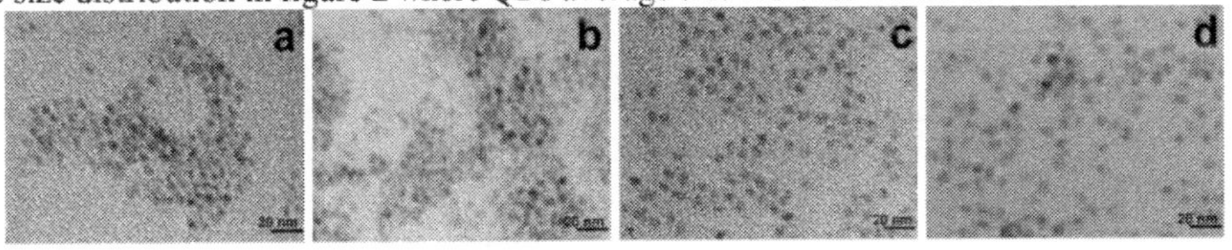

Figure 1. TEM images of the CdTe/CdSe core-shell nanocrystals prepared at the reaction time a) 10s, b) 20s, c) 30s, and d) 40s.

Simultaneously the CdSe shell growth reduces the number of dangling bonds on the CdTe surface, which can act as a trap states for charge carriers and reduce fluorescence Quantum Yield. The CdTe/CdSe nanocrystals are interesting owing to the type II QDs: the conduction band edge of the CdSe shell is located in the band gap of the CdTe core as it is shown in figure 3 for the bulk materials.

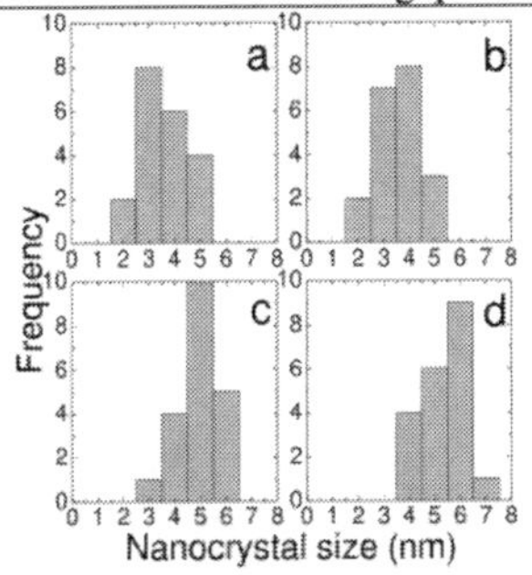

Figure 2. The histogram of nanocrystal size distribution obtained from TEM images in Fig. 1

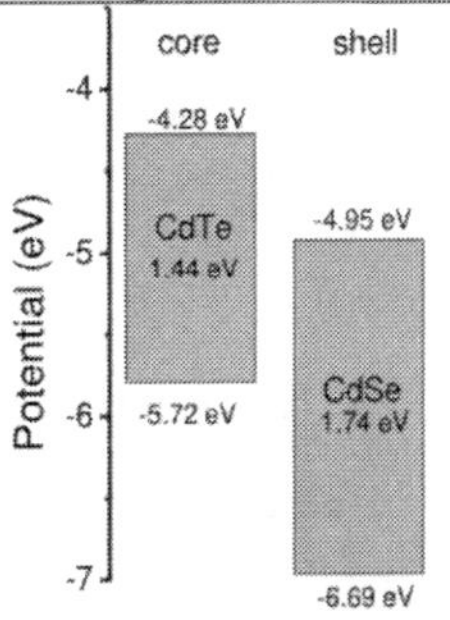

Figure 3. Plot illustrating the CdTe and CdSe band offset alignment in the energy diagram (the bulk values energies were taken from literature).

The staggered band alignment leads to a smaller "effective band gap" than the energy band gap of core and shell materials, that permits obtaining a significant red shift of the emission wavelength. The absorption spectra of four samples are shown in figure 4. The temporal evolution of the UV/Vis absorption spectra of the CdTe core was monitored in as-prepared CdTe/CdSe nanocrystals. This figure shows a continuous red shift in absorption with increasing reaction time at the creation of the CdTe core

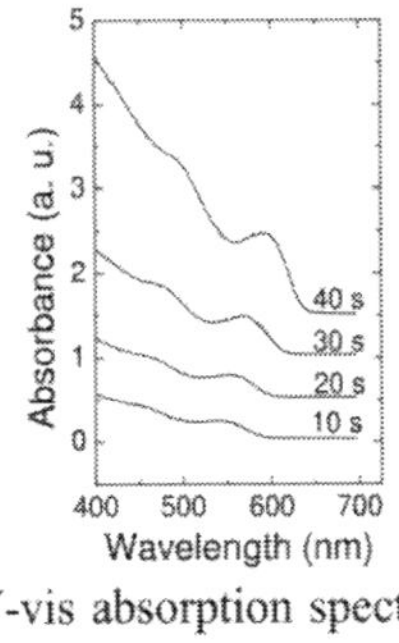

Figure 4. UV-vis absorption spectra for the four studied samples

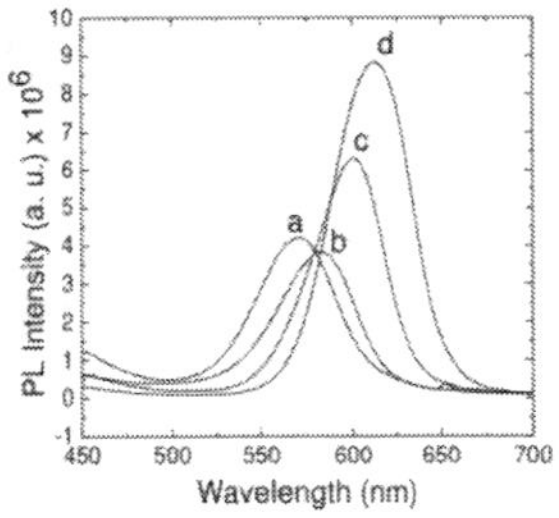

Figure 5. PL spectra corresponding to the samples prepared at the times10, 20, 30 and 40 s presented from left to right.

The first excitonic absorption peaks in samples under study are summarized in table 1. PL spectra of the CdTe/CdSe nanocrystals are presented in figure 5. In the PL spectra we have seen emissions energy maximum localized in 2.17, 2.12, 2.06 and 2.03 eV, in the different four samples, which are higher than the "effective band gap" of the bulk core/shell materials due to the quantum confinement effect. Thus the band gaps of both core and shell increase leading to increasing the "effective band gap" of the CdTe/CdSe nanocrystal. In this case the change of the PL peak position from 570 to 610 nm is related to the enlargement of the nanoparticles size. Note that studied CdTe/CdSe nanocrystal have a size between 3-6 nm that is lower than Borh radii in CdTe equal to 7 nm. The PL intensity rises when the particle size increases, probably, owing to increasing at the same time the concentration of nanocrystals with raising the reaction time. The FWHM values of PL bands are reported in table 1 as well. It is clear that we have obtained a very small FWHM that means that the high-quality CdTe/CdSe core/shell nanocrystals have been produced. Thus using modified hot injection synthesis method, it is possible to obtain samples with 5-10% standard deviation from the mean size.

Table 1

Reaction time (s)	PL peak position (nm)	FEAP* (nm)	Average size (nm)	FWHM (nm)
10	570	540	3	50
20	584	552	4	48
30	600	568	5	46
40	610	591	6	42

*First excitonic absorption peak

4. Conclusions

We have successfully synthesized CdTe/CdSe core/shell QDs with strongly emission from 570 to 610 nm using a modified hot-injection method. This emission can be tuned in the mentioned range in dependence of the reaction time. The presented results show low energy shift and rise of PL intensity with increasing the QD size. An effective suppression on the non-radiative recombination of electrons and holes at the surface of the core has successfully achieved. Obtained core/shell QDS are characterized by the high emission intensity and small size dispersion.

Acknowledgement

The work was supported by project SIP20130424 (Instituto Politecnico Nacional, Mexico). Authors thank to Dr. T.V. Torchynska for the fruitful discussion of presented results.

References

[1] S. Kim, B. Fisher, HJ Eisler, M. Bawendi, J Am. Chem. Soc. 125 (2003) 11466
[2] X. Peng, M. C. Schlamp, A. V. Kadavanich,; A. P. Alivisatos, J. Am. Chem. Soc. 119 (1997) 7019.
[3] LM. Sai and X. Y. Kong, Nanoscale Research Letters 6 (2011) 399.
[4] H. Zhong, Y. Zhou, Y. Yang, C. Yang, Y. Li, J. Phys. Chem. C 111 (2007) 6538.
[5] Y.S. Xia, C.Q. Zhu, Analyst 133 (2008) 928.
[6] M. Kuno, D.P. Fromm, H.F. Hamann, A. Gallagher, D.J. Nesbitt, J. Phys. Chem. 115 (2001) 1028.
[7] T.V. Torchynska, Nanotechnology 20 (2009) 095401.
[8] N. Tessler, V. Medvedev, M. Kazes, S.H. Kan, U. Banin, Science 295 (2002) 1506.
[9] T.V. Torchynska, J.L. Casas Espinola, J.A. Jaramillo Gómez, J. Douda, K. Gazarian, Physica E 51 (2013) 55
[10] (2013) 60
[11] S. Kim, Y. T.Lim, E. G. Soltesz, A. M. De Grand, J. Lee, A. Nakayama, J. A. Parker, T. Mihaljevic, R. G. Laurence, D. M. Dor, L. H. Cohn, M. G. Bawendi, J. V Frangioni, Nat. Biotechnol. 22 (2004) 93.

Photoluminescence observation of electron-hole droplets in a GaAs/AlAs type-II superlattice

Yoshiaki Furukawa and Masaaki Nakayama [1]

Department of Applied Physics, Graduate School of Engineering, Osaka City University, Sugimoto, Sumiyoshi-ku, Osaka 558-8585, Japan

nakayama@a-phys.eng.osaka-cu.ac.jp

Abstract. We have investigated the photoluminescence (PL) properties of a $(GaAs)_{12}/(AlAs)_{12}$ type-II superlattice (SL) at 10 K from the viewpoint of the formation of electron-hole droplets (EHDs). In the type-II SL, the lowest-energy exciton consists of an X-electron and a Γ-heavy-hole confined in the AlAs and GaAs layers, respectively; namely, the optical transition is indirect in momentum and real space. The excitation-power density was widely changed from ~1 mW/cm^2 to ~1 kW/cm^2 for precisely observing changes of PL spectra. It was found that a broad PL band appears with a threshold-like nature in the energy region lower than the biexciton. This suggests the occurrence of the Mott transition. The prominent feature of the broad PL band is the fact that the spectral profile hardly depends on the excitation-power density, which is one of typical properties of EHDs. From the line-shape analysis of the PL spectra, we estimated the stability energy of the EHD to be ~2.9 meV relative to the biexciton energy. The above results consistently demonstrate the formation of the EHD.

1. Introduction

The stability of electron-hole liquid (EHL), which is a liquid phase of electron-hole plasma (EHP) caused by a Mott transition, confined in quantum-well (QW) systems including superlattices (SLs) has attracted much attention as a prominent problem of high-density-excitation phenomena. Note that the EHL is usually observed in experiments as an electron-hole droplet (EHD) because the EHL with a finite size is surrounded by excitons and biexcitons in a coexistence phase [1]. There exist two types of potential structures from an aspect of the spatial configuration: "type I" and "type II". In a type-I potential structure, electrons and holes are confined in the same QW layer. A type-II potential structure corresponds to a staggered lineup of conduction-band (CB) and valence-band (VB) potentials: The QW and barrier layers in the CB turn to the barrier and QW layers in the VB, respectively. Thus, electrons and holes are spatially separated in the type-II potential structure. Kleinman theoretically suggested that the EHL is unstable relative to biexcitons in GaAs type-I QWs with a layer thickness less than 22 nm [2]. In fact, there has been no convincing report on the observation of the EHD in GaAs type-I QWs. In contrast, Hawrylak [3] and Ando *et al.* [4] theoretically suggested the possibility of the EHD formation in type-II QW systems. In GaAs/AlAs QWs and SLs, the GaAs layer is the QW in the CB and VB at the Γ valley, while the AlAs layer is the QW in the CB at the X valley. The lowest-energy exciton state changes from the type-I exciton to the

[1] To whom any correspondence should be addressed.

type-II one with an decrease in GaAs-layer thickness because the energy of the X-electron subband becomes lower than that of the Γ-electron subband, the so-called Γ-X subband crossover at a GaAs thickness of ~13 monolayers [5], where monolayer corresponds to a half of the lattice constant: 0.283 nm in GaAs. As a result, the type-II exciton consists of the X-electron in the AlAs layer and the Γ-heavy-hole in the GaAs layer. The optical transition is indirect in real and momentum space, which results in a long exciton lifetime of the order of μs [5]. This long lifetime is advantageous for the formation of EHDs from an aspect of cooling efficiency of excitons, biexcitons, and carriers. In a GaAs/AlAs type-II SL grown on a (311) GaAs substrate, the structure of which is like a quantum wire due to interface corrugation, the stability of EHDs was confirmed [6, 7]. However, there has been no clear report on the observation of EHDs in usual type-II SLs with flat interfaces.

In the present work, we have investigated the photoluminescence (PL) spectra of a GaAs/AlAs type-II SL, which was grown on a conventional (001) GaAs substrate, at 10 K in a wide excitation-power-density range from ~1 mW/cm^2 to ~1 kW/cm^2 for systematically investigating changes of spectral profiles. The exciton- and biexciton-PL bands were clearly observed. We found that a broad PL band appears with a threshold-like nature in the energy region lower than the biexciton, which suggests the occurrence of the Mott transition for the formation of EHP. The spectral profile of the broad PL band hardly depends on the excitation-power density. This fact is one of typical PL properties of EHDs. From the line-shape analysis of the PL spectra, we revealed the stability energy of the EHD relative to the biexciton energy.

2. Experimental details

The sample used was a $(GaAs)_{12}/(AlAs)_{12}$ type-II SL with 50 periods grown on a (001) GaAs substrate at 660°C by molecular beam epitaxy (MBE). The subscript of the SL structure indicates a number of monolayer in the constituent layers. In the MBE process, individual layer thicknesses were precisely controlled by observing oscillation patterns of reflection-high-energy-electron diffraction [8]. Note that the oscillation period corresponds to a growth time of one monolayer.

In the PL measurements, the excitation light source was a cw-mode Ti:sapphire laser. The excitation energy was fixed to 1.797 eV. We used a beam shaper to homogenize the spatial distribution of the excitation-laser power: The spatial power fluctuation was within ±5%. The PL spectra were detected with a cooled charge couple device attached to a 32-cm single monochromator with a resolution of 0.15 nm. The excitation-power density was widely changed from ~1 mW/cm^2 to ~1 kW/cm^2. All the optical measurements were performed at 10 K.

3. Results and discussion

We first overview the PL spectra as a function of excitation-power density. Figure 1 shows the PL spectra at various excitation-power densities, where the maximum power density is P_0=~1 kW/cm^2, and the intensity of each PL spectrum is normalized by the maximum one. At the lowest power density of $1\times10^{-6}P_0$, a PL band labeled X is attributed to the type-II exciton, the decay time of which was evaluated to be ~2 μs in our previous work [9]. With an increase in excitation-power density, a PL band labeled M continuously grows in the low-energy side of the X-PL band. The M band exhibits a low energy tail, which is a typical feature of the biexciton PL, the so-called inverse Boltzmann line shape. Thus, the M band is assigned to the biexciton PL, which was also confirmed in our previous work [9]. At an excitation-power density of $0.01P_0$, a PL band labeled EHD appears with a threshold-like nature in the energy region lower than the M-PL band. This drastic change of the PL spectra suggests occurrence of the Mott transition. The coexistence of the X-, M-, and EHD-PL bands is a key result in the present work because the coexistence is one of typical characteristics of the EHD [1].

Figure 2 shows the PL spectra at excitation-power densities of $0.5P_0$, $0.7P_0$, and P_0. The intensity of each PL spectrum is scaled by the maximum intensity of the EHD-PL band. Note that the EHD-PL band appears as a shoulder of the M-PL band at excitation-power densities less than $0.5P_0$, so those could not be used for the comparison in Fig. 2. It is evident that the line shape of the EHD-PL band hardly depends on the excitation-power density. This behavior reflects that the carrier density and

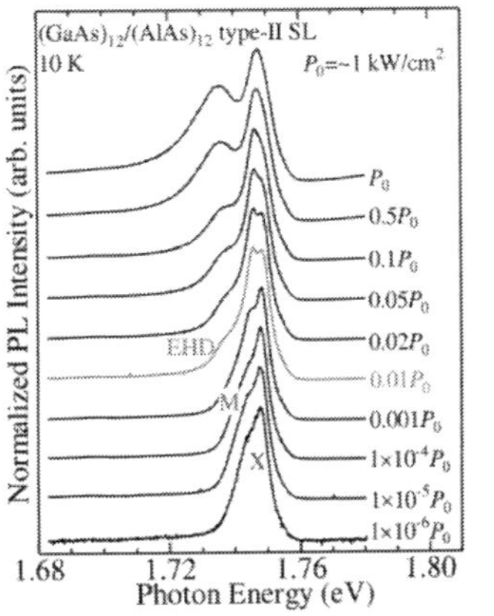

Figure 1. PL spectra at various excitation-power densities at 10 K in the $(GaAs)_{12}/(AlAs)_{12}$ type-II SL, where the maximum power density is P_0=~1 kW/cm^2, and the intensity of each spectrum is normalized by maximum one.

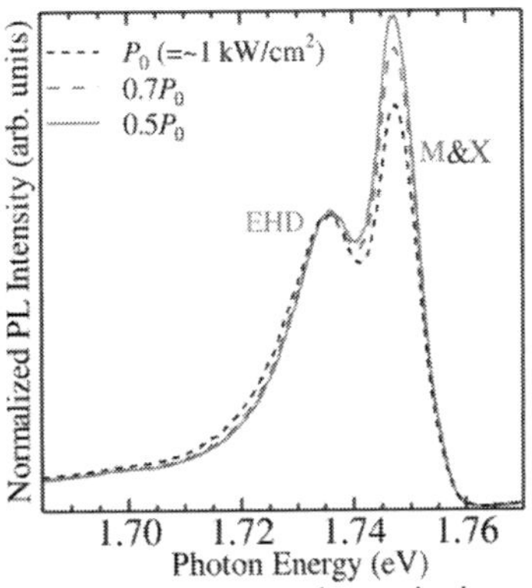

Figure 2. PL spectra at the excitation-power densities of $0.5P_0$, $0.7P_0$, and P_0. Each spectrum is scaled by the maximum intensity of the EHD-PL band.

effective temperature contributing to the EHD formation are independent of the excitation-power density. In other words, the excitons, biexcitons, and EHDs are in quasi-thermal equilibrium.

Hereafter, we perform the line-shape analysis of the PL band to evaluate the parameters of the EHD formation. In the line-shape analysis of the EHD-PL band, we used the following two-dimensional density of states:

$$D_i(E_i)=[m_i^*/(\pi\hbar^2)][1+\exp(-E_i/\Gamma_i)]^{-1}, \qquad (1)$$

where i=e or h, m_e^* and m_h^* are the in-plane effective masses of the X-electron in the AlAs layer and the Γ-heavy-hole in the GaAs layer, respectively. The value of m_h^* was calculated using $m_h^*=m_0/(\gamma_1+\gamma_2)$ [10], where γ_1 and γ_2 are the Luttinger parameters: $m_h^*=0.11m_0^*$ for γ_1=6.8 and γ_1=1.9 [11]. The value of m_e^* is taken from Ref. 12: $m_e^*=0.19m_0^*$. Γ_e and Γ_h are the broadening factors which are fitting parameters. Assuming that the momentum conservation is relaxed, the PL line shape of EHP, which is the EHD in this case, is given by [1]

$$I_{EHD}(\hbar\omega)\propto\int_{-\infty}^{\infty}\int_{-\infty}^{\infty}D_e(E_e)D_h(E_h)f_e(E_e,E_{F,e},T_{eff})f_h(E_h,E_{F,h},T_{eff})dE_e dE_h, \qquad (2)$$

where E_e (E_h) is an electron (hole) energy relative to the conduction-band minimum (valence-band maximum), f_e, (f_h) is the Fermi distribution function of the electron (hole), $E_{F,e}$ ($E_{F,h}$) is a quasi-Fermi energy of the electron (hole), T_{eff} is an effective temperature, and E_g' is a renormalized band-gap energy due to many body effects. The energy conservation of $\hbar\omega=E_g'+E_e+E_h$ was taken into account.

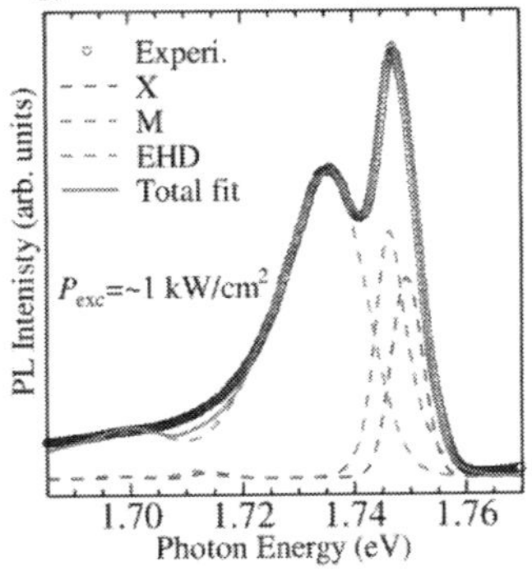

Figure 3. Results of the line-shape analysis of the PL spectrum at an excitation-power density of P_0=~1 kW/cm^2, where the open circles indicate the experimental spectrum, and the solid curve indicates the totally fitted result. The dashed curves depict individually fitted results of the X-, M-, and EHD-PL bands including LO-phonon sidebands of GaAs and AlAs.

Note that Eq. (2) is usually applied to the PL line-shape analysis of EHD and EHP in a gas phase [1] because the EHD and EHP(gas) consist of fermions; therefore, the Fermi distribution functions of electrons and holes dominate the PL line shapes as described by Eq. (2). For the line shape of the M-PL band, we used an inverse-Boltzmann distribution function taking account of the two-dimensional density of states [9]:

$$I_{\mathrm{M}}(\hbar\omega) \propto \frac{\exp[-(E_{\mathrm{X}} - E_{\mathrm{b,M}} - \hbar\omega)/(k_{\mathrm{B}}T_{\mathrm{eff}})]}{1+\exp[-(E_{\mathrm{X}} - E_{\mathrm{b,M}} - \hbar\omega)/\Gamma_{\mathrm{M}}]}, \quad (3)$$

where E_{X} is the exciton energy and $E_{\mathrm{b,M}}$ is the biexciton binding energy. We assumed that the effective temperature of the biexciton is equal to that of the EHD because of the quasi-thermal equilibrium of the EHD and excitonic system. The X-PL band was fitted by the Gaussian function.

Figure 3 shows the results of the line-shape analysis of the PL spectrum at an excitation-power density of P_0. The LO-phonon sidebands of GaAs and AlAs were also taken into account for the line-shape analysis. For the PL bands of the exciton and biexciton, the ratios of the PL intensities of the zero-phonon band to the LO-phonon sidebands were referred to those in the PL spectrum at an excitation-power density below the threshold for the EHD. For the EHD-PL band, however, the PL intensities of the LO-phonon sidebands were treated as fitting parameters because it is expected that the interactions between the LO phonons and EHD are modified by many body effects. The estimated values of the major fitting parameters are as follows: E_g'=1.7280 eV, $E_{\mathrm{F,e}}$=6.0 meV, $E_{\mathrm{F,h}}$=10.6 meV, Γ_e=2.3 meV, Γ_h=6.3 meV, T_{eff}=21 K, $E_{\mathrm{b,M}}$=3.8 meV, Γ_{M}=1.2 meV, and E_{X}=1.7494 eV. These values hardly depend on the excitation-power density. The carrier density was calculated to be 4.9×10^{11} cm^{-2}. The chemical potential of the EHD, which is given by $\mu=E_g'+E_{\mathrm{F,e}}+E_{\mathrm{F,h}}$, was evaluated to be 1.7446 eV. Thus, the stability energy ϕ relative to the biexciton energy, $\phi=E_{\mathrm{M}}/2-\mu=E_{\mathrm{X}}-E_{\mathrm{b,M}}/2-\mu$, was estimated to be 2.9 meV. The stability energy is sufficiently higher than the thermal energy of $k_{\mathrm{B}}T_{\mathrm{eff}}$=1.8 meV. The above results and analysis consistently demonstrate that the type-II EHD is stable.

4. Conclusions

We have investigated the formation of the quantum-confined EHD at 10 K in the $(GaAs)_{12}/(AlAs)_{12}$ type-II superlattice from an aspect of the systematic excitation-power dependence of the PL spectra. It was found that the EHD-PL band appears with a threshold-like nature in the energy region lower than the biexciton-PL band and that the line shape of the EHD-PL band hardly depends on the excitation power. In addition, the EHD-PL band coexists with the exciton- and biexciton-PL bands. The above results phenomenologically indicate the formation of the type-II EHD. From the line-shape analysis of the PL spectrum, we quantitatively demonstrated that the type-II EHD is stable: The stability energy of the EHD relative to the biexciton energy was estimated to be 2.9 meV, which is sufficiently higher than the thermal energy of $k_{\mathrm{B}}T_{\mathrm{eff}}$=1.8 meV.

References

[1] For a review, Kalt H 1995 *Optical properties of III-V semiconductors* (Berlin: Springer) p 41
[2] Kleinman D A 1986 *Phys. Rev. B* **33** 2540
[3] Hawrylak P 1989 *Phys. Rev. B* **39** 6264
[4] Ando T, Nakayama M and Hosoda M 2004 *Phys. Rev. B* **69** 165316
[5] Nakayama M 2000 *J. Lumin.* **87-89** 15 and releted references therein
[6] Kalt H, Nötzel R, Ploog K and Gießen H 1992 *Soild State Commun.* **83** 285
[7] Kalt H 1994 *J. Lumin.* **60&61** 262
[8] Sano N, Kato H, Nakayama M, Chika S and Terauchi H 1984 *Jpn. J. Appl. Phys.* **23** L640
[9] Nakayama M, Suyama K and Nishimura H 1995 *Phys. Rev. B* **51** 7870
[10] Weisbush C 1987 *Semiconductors and Semimetals* ed R Dingle (San Diego, CA: Academic Press) vol 24 p 1
[11] Shanabrook B V, Glembocki O J, Broido D A and Wang W I 1989 *Phys. Rev. B* **39** 3411
[12] Adachi S 1985 *J. Appl. Phys.* **58** R1

Journal of Physics: Conference Series **619** (2015) 012006 doi:10.1088/1742-6596/619/1/012006

Evaluation of crystal structure in $TlInS_2$ by optical second-harmonic generation

K. Wakita[1], M. Hagiwara[1], R. Paucar[1], Y. Shim[2], K. Mimura[2], and N. Mamedov[3]

[1] Chiba Institute of Technology, 2-17-1 Tsudanuma, Narashino, Chiba 275-0016, Japan

[2] Osaka Prefecture University, 1-1Gakuen-cho, Naka-ku, Sakai, Osaka 599-8531, Japan

[3] Institute of Physics, H. Javid 33, Baku AZ-1143, Azerbaijan

Email: kazuki.wakita@it-chiba.ac.jp

Abstract. Second-harmonic generation (SHG) in layered $TlInS_2$ crystals was studied over the temperature range of 77–300 K using a confocal laser microscope system. As expected, the SHG signal was observed in the low temperature ferroelectric phase of the layered compound. In addition, the polarization properties of the SHG signals of $TlInS_2$ were investigated in the 80–180 K range. The results are in good agreement with those of the symmetric space group C_2^3 in the ferroelectric phase.

1. Introduction

$TlMeX_2$ (Me = In, Ga; X = S, Se, Te) incommensurate materials with quasi-one- or two-dimensional lattice structures have attracted much attention because of their interesting electro-optic properties [1] and photo-induced memory effects [2]. Because the phase transition of such compound from the paraelectric to the ferroelectric state occurs via an intermediate incommensurate phase, composed supposedly of nanodomains polarized in opposite directions, the interaction of light waves with such materials is interesting from the viewpoint of basic science and practical applications. Layered $TlMeX_2$, such as $TlInS_2$, has a monoclinic crystal structure with a symmetric space group C_{2h}^6 at room temperature. However, the center of symmetry is lost when the temperature is decreased to that of the ferroelectric phase transition point, where the ferroelectric phase is asymmetric with the space group C_2^3. Second-harmonic generation (SHG) is possible only in materials with no center of symmetry; therefore, it is a very useful and sensitive method for the study of phase transition in $TlMeX_2$ compounds. However, there have been few studies on SHG in these materials [3, 4]. Recently, we reported on the temperature dependence of the SHG properties of $TlMeX_2$ [5]. In the present work, we report the polarization properties of the SHG signal in $TlInS_2$ below the ferroelectric phase transition temperature.

2. Experimental

The samples of $TlInS_2$ used for optical second-harmonic measurements were bulk single crystals grown using the Bridgman-Stockbarger method and then mechanically polished with alumina powder (particle size 0.1 μm). Each sample was mounted in a continuous-flow vibration-free cryostat (JANIS ST-500). Figure 1 is a schematic diagram of our SHG measurement system. A pulsed Ti:sapphire laser

Journal of Physics: Conference Series **619** (2015) 012006 doi:10.1088/1742-6596/619/1/012006

(Spectra Physics Mai Tai VF-T1S), with a repetition rate of 80 MHz, wavelength of 845 nm, and pulse width of 100 fs, was used as the light source. The diameter of the laser beam at the focal point was estimated to be 20 μm. Reflection-type optical second-harmonic measurements were performed using a confocal microspectroscopy system (Tokyo Instruments extended Nanofinder 30) and a charge-coupled device detector (Tokyo Instruments DV420A-OE). The measurements were performed in the temperature range of 77−325 K.

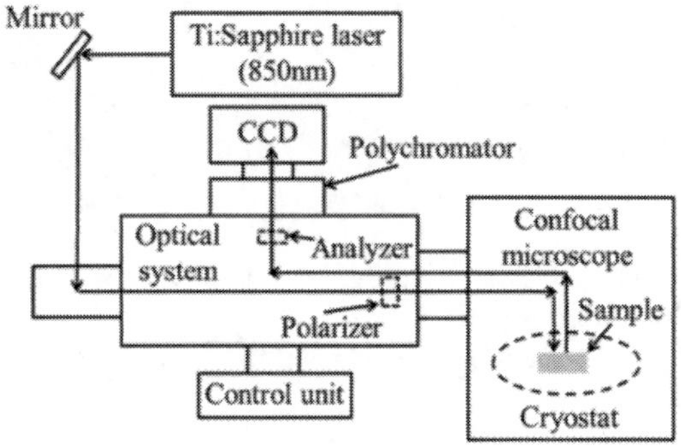

Figure 1. Schematic diagram of our SHG measurement system.

3. Results and Discussion

Figure 2 shows the temperature dependence of the intensity of SHG in the layered $TlInS_2$ crystal. The SHG signal is strong from 77 to 160 K, above which it rapidly decreases and then disappears at 192 K (Fig. 2, inset). The temperature at which $TlInS_2$ transitions from the low-temperature ferroelectric phase to the intermediate incommensurate phase was reported to be ~200 K [2]. Thus, SHG gives the more accurate value of 192 K for the phase-transition-point temperature and verifies that the average structure of the incommensurate phase has a center of symmetry. These results closely coincide with the data of Yamabi and Uchiki [4].

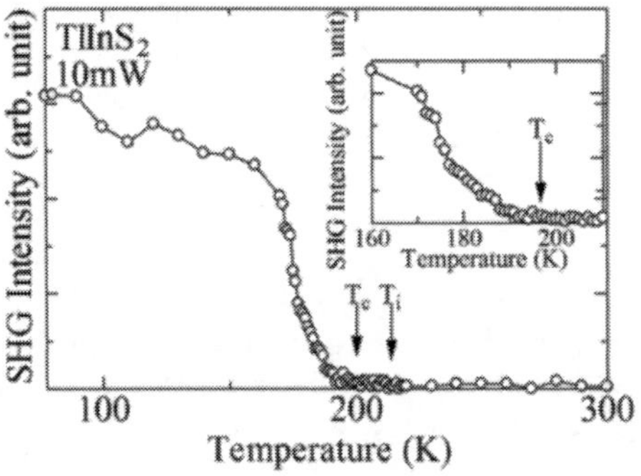

Figure 2. Temperature dependence of SHG intensity in $TlInS_2$.

In addition, we investigated the temperature dependence of the polarization of SHG in $TlInS_2$. Nonlinear polarization at 2ω in the symmetric space group C_2^3 is given by

$$\begin{pmatrix} P_x^{2\omega} \\ P_y^{2\omega} \\ P_z^{2\omega} \end{pmatrix} = \varepsilon_0 \begin{pmatrix} 0 & 0 & 0 & d_{14} & 0 & d_{16} \\ d_{21} & d_{22} & d_{23} & 0 & d_{25} & 0 \\ 0 & 0 & 0 & d_{34} & 0 & d_{36} \end{pmatrix} \begin{pmatrix} E_x^2 \\ E_y^2 \\ E_z^2 \\ 2E_yE_z \\ 2E_zE_x \\ 2E_xE_y \end{pmatrix} \quad (1)$$

The crystallographic a and b axes correspond to the x and y axes, respectively. Therefore, $P_x^{2\omega}$ and $P_y^{2\omega}$ are expressed as

$$P_x^{2\omega} = -\varepsilon_0 d_{16} \sin(2\phi) \quad (2)$$

and

$$P_y^{2\omega} = \varepsilon_0 \left(\frac{1}{2}d_{21} - \frac{1}{2}d_{21}\cos(2\phi) + \frac{1}{2}d_{22} + \frac{1}{2}d_{22}\cos(2\phi)\right) \quad (3)$$

Figures 3 and 4 show the dependence of the intensity of SHG on the fundamental light polarization detected along the a and b axes, respectively, in $TlInS_2$ in the temperature range of 80–180 K, together with results calculated using Eqs. (2) and (3). The observed results for $P_x^{2\omega}$ coincide with the calculated data, while those for $P_y^{2\omega}$ include the other peak. The results agree well with those of the symmetric space group C_2^3 in the ferroelectric phase, although there is the possibility that the samples are polycrystalline.

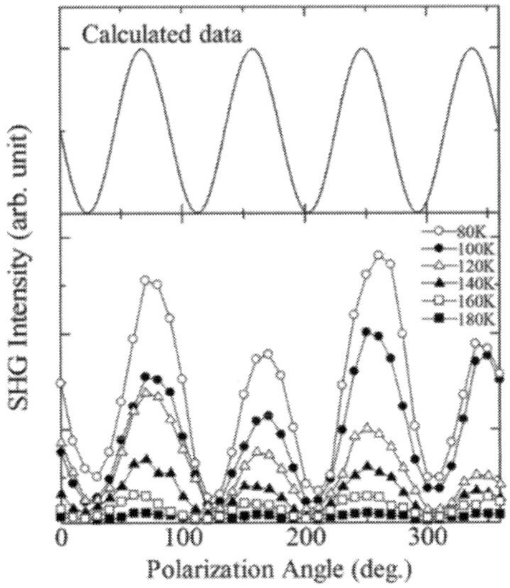

Figure 3. Fundamental light polarization dependence of the SHG intensity detected along a axis in the temperature range of 80 to 180 K in $TlInS_2$ in addition to the result calculated by eq. (2).

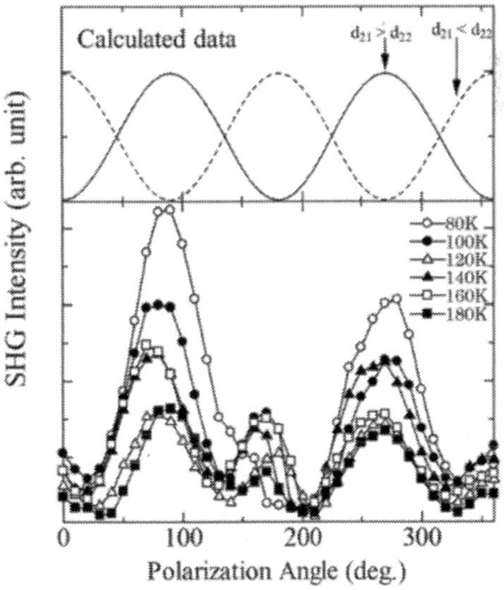

Figure 4. Fundamental light polarization dependence of the SHG intensity detected along b axis in the temperature range of 80 to 180 K in $TlInS_2$ in addition to the result calculated by eq. (3).

4. Conclusion

We performed an SHG experiment on layered $TlInS_2$ crystals in the temperature range of 77−300 K using a confocal laser microscope system. The SHG signal was observed in the low-temperature ferroelectric phase of the layered compounds, as expected. In addition, we investigated the polarization properties of SHG signals on $TlInS_2$ in the 80−180 K range. The results agree well with those of symmetric space group C_2^3 in the ferroelectric phase.

Acknowledgements

This work was supported in part by the Japan Ministry of Education, Culture, Sport, Science, and Technology (grant-in-aid No 23560371) and by the MEXT-supported Program for the Strategic Research Foundation at Private Universities, 2013−2017.

References

[1] L. G. Musaeva et al., *Sov. Phys. Solid State* **18** (1976) 822.
[2] H. Uchiki, D. Kanazawa, N. Mamedov, S. Iida, *J. Luminescence* **87-89** (2000) 664.
[3] K.R. Allakhverdiev et al., *Solid State Commun.* **96** (1995) 827.
[4] R. Yamabi and H. Uchiki, *Accomplishment Report of Professional Group of Ternary and Multinary Compounds* [in Japanese], (2001) 50.
[5] Y. Araki, R. Asaba, K. Wakita, Y.G. Shim, K. Mimura, N. Mamedov, *Phys. Status Solidi* C, **10** (2013) 1136.

Spectroscopic analysis of photo-induced deformation of amorphous carbon nitride films

T Harata, M Aono, N Tamura, N Kitazawa and Y Watanabe

Department of Materials Science and Engineering, National Defense Academy, Hashirimizu 1-10-20, Yokosuka, Kanagawa, 239-8686, Japan

E-mail: aono@nda.ac.jp

Abstract. Reversible photo-induced deformation of amorphous carbon nitride (a-CN_x) films was investigated. The films were deposited at 400 and 600 °C on a rectangular shaped ultrathin SiO_2 substrate by reactive radio frequency magnetron sputtering with graphite target and pure N_2 gas. The amount of deformation change was estimated from bending curvature of a-CN_x/SiO_2 bimorph structure. For the film deposited at 400 °C, it bent toward the film side under illumination. Photo-induced deformation of the film estimated from the bending angle was observed in the excitation energy of 1.77 to 3.35 eV. The maximum deformation was observed under illumination of 2.6 eV which corresponds with π-π^* bond of carbon nitride. In contrast, the deformation was hardly observed in the films deposited at 600 °C. The difference in these two films was probably due to difference in sp^2 bonding and termination structures.

1. Introduction

Amorphous carbon nitride, a-CN_x, is a promising candidate coating material for blood contacting devices because of its good mechanical and biologically compatible properties, such as high hardness [1], low friction coefficient [1, 2] and biocompatibility [3]. Furthermore, a-CN_x films behave as an n-type semiconductors. Thus, electrical and optical properties, such as low dielectric constant [4], electrical conductivity [5, 6], photoconductivity [5, 7, 8] and photoluminescence [9], were studied.

Recently, we revealed a reversible photo-induced deformation of a-CN_x thin films prepared by reactive radio frequency magnetron sputtering [10]. The photo-induced deformation of a-CN_x was observed as a bending of curvature of the ultrathin Si substrate under visible light irradiation [11]. In the future, the photo-induced deformation of a-CN_x films will have great possibility to apply for devices which utilize the photo-mechanical conversion.

Photo-induced phenomena have also been reported on amorphous semiconductor and certain organic materials. In amorphous semiconductors, such as hydrogenated amorphous silicon (a-Si:H) and amorphous arsenic sulfide (a-As_2S_3), photo-induced volume changes and structural changes [12, 13] have been reported during past decades. These phenomena were not reversible. On the other hand, azobenzene-containing materials show the reversible change in shape as the photo-mechanical effect due to the photo-isomerization. Under UV light irradiation, the *trans* configuration of azobenzene is able to change into *cis* configuration. And then, the *cis* form will return to the *trans* form by irradiating visible light [14, 15]. The photo-isomerization of azobenzene also induce the changes of the size from 9.0 Å in the *trans* form to 5.5 Å in the *cis* form [15].

To elucidate the mechanism of the photo-induced deformation of a-CN_x films, it is worthwhile to study the dependence of deformation on the incident photon energy. Thus, in this study, we investigate spectroscopic features of the photo-induced deformation for the a-CN_x films deposited at two different temperatures, which show different structures.

2. Experimantal

Amorphous carbon nitride, a-CN_x, thin films were prepared by reactive radio frequency (RF) magnetron sputtering. The target was graphite plate (purity 99.9999%) of 3 inches in diameter. The reactive gas was N_2 with the purity of 99.99%. N_2 gas flow and the pressure were kept constant at 3 sccm and 0.12 Torr, respectively. The RF power was 85 W. The a-CN_x films were deposited on rectangular shaped ultrathin substrates of SiO_2 glass. The size of the substrate was 30.0 x 2.0 x 0.05 mm^3. The deposition temperature was fixed at 400 and 600 °C.

The amount of deformation change was estimated from the amount of bending for the a-CN_x films/substrate system. One end of the a-CN_x films/substrate was clamped and displacement of the other free end was measured by the optical-lever technique in dark and under illumination with visible light. A He-Ne laser (wavelength 632.8 nm) was used as the probe light, and the laser beam reflected on the free end of the specimen and then the laser spot was observed on the scale. A Xe lamp, 150 W, was used to irradiate the a-CN_x specimen with the spectroscopic light in the wavelength range of 370 to 700 nm, corresponding to the photon energy of 1.77 to 3.35 eV. The absorption coefficient spectra were obtained from the transmittance spectra using UV-vis-NIR spectrophotometer (JASCO, V-570). X-ray photoelectron spectroscopy, Raman spectroscopy and FT-IR spectroscopy were performed on the films to study the nitrogen concentration, the chemical bonding states and the film structures. XPS was operated using Mg K_α radiation under 10^{-9} Pa (Phi, ESCA 1600). The nitrogen concentration was evaluated from the area ratio of N_{1s} to C_{1s} core level peaks. Raman spectra were obtained using Nd-YAG laser with the wavelength of 532 nm (Seishin Trading, RA-07F). Infrared absorption spectra were obtained using Fourier transform infrared spectrometer (Thermo Scientific, NICOLET 6700) in the range of 400 to 4000 cm^{-1} with a resolution of 4 cm^{-1} and averaging over 64 scans.

3. Results and discussion

From the XPS analysis, the nitrogen concentration (N/C) of the films deposited at 400 and 600 °C was evaluated to be 0.50 and 0.32, respectively. Figure 1a shows the XPS N_{1s} spectra from each film. The N_{1s} spectrum can be decomposed into mainly two peaks; N-sp^3C bonding state at 398.5 eV and N-sp^2C bonding state at 400.0 eV. The intensity ratio of N-sp^2C to N-sp^3C bonds (N-sp^2C/N-sp^3C) of the films was calculated to be 0.62 (400 °C) and 0.65 (600 °C). Raman spectra were shown in figure 1b. The obtained spectra were similar to the typical one of amorphous carbon film [16], which indicates the existence of amorphous carbon network in the films. In addition, these two spectra are similar each other. A intensity ratio of D to G peaks (I_D/I_G) was 0.99 (400 °C) and 1.10 (600 °C). The I_D/I_G ratio decreases with decreasing cluster size L_a according to Ferrari [17]. In contrast, FT-IR spectra in 2100 to 2300 cm^{-1} as shown in figure 1c, were different definitely. The absorption band at about 2200 cm^{-1} indicates an existence of -C≡N and -N≡C bondings which are like termination structures.

From these results, it is found that the features of the film deposited at 400 °C are relatively high nitrogen concentration, low sp^2 bonding fraction, small cluster size, and existence of termination structures compared to the film deposited at 600 °C.

Figure 2 shows an absorption coefficient of the a-CN_x films deposited at 400 and 600 °C obtained from the transmittance spectra. The absorption coefficient of the samples prepared at 400 °C decreased slowly from approximately 2.8 eV. On the film deposited at 600 °C, the absorption coefficient gradually decreased in whole region. In the case of amorphous carbon nitride films, the absorption coefficient increases with increasing nitrogen concentration and decreasing sp^2 bonding fraction. We estimated the Urbach energy which corresponds to the width of the band tail and can be evaluated as the width of localization. A plot of ln[α(E)] versus photon energy should be linear, where α(E) is the absorption coefficient. The Urbach energy can be obtained from the slope of the straight line, and was found to be 0.30 (400 °C) and 1.06 eV (600 °C). This means that the film deposited at 600 °C has high disordered structure compared to the film deposited at 400 °C.

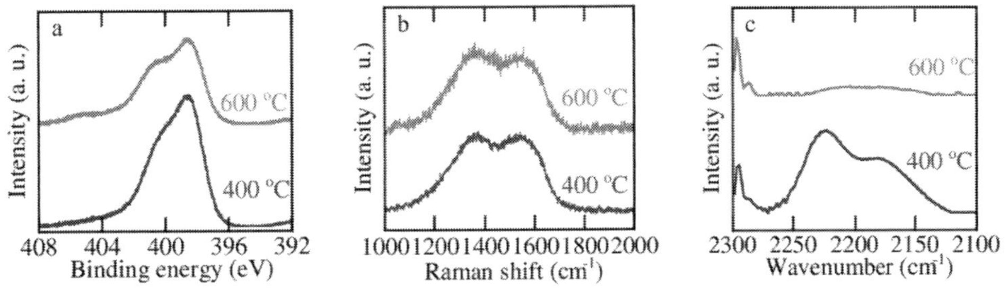

Figure 1. a) XPS N_{1s} spectra; b) Raman spectra; c) FT-IR spectra of the a-CN_x films deposited at 400 and 600 °C.

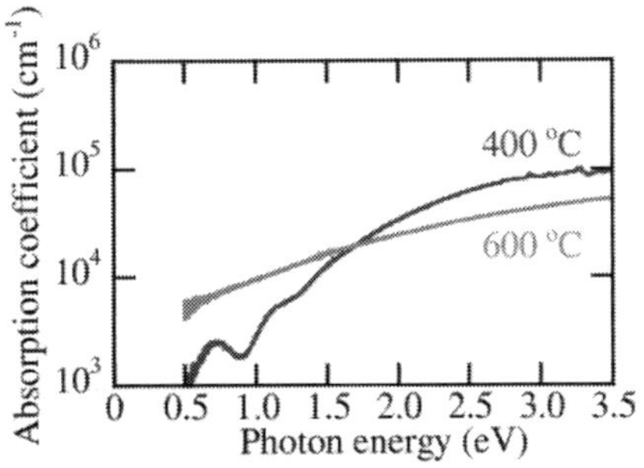

Figure 2. The absorption coefficient of a-CN_x films deposited at 400 and 600 °C.

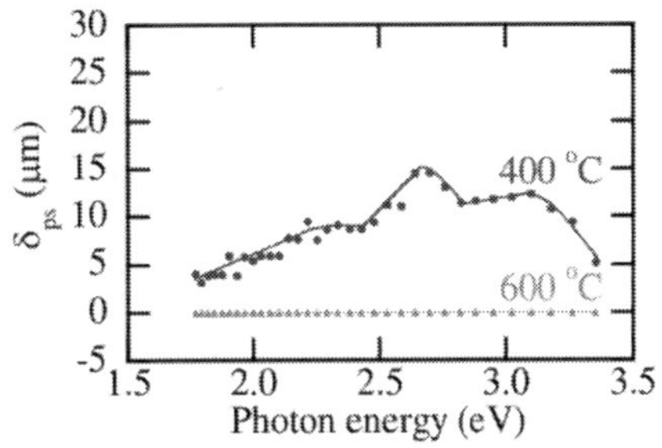

Figure 3. δ_{ps} as a function of incident photon energy. The solid lines are a guide for eyes.

Figure 3 shows a typical spectra of the amount of deformation (δ_{ps}) as a function of incident photon energy. The value of δ_{ps} was normalized by the incident light intensity. For the film deposited at 400 °C, the a-CN_x film/substrate system bent toward the film side when the incident light was irradiated from film side. And this deformation was observed reversibly against turning on and off the light. Moreover, the photo-induced deformation was observed in the whole region of the excitation photon energy (1.77 to 3.35 eV), which is related to the width of the localized states. The maximum value of δ_{ps} was discerned at the photon energy around 2.6 eV, which is close to π-π^* band gap energy in graphitic C_3N_4 [18]. This suggests that the deformation mainly is induced by sp^2C-N bond. On the other hand, δ_{ps} of the specimen deposited at 600 °C was below the detection limit of the present system in the whole region, and a total amount of the deformation under white illumination was almost zero.

Katsuno *et al.* [19] reported the spectrum of photoconductivity of sputtered a-CN_x film. A spectral shape of the photoconductivity at ~3.5 eV is similar to that of deformation obtained in this work. From this result, we surmise that both photoconductivity and photo-induced deformation simultaneously occur by photoexcitation between π and π^* states. We found that the ratio of photo to dark conductivity of the film deposited at 400 °C is 4.7, whereas in the case of the film deposited at 600 °C, it is approximately 100 [20]. We consider that photoconductivity and photo-induced deformation are competitive process and the former process is dominant in the film deposited at 600 °C. As described above, the film deposited at 600 °C has no termination structure, which behaves like traps of conduction electrons. Furthermore, high disordered structure and a large amount of sp^2 bonding can also contribute to high photoconductivity, because the sp^2 chain structures as conducting path increase with increasing the disorder [21]. From these results, we propose that the deformation is induced by mainly *sp*^{2} sites in the films and the existence of termination structure plays an important role for the deformation.

4. Conclusions

We investigated the spectroscopic analysis of photo-induced deformation of a-CN_x films in the range of 1.77 to 3.35 eV using optical-lever technique. The films were prepared on SiO_2 substrate by reactive radio frequency magnetron sputtering at the deposition temperatures of 400 and 600 °C. For the film deposited at 400 °C, the photo-induced deformation was observed in the whole range of the excitation energy. The maximum of the deformation was observed at about 2.6 eV. This energy is close to the π-π^* band gap of graphitic C_3N_4. On the other hand, the displacement of the specimen deposited at 600 °C was observed below detection limit of the measurement system.

Acknowledgements

The authors would like to thank Dr. Hisashi Miyazaki and Dr. Hiroaki Kishimura for useful discussion. This work was supported by JSPS KAKENHI Grant Number 26790054.

References

[1] Hajek V, Rusnak K, Vlcek J, Martinu L and Haw-thorne H M 1997 *Wear* **213** 80
[2] Wang P, Hirose M, Suzuki Y and Adachi K 2013 *Surf. Coat. Technol.* **221** 163
[3] Cui F Z and Li D J 2000 *Surf. Coat. Technol.* **131** 481
[4] Aono M, Naruse Y, Nitta S and Katsuno T 2001 *Diamond Relat. Mater.* **10** 1147
[5] Takada N, Arai K, Nitta S and Nonomura S 1997 *Appl. Surf. Sci.* **113/114** 274
[6] Tamura N, Aono M, Kishimura H, Kitazawa N and Watanabe Y 2012 *Jpn. J. Appl. Phys.* **51** 121401-1
[7] Katsuno T, Nitta S, Habuchi H, Stolojan V and Silva S R P 2004 *Appl. Phys. Lett.* **85** 2803
[8] Aono M, Goto T, Tamura N, Kitazawa N and Watanabe Y 2011 *Diamond Relat. Mater.* **20** 1208
[9] Fanchini G, Ray S C and Tagliaferro A 2003 *Diamond Relat. Mater.* **12** 1084
[10] Aono M, Harata T, Kitazawa N and Watanabe Y 2014 *Diamond Relat. Mater.* **41** 20
[11] Harata T, Aono M, Kitazawa N and Watanabe Y Correlation of Photothermal Conversion on the Photo-iduced Deformation of Amorphous Carbon Nitride Films Prepared by Reactive Sputtering *Appl. Phys. Lett.* under review
[12] Gotoh T, Nonomura S, Nishio M, Masui N, Nitta S, Kondo M and Matsuda A 1998 *J. Non-Cryst. Solids* **227-230** 263
[13] Tanaka K 1975 *Appl. Phys. Lett.* **26** 243
[14] Barrett C J, Mamiya J, Yager K G and Ikeda T 2007 *Soft Matter* **3** 1249
[15] Merio E and Ribagorda M 2012 *Beilstein J. Org. Chem.* **8** 1071
[16] Ferrari A C 2002 *Diamond Relat. Mater.* **11** 1053
[17] Ferrari A C 2000 Phys. Rev. B **61** 14095
[18] Zhang J, Li Y, Zhu P, Huang D, Wu S, Cui Q and Zou G 2011 *Diamond Relat. Mater.* **20** 385
[19] Katsuno T, Nitta S, Habuchi H, Iwasaki T, Itoh T and Nonomura S 2000 *Mat. Res. Soc. Symp. Proc.* **593** 499
[20] Iwasaki C, Aono M, Tamura N, Kitazawa N and Watanabe Y 2012 *Jpn. J. Appl. Phys.* **51** 10NE26
[21] Majumdar A, Das S C, Bogdanowicz R, Shripathi T, Langel W and Hippler R 2013 *Vib. Spectrosc.* **66** 63

Photoelectric conversion characteristics of c-Se-based thin-film photodiodes in imaging device

S Imura[1,2], K Kikuchi[1], K Miyakawa[1], H Ohtake[1] and M. Kubota[1]

[1]NHK Science and Technology Research Laboratories, Kinuta, Setagaya-ku, Tokyo 157-8510, Japan

E-mail: imura.s-la@nhk.or.jp

Abstract. Herein, we report the use of high-efficiency crystalline-selenium-based (c-Se-based) thin-film heterojunction photodiodes in imaging devices. As a novel experiment, we use an image pickup tube with a photoelectric conversion layer consisting of n-gallium oxide (Ga_2O_3)/p-c-Se heterojunction photodiodes to obtain high-resolution images at a relatively low applied voltage. We reduce the thickness of the Ga_2O_3 layer to expand the depletion layer into the c-Se layer at a lower applied voltage. In addition, Sn-doping of the Ga_2O_3 layer effectively increases the carrier concentration, thereby allowing the photodiode to operate at lower voltage.

1. Introduction

Recently, the development of high-resolution high-speed complementary metal-oxide semiconductor-based (CMOS-based) image sensors has rapidly increased the demand for highly sensitive visible-light photodetectors. Satisfying this demand requires a significant enhancement of the sensitivity of photodetectors beyond what is possible with conventional silicon-based photodiodes. Motivated by this demand, we are developing a CMOS image sensor overlaid with a photoelectric conversion layer, which would allow satisfying the various requirements of high sensitivity, low crosstalk, high dynamic range, and so on.

Because of the excellent spectral response of crystalline-selenium-based (c-Se-based) photodiodes over almost the entire visible spectrum, they are promising candidates for fabricating highly sensitive visible-light photodetectors. In addition, because of the low crystallization temperature of selenium [1], all fabrication processes may be done below 200°C, which is advantageous for CMOS manufacturing.

In this study, n-Ga_2O_3/p-c-Se heterojunction photodiodes are used to demonstrate the possibility of high-efficiency visible-light photodetectors operating at a low applied voltage. Furthermore, in a novel experiment using a pickup tube with a photosensitive layer consisting of Ga_2O_3/c-Se photodiodes, we obtain high-resolution images at a relatively low voltage.

2. Experimental procedure

Figures 1(a) and 1(b) show the cross section of a pickup-tube structure and a cell structure, respectively. For the pickup tube, an n-type hole barrier was made by depositing 20-nm-thick Ga_2O_3 films over indium-tin-oxide (ITO) films on glass substrates by room-temperature radio-frequency

[2] To whom any correspondence should be addressed.

Journal of Physics: Conference Series **619** (2015) 012008 doi:10.1088/1742-6596/619/1/012008

magnetron sputtering. Next, sub-nanometer-thick tellurium (Te) films and 500-nm-thick a-Se films were successively deposited by room-temperature vacuum evaporation. The amorphous selenium (a-Se) was completely converted to c-Se by annealing at 200°C. Finally, beam-landing layers made of 10-nm-thick di-antimony tri-sulfide (Sb_2S_3) porous films deposited by vacuum evaporation served to prevent the secondary-electron emission.

The cells had the same structure and materials as the pickup tubes except that the di-antimony tri-sulfide films were replaced by 30-nm-thick ITO films deposited by direct-current magnetron sputtering to serve as transparent top electrodes.

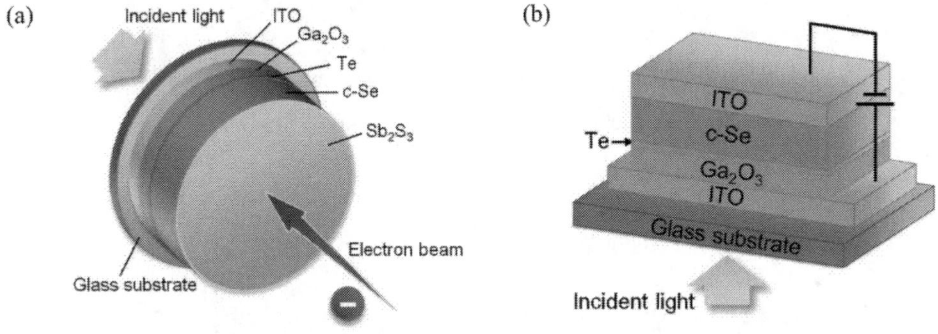

Figure 1. Cross-sectional schematic of (a) pickup tube structure and (b) cell structure.

3. Results and Discussion

3.1. Imaging experiment with pickup tube

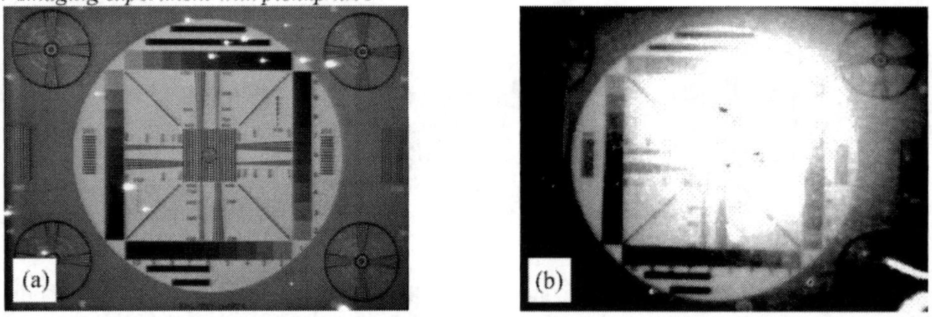

Figure 2. Images from pickup tube incorporating (a) Ga_2O_3/c-Se pn photodiode and (b) ITO/c-Se Schottky photodiode at an applied voltage of 30 V.

Figures 2(a) and 2(b) compare an image acquired with a pickup tube containing Ga_2O_3/c-Se pn photodiodes and that containing ITO/c-Se Schottky photodiodes, respectively. This comparison clearly shows that a lower dark current and a higher resolution results from using Ga_2O_3/c-Se pn photodiodes rather than ITO/c-Se Schottky photodiodes. To obtain a resolved image with the pickup tube, the applied voltage must be increased to spread the depletion layer from the Ga_2O_3/c-Se interface into the c-Se layer, which is scanned by the electron beam. However, when using Schottky photodiodes, the potential barrier between ITO and c-Se is too low to prevent carriers from being injected into the c-Se from the ITO. At high electric fields, these carriers are the main contributors for the dark current. As a result, the image of Figure 2(b) is not clearly resolved. In contrast, when using Ga_2O_3/c-Se pn photodiodes, the potential barrier between ITO and Ga_2O_3 is extremely high because of the large band gap of Ga_2O_3 (4.9 eV) [2], leading to low dark current at high electric fields.

However, these devices still require a relatively high operation voltage (~30 V). In the next section, we discuss how to reduce this voltage to a level acceptable for CMOS circuits.

3.2. Influence of thickness and carrier concentration on operating voltage

We prepared the cell structures consisting of Ga_2O_3 layers with varying thickness (5, 10, and 20 nm) and Sn concentration (0, 5, and 5.6 mol%) to study how the thickness and carrier concentration of the Ga_2O_3 layer affects the threshold voltage for saturation [see Figure 1(b)]. Figures 3(a) and 3(b) show the current–voltage (I–V) characteristics of an illuminated Ga_2O_3/c-Se heterojunction photodiode. The results show that the threshold voltage for saturation decreases with decreasing thickness of the Ga_2O_3 layer and with increasing Sn concentration in the Ga_2O_3 layer. Because of the low carrier concentration in the Ga_2O_3 layer, the depletion layer mainly spreads through the Ga_2O_3 layer at a lower voltage. Decreasing the thickness of the Ga_2O_3 layer and increasing the Sn concentration leads to an increased carrier concentration in this layer [3], thus requiring a smaller applied voltage to spread the depletion layer into the c-Se layer.

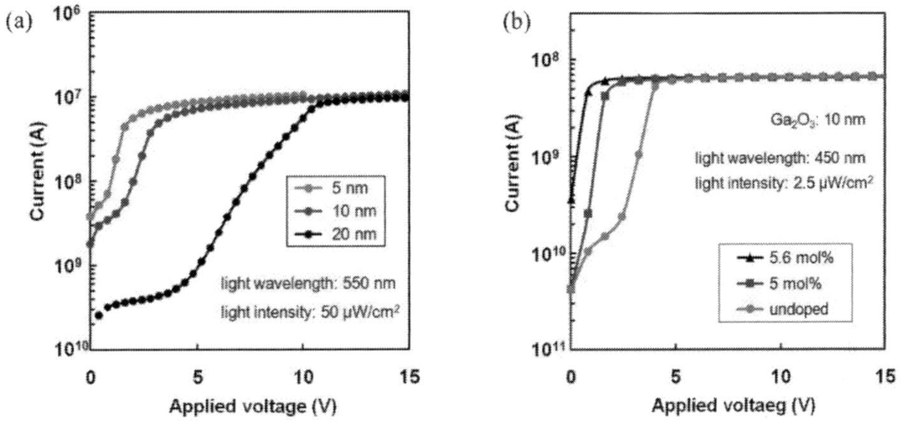

Figure 3. Photocurrent as a function of applied voltage for (a) Ga_2O_3/c-Se photodiode with varying Ga_2O_3-layer thickness (5, 10, and 20 nm) and (b) Sn-doped Ga_2O_3/c-Se photodiode with different Sn-doping concentrations (0, 5, and 5.6 mol%).

4. Conclusion

We report high-efficiency c-Se-based heterojunction photodiodes. By using a pickup tube with a photosensitive layer consisting of n-Ga_2O_3/p-c-Se heterojunction photodiodes, we obtained novel high-resolution images for such a technique. In addition, by decreasing the thickness of the Ga_2O_3 layer and increasing the Sn concentration, we made photodetectors that operate at lower voltages, which is advantageous for CMOS-based solid-state imaging devices.

References

[1] Stuhl D R and Sinke G C *Advances in Chemistry Series No. 18* 1965 (Washington, DC: American Chemical Society).
[2] Tippins H H 1965 *Phys. Rev.* **140** A316.
[3] Ueda N, Hosono H, Waseda R and Kawazoe H 1997 *Appl. Phys. Lett.* **70** 3561.

Properties of Ga_2O_3/Ga_2O_3:Sn/CIGS for visible light sensors

K Kikuchi[1], S Imura, K Miyakawa, H Ohtake, M Kubota

NHK Science and Technology Research Laboratories
1-10-11, Kinuta, Setagaya-ku, Tokyo, 157–8510, Japan

E-mail: kikuchi.k-oa@nhk.or.jp

Abstract. A Ga_2O_3/Ga_2O_3:Sn/$CuIn_{1-x}Ga_x(Se_{1-y}S_y)_2$(CIGS) structure is proposed for use in visible light sensors. CIGS thin films have been investigated for high efficiency thin-film solar cells. The typical structure of CIGS solar cells consists of ZnO/CdS/CIGS. However, the quantum efficiency of this structure at short wavelengths is decreased due to optical absorption loss from the CdS layer. In this study, a Ga_2O_3 layer was adopted instead of a CdS layer. Ga_2O_3 has a wide band gap and high transmittance in the visible region. Furthermore, it was assumed that it could function as a hole-blocking layer to suppress hole injection from the anode. Additionally, a Ga_2O_3:Sn thin layer was laid on the CIGS layer to spread the depletion region to the CIGS layer, because the carrier density of the undoped Ga_2O_3 layer was much lower than that of the CIGS layer. In this structure, signal current multiplication and quantum efficiency greater than unity were observed at applied voltages over 3.5 V.

1. Introduction

$CuIn_{1-x}Ga_x(Se_{1-y}S_y)_2$ (CIGS) chalcopyrite thin films have been used for thin film solar cells with very high energy conversion efficiency of over 20 % [1, 2]. They have high quantum efficiencies and high absorption coefficients. Moreover, the band gap is adjustable by varying the composition ratios [3, 4]. CIGS films have great potential for improving the sensitivity of image sensors. For solar cells, a cadmium sulfide (CdS) buffer layer is usually applied to a CIGS film. However, cadmium is undesirable from the viewpoint of environmental safety. Also, this structure has low sensitivity to blue light, because the band gap of CdS is 2.4 eV [5]. This is especially a problem for solar cells and visible light sensors. For these reasons, much effort has been expended on finding a Cd-free replacement material for solar cells [6, 7]. Another problem is that the dark currents of CIGS films are high because of their low resistivities [8]. We used gallium oxide (Ga_2O_3) instead of CdS, as Ga_2O_3 has a wide band gap of 4.9 eV and high transmittance of visible light [9]. Higher short-wavelength quantum efficiency can be expected by widening the band gap beyond that of CdS. Moreover, we assumed that the Ga_2O_3 thin film would function as a hole-blocking layer for CIGS films and suppress the injection of holes from the electrode [10]. A Sn-doped Ga_2O_3 (Ga_2O_3:Sn) layer was layered between the CIGS and Ga_2O_3 layers to spread the depletion region to the CIGS layer, because the carrier density of the undoped Ga_2O_3 layer was much lower than that of CIGS layer. In this study, the electric characteristics of Ga_2O_3/Ga_2O_3:Sn/CIGS structure were investigated.

[1] To whom any correspondence should be addressed.

2. Experimental

The Ga_2O_3:Sn (50 nm)/CIGS (1µm) structures were formed on Mo (1µm)-coated alkali-free Corning glass substrates (EAGLE-XG) in a procedure similar to those in our previous studies [11, 12]. Then, Ga_2O_3 films (100 nm) were deposited by pulsed laser deposition (PLD) from a Ga_2O_3 target using an Nd:YAG laser (wavelength of 213 nm; repetition rate of 1.0 Hz). The substrate temperature during PLD was 295 K. Finally, a front contact 30-nm ITO film was deposited by direct current (DC) magnetron sputtering onto each sample in Ar/O2 1% atmosphere. The optical transmittance and reflectance of the CIGS and Ga_2O_3 films were measured with a spectrophotometer to estimate their optical band gaps. The energy difference between the Fermi level and valence band for each film was analyzed by X-ray photoelectron spectroscopy (XPS).

3. Results and discussion

Figure 1 shows the energy band diagram of Ga_2O_3 and $CuIn_{0.5}Ga_{0.5}S_2$. The work function of Ga_2O_3 and electron affinity of CIGS was taken from other studies [13, 14]. As observed in the diagram, Ga_2O_3 should have a high potential barrier against holes from the anode.

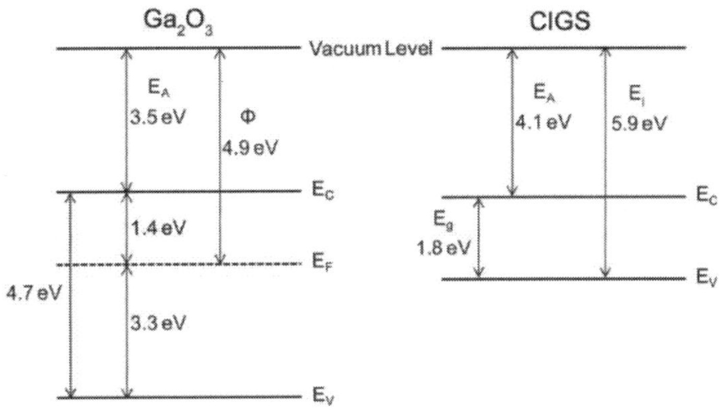

Figure 1. Energy band diagram of Ga_2O_3 and CIGS.

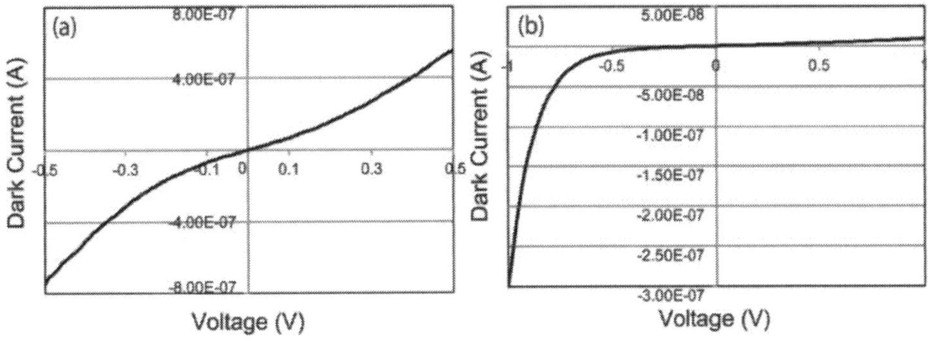

Figure 2. The dark current–voltage characteristics for the (a) Ga_2O_3:Sn/CIGS and (b) Ga_2O_3/Ga_2O_3:Sn/CIGS structures.

Journal of Physics: Conference Series **619** (2015) 012009 doi:10.1088/1742-6596/619/1/012009

Figure 2 plots the dependence of the dark current on the voltage for the Ga_2O_3:Sn/CIGS and Ga_2O_3/Ga_2O_3:Sn/CIGS structures. These results indicate that the Ga_2O_3 layer improved the diode characteristics, because the Ga_2O_3 layer increased the resistance in the shunt path and decreased the leakage current.

The current–voltage characteristics and quantum efficiency of the Ga_2O_3/Ga_2O_3:Sn/CIGS structure is shown in Figure 3. The signal current from irradiated visible light (450 nm, 550 nm, and 650 nm) significantly increased at applied voltages over ~3.5 V. We consider this carrier multiplication was due to avalanche current or tunneling current. The quantum efficiency exceeded unity due to carrier multiplication at applied voltages over 3.5 V. These results indicate that the Ga_2O_3/Ga_2O_3:Sn/CIGS structure had a very high sensitivity to visible light. We deposited our CIGS films by one-step RF sputtering, which increased the number of sulfur vacancies, because the vapor pressure of sulfur is very high. We confirmed that there were many sulfur deficiencies in our CIGS film in a previous study [11]. This means that this structure could have a higher quantum efficiency and lower dark current if a higher quality CIGS film is used.

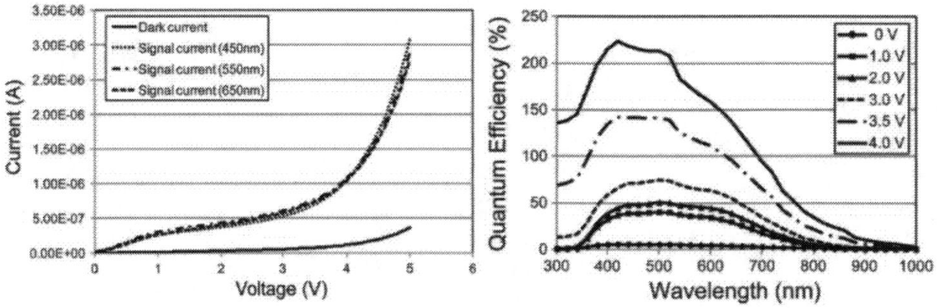

Figure 3. The current–voltage characteristics and quantum efficiency of the Ga_2O_3/Ga_2O_3:Sn/CIGS structure.

4. Conclusion

The electrical properties of Ga_2O_3/Ga_2O_3:Sn/CIGS heterojunctions were investigated. We succeeded in improving the pn characteristics by applying Ga_2O_3 onto the Ga_2O_3:Sn layer. The dark current of this structure was very low because of the high potential barrier of Ga_2O_3 against holes. The carrier multiplication phenomenon was observed, which is beneficial for high-sensitivity sensors. Further optimization in the fabrication of the CIGS film is needed for higher sensitivity.

5. Acknowledgments

This work was partly supported by the Foundation for Promotion of Material Science and Technology of Japan (MST Foundation).

6. References

[1] M. A. Green, K. Emery, Y. Hishikawa and W. Warta 2009 Prog. Photovolt. **17** 320

[2] N. Nakamura, N. Yoneyama, K. Horiguchi, Y. Iwata, K. Yamaguchi, H. Sugimoto and T. Kato 2014 Proc. 40th IEEE Photovolt. Specialist Conf. 0107

[3] Turcu M, Kotschau I M and Rau U 2001 *Appl. Phys.* A **73** 769

[4] Chirila A et al 2013 *Nature Materials* 12 1107

[5] Pudov A, Sites J, Nakada T 2002 *Jpn. K. Appl. Phys.* **41** L672

[6] Nakada T, Furumi K and Kunioka A 1999 *IEEE T. Electron Dev.* **46** 2093

[7] Hultqvist A, Platzer-Bjorkman C, Coronel E and Edoff M 2011 *Sol. Energy Mater. Sol. Cells* **95** 497
[8] Tanaka K, Kosugi M, Ando F, Ushiki T, Usui H and Sato K 1993 *Jpn. J. Appl. Phys. Suppl.* **32** 113
[9] Varley J B, Weber J R, Janotti A and Van de Walle C G 2010 *Appl. Phys. Lett. 97* 142106
[10] Kikuchi K, Ohkawa Y, Miyakawa K, Matsubara T, Tanioka K, Kubota M, Egami N 2011 *Phys. Stat. Sol.* C **8** 2800
[11] Kikuchi K, Imura S, Miyakawa K, Kubota M and Ohta E 2014 *Thin Solid Films* **550** 635
[12] Kikuchi K, Imura S, Miyakawa K, Ohtake H, Kubota M and Ohta E 2014 *2013 MRS Fall Meeting* 1635
[13] Minami T, Miyata T and Yamamoto T 1998 *Surf. Coat. Technol.* **108** 583
[14] Sugiyama M, Nakanishi H and Chichibu S 2001 *Jpn. J. Appl. Phys.* **40** L428

Radiative coupling between multi-component excitons in ZnO thin films

Takashi Kinoshita and Hajime Ishihara

Department of Physics and Electronics, Osaka Prefecture University, 1-1 Gakuen-cho, Naka-ku, Sakai, Osaka 599-8531, Japan

E-mail: kinoshita@pe.osakafu-u.ac.jp

Abstract. We study the linear optical responses of of ZnO, in which the degeneracy of the p-like states at the Γ point is lifted by the crystal field and spin-orbit interaction, and the $n = 1$ (A- and B-) excitons are optically allowed under the condition of $\boldsymbol{E} \perp \boldsymbol{c}$. The coupled system of the light and multi-component excitons forms the three branches, i. e., lower, middle, and upper branches through the radiative coupling between A- and B-excitons. We theoretically reveal the ratio of the components of A- and B-excitons in the respective branches which is determined from diagonalization of self-consistent equation of the nonlocal induced-polarization and Maxwell electric field.

1. Introduction

Wurtzite ZnO is one of the promising materials for opto-electronic devices because of its high excitonic stability due to the large exciton binding energy. In wz-ZnO, the crystal field and spin-orbit interaction lifts the degeneracy of the p-like states at the Γ point of valence bands. We present a theoretical study to reveal that the main quantum number $n = 1$ (namely A- and B-) excitons are coupled via radiation in bulk system and thin films. In particular for thin films, radiative properties, such as the radiative shift, and the radiative width of diagonalized eigenstates, which appear in the optical spectra, vary with the film thickness according to the thickness-dependent ratio of the components of A- and B-excitons. This situation is in striking contrast with the picture in conventional analysis to identify a single component solely contributing to each spectral peak. In this study, we consider the film thickness of weak confinement regime where the centre-of-mass (c. m.) motions of excitons are confined. Especially, we focus on the regime beyond the long wavelength approximation (LWA) where the microscopic nonlocal response plays an important role. In such regime, A- and B-exciton are coupled in a drastic way through the wave-wave coupling between radiation and excitons [1], forming upper branches (UBs), middle branches (MBs), and lower branches (LBs) as observed in cavity polaritons [2]. Further, the phase-matched modes exhibit ultrafast radiative decay time. This study will change the conventional view of ZnO excitons in film samples, which will be definitely helpful for future development of optical devices of this material.

2. Theoretical framework

As a model system, we considered c. m. motions of excitons are along the z-axis (the optical axis). In the case that relative motions of electron and hole are regarded as the same with that in bulk system, the transition matrix element of the polarization operator $\rho_{\sigma\lambda}(z)$ is written as $\rho_{\sigma\lambda}(z) = \sqrt{\frac{E_\sigma f_\sigma}{8\pi}} g_{\sigma\lambda}(z)$, where σ is a band index, E_σ is the bulk transverse energy of σ exciton, f_σ is the oscillator strength, and $g_{\sigma\lambda}(z)$ is the wave function of excitons' c. m. motion. According to the Kubo formula in the site representation, the induced polarization $P(z, \omega)$ is described as $P(z, \omega) = \int dz' \chi(z, z'; \omega) E(z', \omega)$, where the nonlocal susceptibility $\chi(z, z'; \omega)$ is given as

$$\chi(z, z'; \omega) = \sum_\sigma \sum_\lambda \frac{\rho_{\sigma\lambda}(z)\rho^*_{\sigma\lambda}(z')}{E_{\sigma\lambda} - \hbar\omega - i\Gamma_\sigma}. \tag{1}$$

In this expression, Γ_σ is a dephasing constant, and $E_{\sigma\lambda}$ is the energy of excitons in the framework of standard effective-mass approximation written as $E_{\sigma\lambda} = E_\sigma + \frac{\hbar^2 k^2_{\sigma\lambda}}{2M_\sigma}$, where $k_{\sigma\lambda}$ is the λth quantized wave vector, and M_σ is the total mass of the exciton parallel to z-axis. We can self-consistently solve this constitutive equation and the Maxwell equation, given by $-\frac{\partial^2}{\partial z^2}E(z, \omega) - q^2\epsilon_b E(z, \omega) = 4\pi q^2 P(z, \omega)$, where q ($= \omega/c$) is the wavenumber of light and ϵ_b is the background dielectric constant. This solution is based on the nonlocal response theory [3], and in particular, we extended the method in Ref. [4] to the multi-component excitons. Using Green's function $G(z, z'; \omega)$ [5], the electric field $E(z, \omega)$ is written as follows: $E(z, \omega) = E^0(z, \omega) + 4\pi q^2 \int dz G(z, z'; \omega) P(z', \omega)$, where $E^0(z, \omega)$ is the electric field of the incident light. We derived a simultaneous system of linear equations [4]:

$$(E_{\sigma'\lambda'} - \hbar\omega - i\Gamma_{\sigma'})H_{\sigma'\lambda'} + \sum_\sigma \sum_\lambda A_{\sigma'\sigma\lambda'\lambda} H_{\sigma\lambda} = F^{(0)}_{\sigma'\lambda'}, \tag{2}$$

where $H_{\sigma\lambda} \equiv (E_{\sigma\lambda} - \hbar\omega - i\Gamma_\sigma)^{-1} \int dz \rho^*_{\sigma\lambda}(z) E(z, \omega)$ and $F^{(0)}_{\sigma\lambda} \equiv \int dz \rho^*_{\sigma\lambda}(z) E^0(z, \omega)$. In Eq. (2), $A_{\sigma'\sigma\lambda'\lambda}$ indicates the interaction between the induced polarizations originating from the λ'th states of the σ' exciton and the λth states of the σ exciton.

Equation (2) can be rewritten in a matrix form as $\boldsymbol{SH} = \boldsymbol{F}^{(0)}$. Det[$\boldsymbol{S}$] provides the complex eigenmodes Ω of the exciton-radiation coupled system. The real part of the eigenmodes Re[$\hbar\Omega$] provides the eigenenergies of the excitons including the radiative shift. The imaginary part of the eigenmodes Im[$\hbar\Omega$] indicates the radiative width. We defined the diagonalized density $D_{\sigma\lambda}$ as

$$D_{\sigma'\lambda'} = \frac{|H_{\sigma'\lambda'}|^2}{\sum_{\sigma\lambda} |H_{\sigma\lambda}|^2}. \tag{3}$$

This quantity indicates the ratio of the λ'th σ' exciton component.

3. Exciton-polaritons in bulk system

In bulk system, the wavefunction of c. m. motion of free excitons is written as $g_{\sigma\lambda}(z) = \sqrt{\frac{1}{L}} e^{ik_\lambda z}$ where L is normalization factor, and k_λ satisfies the periodic boundary condition. We use Green's function of one-dimension free space in the wavenumber representation as $\tilde{G}(k, k'; \omega) = -\frac{\delta(k-k')}{\epsilon_b q^2 - k^2}$. Radiative correction $A_{\sigma'\sigma\lambda'\lambda}$ is written as

$$A_{\sigma'\sigma\lambda'\lambda} = \frac{q^2}{2} \frac{\sqrt{f_{\sigma'} f_\sigma E_{\sigma'} E_\sigma}}{\epsilon_b q^2 - k^2} \delta_{\lambda', \lambda}. \tag{4}$$

$A_{\sigma'\sigma\lambda'\lambda}$ has the finite value under the condition that both σ' and σ exciton have the same wavenumber ($\lambda' = \lambda$), even though σ' is different from σ. Therefore, A- and B-excitons couple with each other via radiation even in bulk system. Det[$\boldsymbol{S}$] provides the exciton-polariton dispersion relation in bulk system as follows

$$\frac{c^2k^2}{\omega^2} = \epsilon_b + \frac{\frac{1}{2}f_A E_A}{E_A + \frac{\hbar^2k^2}{2M_A} - \hbar\omega - i\Gamma_A} + \frac{\frac{1}{2}f_B E_B}{E_B + \frac{\hbar^2k^2}{2M_B} - \hbar\omega - i\Gamma_B}. \quad (5)$$

Assuming a ZnO sample, we use the following parameters, $E_A = 3.3758$ eV, $E_B = 3.3810$ eV [6], $\epsilon_b = 3.75$, $f_A = 0.0125$, $f_B = 0.0125$, $\Gamma_A = \Gamma_B = 0$ meV, $M_A = 1.0m_0$, $M_B = 0.8m_0$, where m_0 is the electron rest mass.

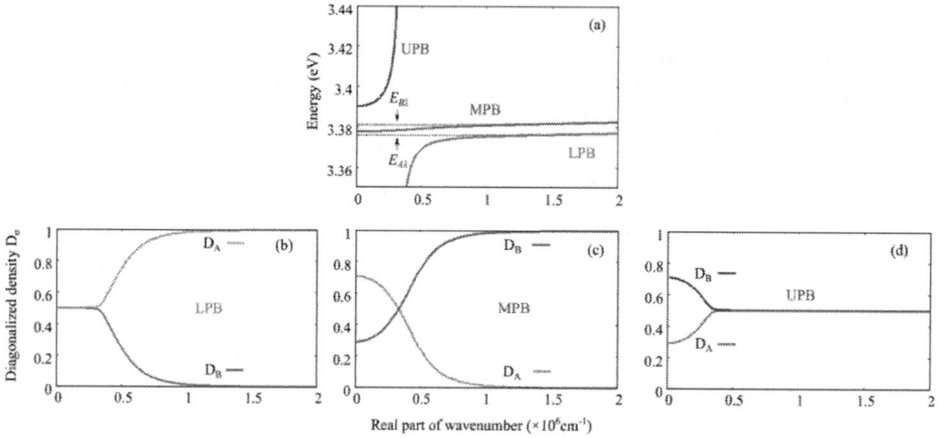

Figure 1. (a) Exciton-polariton dispersion relation obtained from Eq. (5), and wavenumber dependence of the diagonalized density D_σ for (b) lower polariton branch (LPB), (c) middle polariton branch (MPB), (d) upper polariton branch (UPB).

Three exciton-polariton branches, namely LPB, MPB, and UPB, appear in the dispersion relation in Figs. 1(a). The diagonalized density $D_{\sigma\lambda}$ that indicates the ratio of A- and B-component for each branch is shown in Figs. 1(b), (c), and (d). We found that $D_A : D_B \approx 0.7 : 0.3$ for MPB and $D_A : D_B \approx 0.3 : 0.7$ for UPB at $k = 0$. This means that the radiative coupling effect between A- and B-exciton is significant even in bulk system. We should remark that LT-splitting energy of σ exciton is not proportional to oscillator strength f_σ due to this effect.

4. A and B excitons including radiative correction in thin films

We considered a thin film whose film thickness is d. In order to address the distortion of wavefunctions near the surface of the film, we employed the Transition Layer Model (TLM) [7]. In this model, $k_{\sigma\lambda}$ satisfies the following quantization condition: $k_{\sigma\lambda}d - 2\tan^{-1}(k_{\sigma\lambda}/P_\sigma) = \lambda\pi$ $(\lambda = 1, 2, \cdots)$, where P_σ ($P_A = 1/1.6$ cm^{-1}, $P_B = 1/1.8$ cm^{-1}) is a decay constant of evanescent wave of c. m. motion (see Ref [7]). Radiative correction $A_{\sigma'\sigma\lambda'\lambda}$ is written as

$$A_{\sigma'\sigma\lambda'\lambda} = -4\pi q^2 \int\int \rho^*_{\sigma'\lambda'}(z)G(z, z'; \omega)\rho_{\sigma\lambda}(z')dzdz', \quad (6)$$

where $G(z, z'; \omega)$ is Green's function for planar structure in the site representation [5].

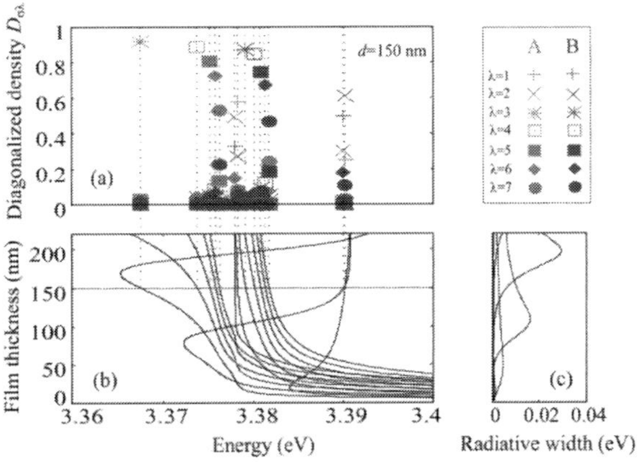

Figure 2. (a) Diagonalized density $D_{\sigma\lambda}$ for the film thickness $d = 150$ nm, which indicates the ratio of λth σ exciton component. Red (blue) dots means A- (B-) exciton components. (b) Eigen energies of excitons including radiative correction obtained from Det[$\boldsymbol{S}$]. Horizontal dashed line means the film thickness 150 nm. (c) Radiative widths.

Figures 2 show (a) diagonalized density $D_{\sigma\lambda}$ for the film thickness $d = 150$ nm, which indicates the ratio of λth σ exciton component, (b) eigen energies of excitons including radiative correction obtained from Det[$\boldsymbol{S}$], and (c) radiative widths. Each exciton mode shows a large and strongly size-dependent radiative shift, depending on the radiation phase matching [1]. Additionally, the radiative width of each mode attains a maximum value at the best radiation phase matching (Figs. 2(c)), which leads to ultrafast radiative decays. In regard to the ratio of the component (Figs. 2 (a)), $\lambda = 1, 2$th A and B exciton have significant values in MBs and UBs, though A-exciton components are dominant in LBs.

5. Conclusion

Eigenmodes and its ratio of each (A- and B-) exciton component in bulk system and thin film structures for ZnO are theoretically revealed by the nonlocal response theory. In bulk system, both A- and B-exciton components have significant density D_σ in three branches, namely LPB, MPB, and UPB, even for $k = 0$. In thin film structures, the radiative coupling between A and B components strongly affects the formation of mode structures. Due to this effect, a very large radiative shift and width are realized. From the aspect of ratio of the components, both A- and B-exciton components have significantly high rate in MBs and UBs.

These results are also expected to be important in the discussion of the valence-band-ordering of ZnO [8] because the density of each A- and B-exciton components through the radiative coupling determine the ratio of the components in the optical spectra.

References

[1] M. Ichimiya, M. Ashida, H. Yasuda, H. Ishihara, and T. Itoh, 2009 Phys. Rev. Lett. **103**, 257401
[2] T. Kawase, D. Kim, K. Miyazaki, and M. Nakayama, 2011 Phys. Stat. Sol. B **248**, No.2, 460
[3] K. Cho, 1986 J. Phys. Soc. Jpn. **55**, 4113
[4] H. Ishihara, J. Kishimoto,and K. Sugihara, 2004 J. Lumin. **108**, 343
[5] W. C. Chew: Waves and Fields in Inhomogeneous Media (Wiley-IEEE PRESS, 1995).
[6] J. Lagois, 1981 Phys. Rev. B **23**, 5511
[7] A. D'Andrea and R. Del Sole, 1982 Phys. Rev. B **25**, 3714
[8] K. Hazu, S. F. Chichibu, S. Adachi, and T. Sota, 2012 J. Appl. Phys. **111**, 093522

Optimisation of 1.3-μm InAs/GaAs Quantum-Dot Lasers Monolithically Grown on Si Substrates

Mingchu Tang,* Siming Chen, Jiang Wu, Qi Jiang, Dongyoung Kim, Alwyn Seeds, Huiyun Liu

Department of Electronic and Electrical Engineering, University College London, London WC1E 7JE, United Kingdom

*mingchu.tang.11@ucl.ac.uk

Abstract. We present a study of 1.3-μm InAs/GaAs quantum dot lasers monolithically grown on Si substrates by molecular beam epitaxy. We focused on the optimization of III-V buffer layers epitaxy grown on Si substrates, which includes the nucleation layers and the dislocation filter layers. The effect of growth temperature of GaAs nucleation layer has been investigated. Additionally, InAlAs/GaAs and InGaAs/GaAs strained layer superlattices (SLSs) are compared as dislocation filter layers. Our results show the optimization of III-V buffer layers grown on Si is critical to achieve high performance quantum-dot lasers. An optimised 1.3-μm board-area laser has been demonstrated with a low threshold current density of 194 A/cm^2 and output power of 77 mW at room temperature.

1. Introduction

Si photonics has attracted much attention currently. Although in the last 20 years, intensive studies have focused on Si-based laser, which is the last missing element of Si photonics, efficient Si laser is not yet developed due to the indirect band-gap of Si [1]. Si Raman lasers, Si hybrid lasers, and III-V and Ge lasers monolithically grown on Si substrates have been demonstrated [2-6]. Among these efforts, the integration of III-V lasers monolithically on Si substrate is considered to be a promising solution to obtain efficient light sources on the Si platform in the near future.

The major issues of direct epitaxy of III-V materials on Si substrates are the formation of anti-phase boundary (APB) due to the polar/non-polar nature of III-V and group IV materials, and the threading dislocations (TDs) caused by lattice mismatch between GaAs and Si (4.1%). Also the difference in thermal expansion coefficient between III-V materials and Si substrates presents further challenges in the direct combination of these two material systems. The design of buffer layers has been proved to play a key role in obtaining high quality III-V materials on Si substrates [7]. The two-step GaAs buffer layer grown at low temperature could eliminate the APB and stop most of the TDs [7]. SLSs is also an effective method to reduce the TDs propagating in the semiconductor materials including AlGaN/GaN SLSs on sapphire [8], InGaAs/GaAs SLSs on Si [9] etc. III-V Quantum dot (QD) lasers have been demonstrated with very low threshold current, high operating temperature, high characteristic temperature, and high output power lasers in communication wavelengths (1.3 and 1.5 μm) due to its delta-function density of states [10, 11]. Furthermore, QD device is less sensitive to defects [12]. The recent demonstrations of high-performance 1.3-μm InAs/GaAs QD lasers on Ge and Ge/Si substrates have made a significant progress towards solving monolithic integration of III-V emitters on Si substrates. However, the direct epitaxy of III-V on Si substrates still needs to be further improved [3, 12, 13]. In this paper, the III-V buffer growth on Si substrates has been studied, including the optimization of GaAs nucleation layers and the investigation of InGaAs/GaAs and InAlAs/GaAs SLSs as dislocation filter layers (DFLs). 1.3-μm InAs/GaAs quantum dot lasers directly grown on Si substrates have been demonstrated with a low threshold current density of 725 A/cm^2 with InGaAs/GaAs DFLs and 194 A/cm^2 with InAlAs DFLs, respectively.

2. Optimisation of GaAs nucleation layer grown on Si substrate

The Si substrate has been de-oxidized by holding it at 900 °C for 10 minutes and then cooled down to grow the first 30-nm GaAs nucleation layer. The GaAs nucleation layer has been investigated at temperatures of 380, 400 and 420 °C, and additional 970-nm GaAs buffer layer has been grown with higher temperatures. Test samples with 5 layers of InAs/GaAs quantum dots-in-a-well (DWELL) [14] have been grown to investigate the effects of GaAs nucleation temperatures.

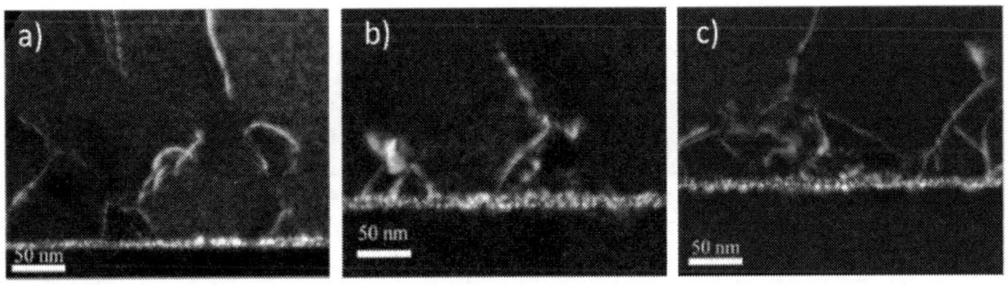

Figure 1. Cross-sectional TEM images of GaAs grown on Si substrates with different nucleation temperatures: (a) 380 °C, (b) 400 °C, and (c) 420 °C.

Transmission electron microscopy (TEM) studies were used to study the effects of the GaAs nucleation temperature on Si substrates. It clearly shows the GaAs with 400°C has the lowest threading dislocation density, as shown in Figure 1. The defects, mostly TDs, are mainly due to the misfit between GaAs and Si (~4%). The TDs easily propagate through buffer layer into active region, which could be non-radiative combination centers and increase threshold current density [15]. In Figure 1, the first 30nm GaAs nucleation layer is full of defects which generated from the GaAs/Si interface. However, most of defects are strongly confined in the nucleation layer and only few are able to propagate to the subsequent buffer layer. The effects of GaAs nucleation layer was further studied by photoluminescence (PL) study, as shown in Figure 2. The samples have been excited from a 532-nm diode-pumped solid-state laser with 94-mW power at room temperature. Figure 2 shows the one grown at 400°C has the strongest PL emission at room temperature, which is consistent with TEM results shown in Figure 1.

3. The reduction of threading dislocation by dislocation filter layers

In order to reduce the propagation of TDs into active region, InAlAs/GaAs and InGaAs/GaAs DFLs have been both studied in this work. Five layers of InAlAs/GaAs SLSs or InGaAs/GaAs SLSs with 3 repeats separated by GaAs spacing layers have been grown on an optimised GaAs buffer layer. 5-layers of typical InAs/InGaAs dot-in-a-well (DWELL) structures have been grown to compare the effectiveness of DFLs [16]. The room-temperature PL shown in Figure 3, shows a stronger emission for the sample with InAlAs/GaAs SLSs around 1300 nm. The relative lower emission for the sample with InGaAs/GaAs SLSs indicates the worse crystal quality than the one with InAlAs/GaAs SLSs. The spontaneously emission of quantum dots is high relied on the DFLs effectiveness, because the threading dislocation could kill the minority-carriers easily [17].

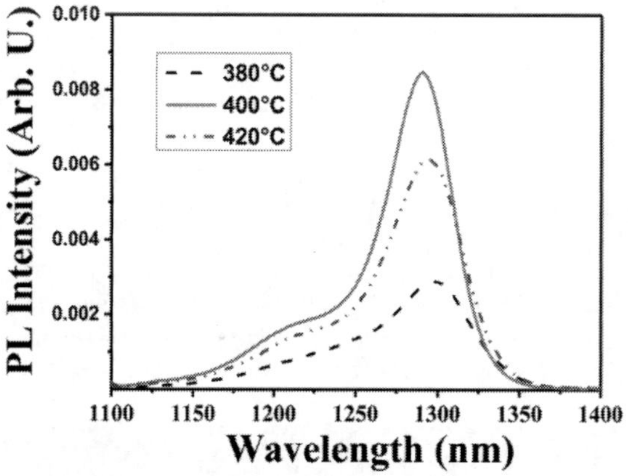

Figure 2. Room-temperature PL spectra of InAs/GaAs QDs grown on Si substrates with different initial GaAs nucleation temperatures from 380 to 420°C.

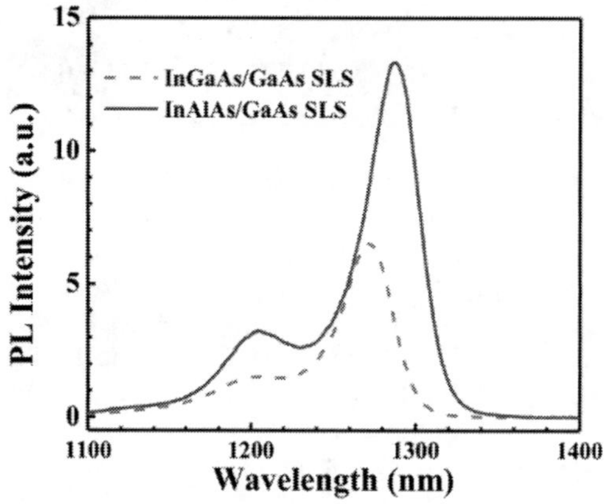

Figure 3. Room temperature PL spectra of InAs/GaAs quantum dots grown on Si substrate with InGaAs/GaAs SLSs and InAlAs/GaAs SLSs as DFLs.

Atomic force microscopy (AFM) and TEM have been used to examine the quality of quantum dots. Figure 4(a) and (b) show the uncapped InAs/GaAs QDs grown on Si substrate with InGaAs/GaAs and InAlAs/GaAs DFLs separately. It obviously shows the InAlAs/GaAs DFLs provide better dot size distribution than InGaAs/GaAs DFLs. In addition, Figure 4(b) is free of quantum dot coalescence but Figure 4(a) is not, which means the InAlAs/GaAs DFLs provides better GaAs matrix for InAs/InGaAs DWELL structures than the InGaAs/GaAs DFLs. Transmission electron microscopy has been used to examine the defects propagating within GaAs buffer and effects of DFLs. Figure 5(a) shows most of the TDs have been stopped by two sets of DFL layers. Most of TDs are confined below the first layer of SLSs. Only a few are able to be survived after the second set of InAlAs/GaAs SLSs. InAlAs/GaAs SLSs are shown to be effective enough to eliminate TDs and there is free of defect in the active region, as shown in Figure 5(b).

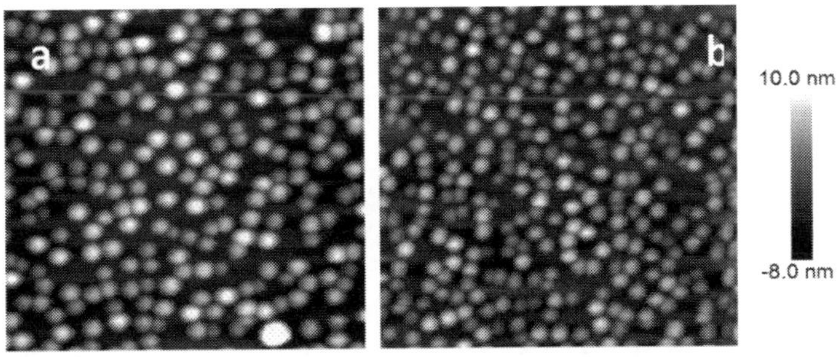

Figure 4. AFM images (1×1 μm^2) of InAs/GaAs QDs grown on Si substrates with (a) InGaAs/GaAs DFLs; (b) InAlAs/GaAs DFLs.

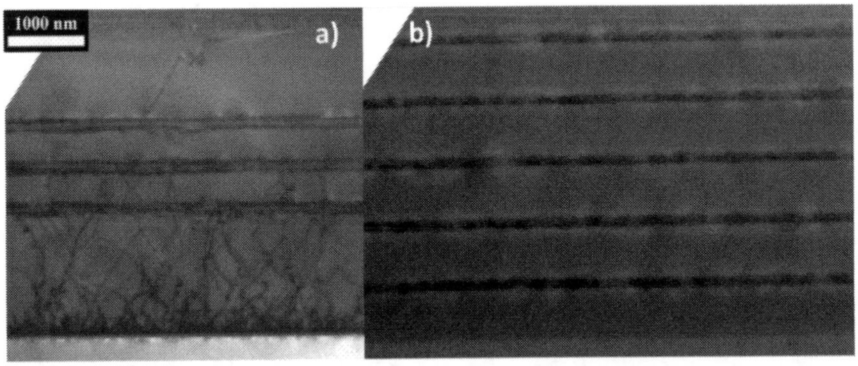

Figure 5. Cross-sectional TEM images of (a) 3 repeats of 20-layers InAlAs/GaAs SLSs and (b) active region which consists of 5layers of InAs/GaAs QDs capped with InGaAs.

4. Result and discussions

The laser devices have been studied. 1-μm GaAs buffer layer has been deposited on Si substrates similar to the test samples. InGaAs/GaAs and InAlAs/GaAs DFLs have both tested in laser structures for comparison. Both DFLs have 4 repeats of SLSs separated by 400-nm GaAs space layer. 1.5-μm $Al_{0.4}Ga_{0.6}As$ cladding layers have been grown on top and bottom of the active region with n and p doping separately. The active region is a typical DWELL structure with five layers of InAs QDs. 300-nm p-type GaAs contact layer is deposited on the top of laser structure. The samples have been fabricated into board-area lasers with 3.0-mm cavity. The laser with InGaAs/GaAs DFLs structure has a maximum operating temperature of 42°C. Figure 6 shows that the threshold current density of InAs/GaAs quantum dot lasers with InGaAs/GaAs SLSs DFLs is 725 A/cm^2 at 20°C. In comparison, the laser with InAlAs/GaAs DFLs shows a relative low threshold current density of 196 A/cm^2 at room temperature and the maximum operating temperature is 85°C (Figure 7). The characteristic temperature (T_0) is 44 K between operating temperature from 20 to 45°C for the laser with InGaAs/GaAs DFLs and is 46 K between 20 to 85°C for the one with InAlAs/GaAs DFLs. The poor T_0 observed for both devices is due to the carrier escape in the heterostructure when the laser is heated up. The future work will be the optimisation of characteristic temperature by using p-type modulations.

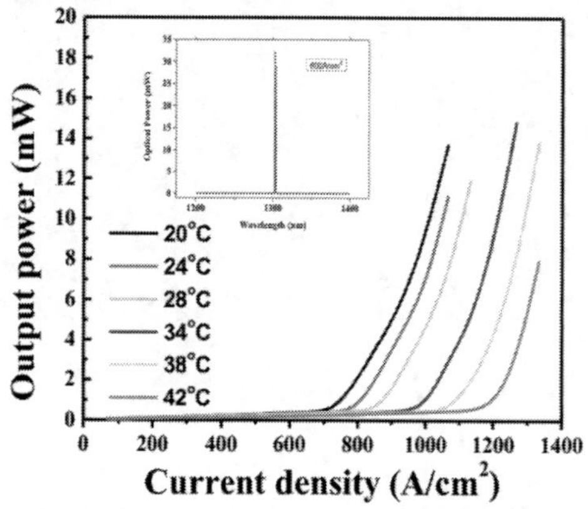

Figure 6. InAs/GaAs quantum dot lasers grown on Si substrate with InGaAs/GaAs DFLs: Output power against current density at various temperatures from 20 to 42°C.

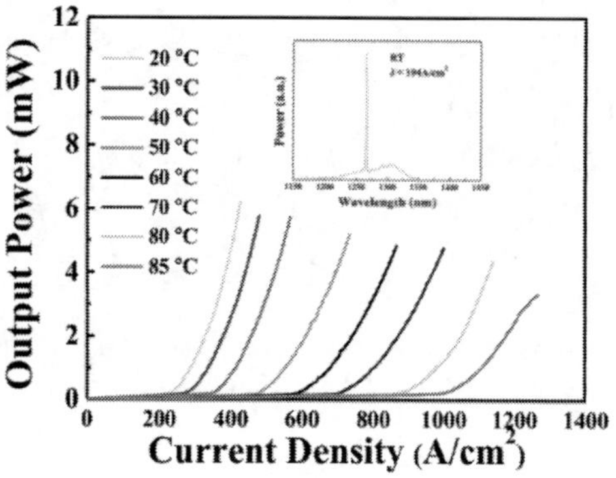

Figure 7. InAs/GaAs quantum dot lasers grown on Si substrate with InAlAs/GaAs DFLs: Output power against current density at various temperatures from 20 to 85°C.

5. Conclusion

We reported the optimisation of growing InAs/GaAs QD lasers on Si substrates. It has been shown that the growth temperature of the GaAs nucleation layer plays a critical role in material quality of the arsenide layers grown on Si substrates. We also compared the effectiveness of InAlAs/GaAs and InGaAs/GaAs SLSs as DFLs and show that InAlAs/GaAs SLSs provides better crystal quality as confirmed by PL and TEM measurements. 1.3-μm InAs/GaAs QD lasers directly grown on Si substrate with both InGaAs/GaAs and InAlAs/GaAs DFLs have been successfully demonstrated and compared. The significant improvement in the operating temperature and threshold current density has been demonstrated fro the lasers with InAsAs/GaAs DFLs. The laser device shows high output power around 80 mW and low threshold current density of 196 A/cm^2. As-cleaved laser is able to operate at maximum temperature of 85°C under pulsed mode.

Acknowledgement
The authors would like to acknowledge the funding support by EPSRC under Grant No. EP/J012904/1. H. Liu would like to thank The Royal Society for funding his University Research Fellowship.

References
[1] Jalali S and Fathpour S 2006 Silicon photonics *J. Lightw Technol.* **24**, 4600-4615
[2] Liu J, Sun X, Camacho-Aguilera R, Kimerling L C and Michel J 2010 Ge-on-Si laser operating at room temperature *Opt. Lett* **35**, 679-681
[3] Wang T, Liu H, Lee A, Pozzi F and Seeds A 2011 1.3-μm InAs/GaAs quantum-dot lasers monolithically grown on Si substrates *Opt. Express* **19**, 11381-11386
[4] Chen X, Li C, and Tsang H K 2011 Device engineering for silicon photonics *NPG Asia Materials* **3**, 34-40.
[5] Liang D and Bowers J E 2010 Recent progress in lasers on Silicon *Nature Photon.* **4**, 511-517
[6] Liu H, Wang T, Jiang Q, Hogg R, Tutu F, Pozzi F and Seeds A, 2011 Long-wavelength InAs/GaAs quantum-dot laser diode monolithically grown on Ge substrate *Nature Photon.* **5**, 416-419
[7] Wang T, Lee A, Tutu F, Seeds A, Liu H, Jiang Q, Groom K and Hogg R 2012 The effect of growth temperature of GaAs nucleation layer on InAs/GaAs quatum dots monolithically grown on Ge substrate *Appl. Phys. Lett.* **100**, 052113
[8] Nakamura S, Senoh M, Nagahama S, Iwasa N, Yamada T, Matsushita T, Kiyoku H, Sugimoto H, Kozaki T, Umemoto H, Sano M and Chocho K 1998 InGaN/GaN/AlGaN-based laser diodes with modulation-doped strained-layersuperlattices grown on an epitaxially laterally overgrown GaN substrate *Appl. Phys. Lett.* **72**, 211
[9] Nozawa K and Horikoshi Y 1991 Low Threading Dislocation Density GaAs on Si(100) with InGaAs/GaAs Strained-Layer Superlattice Grown by Migration-Enhanced Epitaxy *Jpn. J. Appl. Phys* **30** L668-L671
[10] Liu H 2013 III–V Quantum-Dot Materials and Devices Monolithically Grown on Si Substrates *Silicon-based Nanomaterials*, H. Li, J. Wu, and Z. M. Wang, eds. (Springer New York, 2013), pp. 357-380.
[11] Wu J, Lee A, Jiang Q, Tang M, Seeds A J, and Liu H 2014 Electrically pumped continuous-wave 1.3-μm InAs/GaAs quantum dot lasers monolithically grown on Si substrates *IET Optoelectronics* **8**, p20-24
[12] Liu H, Wang T, Jiang Q, Hogg R, Tutu F, Pozzi F and Seeds A 2011 Long-wavelength InAs/GaAs quantum-dot laser diode monolithically grown Ge substrate *Nat. Photon*, **5** 416-419
[13] Lee A, Jiang Q, Tang M, Seeds A and Liu H 2012 Continuous-wave InAs/GaAs quantum-dot laser diodes monolithically grown on Si substrate with low threshold current densities, *Opt. Express* **20**, 22181-22187
[14] Liu H, Hopkinson M, Harrison C, Steer M, Frith R, Sellers I, Mowbray D, and Skolnick M 2003 Optimizing the growth of 1.3 μm InAs/In,GaAs dots-in-a-well structure *J. Appl. Phys.* **93**, 2931-2936
[15] Fischer R, Kopp W, Morkoc H, Pion M, Specht A, Burkhart G, Appelman H, McGougan D and Rice R 1986 Low threshold laser operation at room temperature in GaAs/(Al,Ga)As structures grown directly on (100)Si *Appl. Phys. Lett.* **48**, 1360-1361
[16] Wang T, Liu H, Lee A, Pozzi F and Seeds A 2011 1.3-μm InAs.GaAs quantum-dots lasers monolithically grown on Si substrates, *Opt. Express* **19**, 11381-11386
[17] Sieg R. M, Carlin J. A, Boeckl J. J, Ringel S. A, Currie M. T, Ting S. M, Langdo T. A, Taraschi G, Fitzgerald E. A., and Keyes B. M 1998 High minority-carrier lifetimes in GaAs grown on low-defect-density Ge/GeSi/Si substrates, *Appl. Phys. Lett.* **73**, 3111-3113

Anomalous Photoluminescence of Weakly Confined Excitons including Radiative Correction in Nano-to-Bulk Crossover Regime

Takuya Matsuda, Nobuhiko Yokoshi, Hajime Ishihara

Department of Physics and Electronics, Osaka Prefecture University, 1-1, Gakuen-cho, Naka-ku, Sakai, Osaka, 599-8531, Japan

E-mail: matsuda@pe.osakafu-u.ac.jp

Abstract. We develop a theoretical formalism to calculate photoluminescence (PL) spectrum of weakly confined excitons incorporating the microscopic nonlocal optical response. The nonlocality is caused by the center-of-mass (c. m.) motion of exciton and becomes remarkable in nano-to-bulk crossover regime. The theory successfully explains the characteristics of recently observed peculiar PL spectra in high quality CuCl films [5], wherein the signals appear at the exciton states with the very large radiative corrections not only for the lowest level but also for the higher ones including non-dipole types of excitons.

1. Introduction

The exciton-radiation coupling has attracted a great deal of interest from viewpoint of the remarkable sample size dependence [1-4]. If the c. m. wavefunctions of exciton coherently spread across the entire crystal with the size beyond the long-wavelength approximation (LWA), the spatial interplay between the exciton wave and light wave greatly affects the sample size dependence, so the nonlocal optical response plays an important role therein. Such a wave-wave interplay causes the large exciton-radiation interaction volume which leads to the anomalous optical response. For example, it has been theoretically predicted that the radiative decay becomes exceptionally fast with the increase of the sample size [5], and actually the experiment with the quite high quality CuCl films revealed that the observed radiative decay time reaches the order of 100 [fs] in degenerate four-wave mixing [6].

Recently, the above-mentioned anomalous aspects have also been found in the PL processes [7]. In a contrast of a standard picture of PL processes in solids, in the experiment in Ref. [7], it has been succeeded in systematically observing the PL signals from multiple exciton-radiation coupled modes in the high quality CuCl films with thickness of a few hundred nanometers. The observed PL spectra exhibit following unconventional aspects: (1) PL signals appear from multiple higher excited states. (2)The signals appear even from "optically forbidden" exciton states such as that with odd parity c. m. wavefunctions. (3)The signals appear at the energies of excitons including the remarkably large radiative corrections. The key to understand these features is the nonlocal wave-wave coupling established in the size beyond LWA that causes the extremely large exciton-radiation interaction volume. This leads to the ultrafast radiative decay of excitons even including non-dipole type states. In order to elucidate the PL processes of wave-wave coupled system, the PL theory taking account of the microscopic nonlocal effect is necessary. The purpose of this paper is to develop such a theoretical formalism and demonstrate PL spectra which exhibit above-mentioned anomalous features.

2. Theory

We explain the calculation method of PL spectra for weakly confined excitons including radiative correction. We consider the whole Hamiltonian as $\hat{H} = \hat{H}_{\rm ex} + \hat{H}_{\rm rad} + \hat{H}_{\rm int} + \hat{H}_{\rm res} + \hat{H}_{\rm er}$, where $\hat{H}_{\rm ex}$, $\hat{H}_{\rm rad}$, $\hat{H}_{\rm int}$, $\hat{H}_{\rm res}$ and $\hat{H}_{\rm er}$ represent, respectively, the excitons, the radiation field, the interaction between excitons and radiation field, the reservoirs, and the interaction between excitons and reservoirs . From the whole Hamiltonian, we derive the Markovian master equation as $(\partial/\partial t)\hat{\rho}(t) = (1/{\rm i}\hbar)[\hat{H}_{\rm ex} + \hat{H}_{\rm rad} + \hat{H}_{\rm int},\ \hat{\rho}(t)] + \hat{\mathcal{L}}_{\rm damp}\hat{\rho}(t) + \hat{\mathcal{L}}_{\rm phase}\hat{\rho}(t)$, where $\hat{\mathcal{L}}_{\rm damp}\hat{\rho}(t) = \sum_\lambda \gamma_{\rm ex}^\lambda/2(2\hat{b}_\lambda\hat{\rho}(t)\hat{b}_\lambda^\dagger - \{\hat{b}_\lambda^\dagger\hat{b}_\lambda,\ \hat{\rho}(t)\})$, $\hat{\mathcal{L}}_{\rm phase}\hat{\rho}(t) = \sum_\lambda \Gamma_{\rm ex}^\lambda/2[\hat{b}_\lambda^\dagger\hat{b}_\lambda\ [\hat{b}_\lambda^\dagger\hat{b}_\lambda\ \hat{\rho}(t)]]$. Here, $\hat{\rho}(t)$ is the density matrix, $\gamma_{\rm ex}^\lambda$, $\Gamma_{\rm ex}^\lambda$ and $\hat{b}_\lambda$, respectively, show the nonradiative decay rate, the dephasing rate and the annihilation operator of the λth exciton state. From the master equation, we derive the motion equations of the λth excitonic polarization $\langle\hat{b}_\lambda(t)\rangle$, the coherence between the λth and λ'th exciton states $\langle\hat{C}_{\lambda,\lambda'}(t)\rangle = \langle\hat{b}_\lambda^\dagger(t)\hat{b}_{\lambda'}(t)\rangle$ $(\lambda \neq \lambda')$, and the λth excitonic population $\langle\hat{N}_\lambda(t)\rangle = \langle\hat{b}_\lambda^\dagger(t)\hat{b}_\lambda(t)\rangle$ as

$$\frac{\partial}{\partial t}\langle\hat{b}_\lambda(t)\rangle = -\left({\rm i}\Omega_\lambda^{\rm ex} + \gamma_{\rm ex}^\lambda/2 + \Gamma_{\rm ex}^\lambda/2\right)\langle\hat{b}_\lambda(t)\rangle + \frac{\rm i}{\hbar}\int {\rm d}\boldsymbol{r}\ \boldsymbol{\mathcal{P}}_\lambda^*(\boldsymbol{r})\cdot\boldsymbol{E}(\boldsymbol{r},t), \tag{1}$$

$$\begin{aligned}\frac{\partial}{\partial t}\langle\hat{C}_{\lambda,\lambda'}(t)\rangle = &\left[{\rm i}(\Omega_\lambda^{\rm ex} - \Omega_{\lambda'}^{\rm ex}) - \frac{\gamma_{\rm ex}^\lambda + \gamma_{\rm ex}^{\lambda'} + \Gamma_{\rm ex}^\lambda + \Gamma_{\rm ex}^{\lambda'}}{2}\right]\langle\hat{C}_{\lambda,\lambda'}(t)\rangle \\ &+ \frac{\rm i}{\hbar}\int {\rm d}\boldsymbol{r}\ \boldsymbol{\mathcal{P}}_{\lambda'}^*(\boldsymbol{r})\cdot\boldsymbol{E}(\boldsymbol{r},t)\langle\hat{b}_\lambda^\dagger(t)\rangle - \frac{\rm i}{\hbar}\int {\rm d}\boldsymbol{r}\ \boldsymbol{\mathcal{P}}_\lambda(\boldsymbol{r})\cdot\boldsymbol{E}^*(\boldsymbol{r},t)\langle\hat{b}_{\lambda'}(t)\rangle,\end{aligned} \tag{2}$$

$$\begin{aligned}\frac{\partial}{\partial t}\langle\hat{N}_m(t)\rangle = &-\sum_{m(<\lambda)}\delta^{m,\lambda}\langle\hat{N}_\lambda(t)\rangle + \sum_{n(>\lambda)}\delta^{\lambda,n}\langle\hat{N}_n(t)\rangle - \gamma_{\rm ex}^\lambda\langle\hat{N}_\lambda(t)\rangle \\ &+ \frac{\rm i}{\hbar}\int {\rm d}\boldsymbol{r}\ \left(\boldsymbol{\mathcal{P}}_\lambda^*(\boldsymbol{r})\cdot\boldsymbol{E}(\boldsymbol{r},t)\langle\hat{b}_\lambda^\dagger(t)\rangle - \boldsymbol{\mathcal{P}}_\lambda(\boldsymbol{r})\cdot\boldsymbol{E}^*(\boldsymbol{r},t)\langle\hat{b}_\lambda(t)\rangle\right).\end{aligned} \tag{3}$$

Here, we introduce phenomenologically the level-to-level nonradiative decay rate $\delta^{\lambda,\lambda'}$ from λ'th exciton state to λth exciton one into the motion equation of population Eq.(3), and $\boldsymbol{E}(\boldsymbol{r},t)$ is the classical Maxwell electric field. $\Omega_\lambda^{\rm ex}$ means the eigenfrequency of λth exciton, and the expansion coefficient of the polarization $\boldsymbol{\mathcal{P}}_\lambda(\boldsymbol{r})$ is written as $\boldsymbol{\mathcal{P}}_\lambda(\boldsymbol{r}) = \boldsymbol{\mathcal{M}}\psi_\lambda(\boldsymbol{r})$. $\boldsymbol{\mathcal{M}}$ is the transition dipole moment and $\psi_\lambda(\boldsymbol{r})$ is proportional to the c. m. wavefunction of λth exciton. Solving the motion equations Eqs.(1)-(3), and Maxwell's equation self-consistently, we can obtain the mean values of the polarization $\langle\hat{b}_\lambda\rangle$, the coherence $\langle\hat{C}_{\lambda,\lambda'}\rangle$, and the population $\langle\hat{N}_\lambda\rangle$ in the steady state. We calculate PL spectrum from the quantized output radiation field expressed in frequency component by using the photon Green function as $\hat{\boldsymbol{E}}_{\rm out}(\boldsymbol{r}_0,\omega) = \mu_0\omega^2\int {\rm d}\boldsymbol{r}'\ \mathbf{G}_{\rm out}(\boldsymbol{r}_0,\boldsymbol{r}',\omega)\cdot\hat{\boldsymbol{P}}_{\rm ex}(\boldsymbol{r}',\omega)$, where $\boldsymbol{r}_0$ is an arbitrary position outside the material. Here, under the rotating-wave approximation (RWA), the excitonic polarization operator $\hat{\boldsymbol{P}}_{\rm ex}(\boldsymbol{r},\omega)$ is represented in terms of the annihilation operator of excitons $\hat{b}_\lambda(\omega)$ in frequency component as $\hat{\boldsymbol{P}}_{\rm ex}(\boldsymbol{r},\omega) = \sum_\lambda \boldsymbol{\mathcal{P}}_\lambda(\boldsymbol{r})\hat{b}_\lambda(\omega)$. The photon Green function $\mathbf{G}_{\rm out}(\boldsymbol{r}_0,\boldsymbol{r}',\omega)$ includes information of the spatial structure of the background dielectric function, and thus, this formulation can be applied to an arbitrary sample shape. From the first-order correlation function, the spectrum is defined as $S(\boldsymbol{r}_0,\omega) \equiv \frac{1}{(2\pi)^2}\int_0^\infty {\rm d}t_1\int_0^\infty {\rm d}t_2\ {\rm e}^{{\rm i}\omega(t_2-t_1)}\ \langle\hat{\boldsymbol{E}}_{\rm out}^\dagger(\boldsymbol{r}_0,t_1)\cdot\hat{\boldsymbol{E}}_{\rm out}(\boldsymbol{r}_0,t_2)\rangle$.

Journal of Physics: Conference Series **619** (2015) 012012 doi:10.1088/1742-6596/619/1/012012

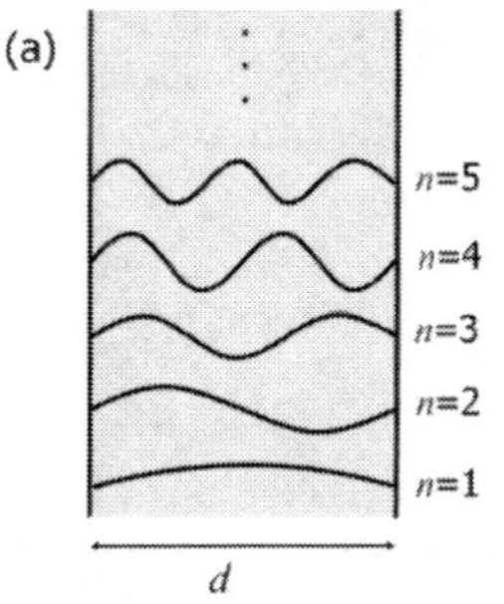

(b)

n	κ_n [meV]
1	$\kappa_1 = 1.12$
2	$\kappa_2 = 3.10$
3	$\kappa_3 = 6.60$
4	$\kappa_4 = 55.6$
5	$\kappa_5 = 3.29$
6	$\kappa_6 = 0.635$
7	$\kappa_7 = 0.215$
8	$\kappa_8 = 0.100$
9	$\kappa_9 = 0.052$
10	$\kappa_{10} = 0.033$
11	$\kappa_{11} = 0.019$
12	$\kappa_{12} = 0.015$

Figure 1. (a) Sketch of the c. m. wavefunctions of confined excitons. (b)Table shows the radiative rates of the respective coupled modes.

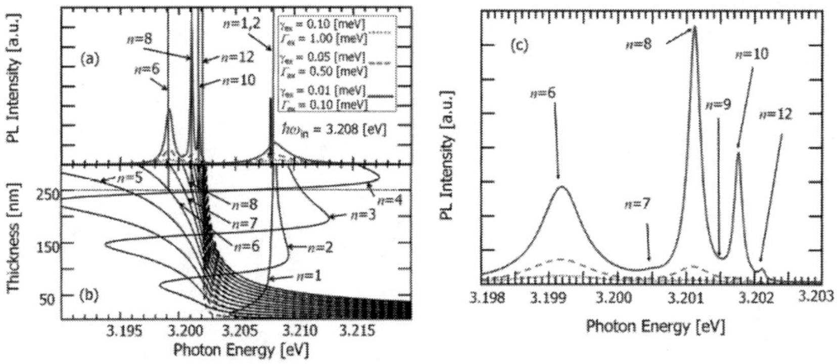

Figure 2. (a) Calculated PL spectra of a CuCl film with thickness of 250 [nm] for different values of nonradiative population decay rates. The assumed values of $\gamma_{\rm ex}$ and $\mathit{\Gamma}_{\rm ex}$ are indicated. The equal values of $\gamma_{\rm ex}$ and $\mathit{\Gamma}_{\rm ex}$, respectively, are taken for the different states. (b) The real parts of the eigenenergies of the coupled modes. Those for non-coupled excitons are also plotted by dashed lines. (c) An enlarged view of the spectra in Fig.2 (a) from 3.198 [eV] to 3.203 [eV].

3. Model and Results

In the numerical demonstrations, we consider a CuCl film with thickness d, where the c. m. motion is confined in the direction of the film thickness. With regard to the c. m. wavefunctions, we consider the simple sinusoidal waves without distortions near the surfaces as illustrated in Fig.1 (a). The excitation light is the monochromatic cw laser whose wavevector is perpendicular to the film surface. We divide the calculation process into the two steps. In the first step, we calculate the mean values of the polarization $\langle \hat{b}_\lambda \rangle$, the coherence $\langle \hat{C}_{\lambda,\lambda'} \rangle$, and the population $\langle \hat{N}_\lambda \rangle$ in the steady state in the presence of the incident light. In the second step, we cut off the incident light, and calculate PL spectrum by using the above mean values as initial ones. This procedure is valid in the presence of the sufficient excitonic dephasing that disconnects the initial coherent process from the incoherent emission one.

Figure 2 (a) shows calculated PL spectra, where we choose the thickness (d=250 [nm]) different from those in the experiment [7] for the sake of the clearer view of respective contributions of the coupled modes, which does not change the essential features of observed PL spectra. The excitation light is assumed to be tuned to 3.208 [eV]. In Fig. 2 (c), we show an enlarged view of the spectra in Fig. 2 (a) from 3.198 [eV] to 3.203 [eV]. The table in Fig. 1 (b) shows the imaginary parts of the complex eigenenergies of exciton-radiation coupled modes corresponding to the radiative decay rates, and Fig. 2 (b) shows the real parts corresponding to the eigenenergies including radiative shifts. Those values can be obtained from the self-consistent solutions of Eq. (1) and the Maxwell's equations. In Fig. 2 (b), the eigenenergies of non-coupled excitons are also plotted (dashed line). Comparing the exciton-radiation coupled modes with the non-coupled excitons, we can see the large radiative shift and their drastic behavior with the change of the film thickness, where exciton states $m = 1, 2, \cdots$, in turn, changes from the lower branch modes to the upper branch ones with the increase of film thickness [5][8]. By comparing Fig.2 (a) with Fig.2 (b), we find that the respective peaks in PL spectrum appear at the eigenenergies of the coupled modes. In the experiment in Ref. [7], the observed PL signals for a CuCl film with the thickness of 325 [nm] are from the quantum states of n=1,2,3,4,7-12, where the even numbered states are optical forbidden if the LWA is satisfied because their wavefunctions have odd parity. The state n=5 is not detectable in the above experiment because its peak is too wide due to the very large radiative width. Our calculated result essentially reproduces similar features though the quantum states which appear in PL spectrum are slightly different owing to the different thickness. The remarkable point for this thickness is that the even numbered states (odd parity of states) clearly appear. This means that the parity of the light wave is odd like in the present case, and the selection rule is almost inverted from that under normal LWA. In this case, the states of n=4 has very large radiative width as seen in the table in Fig. 1(b), and hence, it is undetectable as n=5 in the experiment. The signals from the level of n=3 is not seen in Fig. 2(b) because its eigenenergy is higher than the excitation energy. The radiative decay rates of the levels of n=8-12 are of the order of 0.1 [meV]-0.01 [meV]. Therefore, the peaks of these states clearly appear when the nonradiative population decay rate γ_{ex} is smaller than them (blue solid line) in Fig. 2(a) and (c)). In the present case, we assume the level-to-level nonradiative decay rate smaller than the radiative decay rate of each quantum level. That is why their effect does not clearly appear in the present demonstration. The explicit effect of the level-to-level nonradiative decay will be studied in the next publication.

4. Summary

In summary, we develop a theoretical formalism to calculate PL spectra of weakly confined excitons beyond the LWA regime. By using this calculation method, we compute the PL spectra, and reproduce the essential features of the recent reported experimental results which show unconventional aspects: (1)The PL signals appear from multiple higher excited states. (2)The signals appear even from optically forbidden exciton states such as that with odd parity c. m. wavefunctions. (3)The signals appear at the energies of excitons with the radiative shifts. Although the quantitative comparison between the theory and experiment was not done because of the different incident conditions (The experiment was made under the pulse incidence), the study with more detailed comparison by further generalized theory would reveal the PL processes comprehensively for nano-to-bulk crossover regime, which is our future subject.

References

[1] Takagahara T, 1987 Phys. Rev. B. **36** 9293
[2] Hanamura E, 1988 Phys. Rev. B. **37** 1273
[3] Gaponenko S V 1998 Optical Properties of Semiconductor Nanocrystals (Cambridge: Cambridge University)
[4] Ishihara H, J. Phys.: Condens. Matter **16** 2004 R247-R273
[5] Ishihara H, Kishimoto J, and Sugihara K, 2004 J. lumin. **108** 343-346
[6] Ichimiya M, Ashida M, Yasuda H, Ishihara, and H, Itoh T, 2009 Phys. Rev. Lett. **103** 257401
[7] Phuong L.Q, Ichimiya M, Ishihara H, Ashida M, 2012 Phys. Rev. B. **86** 235449

Model-simulation of light-induced defect creation in hydrogenated amorphous silicon

K Morigaki[1*] **and H Hikita**[2]
[1] Department of Electrical-System Engineering, Hiroshima Institute of Technology, Miyake, Saeki-ku, Hiroshima 731-5193, Japan
* Present address: C-305, Wakabadai 2-12, Inagi, Tokyo 206-0824, Japan
[2] Physics Laboratory, Meikai University, Urayasu, Chiba 279-8550, Japan

E-mail: k.morigaki@yacht.ocn.ne.jp

Abstract. The processes of light-induced defect creation in hydrogenated amorphous silicon consist of self-trapping of holes in a weak Si-Si bond adjacent to a Si-H bond, nonradiative recombination of self-trapped holes with electrons, the Si-H bond switching towards the weak Si-Si bond, and hydrogen-hopping or –tunneling along the weak Si-Si bond. The Si-H bond switching and hydrogen movements are treated by Monte-Carlo computer simulation in a simple cubic lattice. From the result of $<r^2>$ vs t in which r and t designate diffusion distance of hydrogen and diffusion time, respectively, the density of light-induced defects is estimated in a good agreement with the observed density.

1. Introduction

The mechanism of light-induced defect creation in hydrogenated amorphous silicon (a-Si:H) has received much attention from viewpoints of physics and applications such as solar cells. Prolonged illumination causes Si dangling bonds (DBs) to be created, so that the conversion efficiency of solar cells decreases. So far the consensus has not yet been obtained for the mechanisms of light-induced defect creation although many models have been proposed. We have already proposed a mechanism of light-induced defect creation [1], in which the breaking of a weak Si-Si bond adjacent to a Si-H bond (specific weak Si – Si bond) and hydrogen movement are key factors, as mentioned below. In this paper, we present model-simulation for the light-induced defect creation process in the following way: A Si-H bond adjacent to a weak Si-Si bond is switched towards the weak Si-Si bond, i.e., hydrogen occupies a central site of the weak Si-Si bond by use of an energy associated with nonradiative recombination of an electron with a self-trapped hole in the weak Si-Si bond under illumination, and then a normal DB is created behind. The switched hydrogen at the central site is bonded to a neighboring Si atom, and then this Si-H bond is switched towards its neighboring Si-Si bond in the same way as before. Such a switching of the Si-H bond is repeated, and a DB having hydrogen at the nearby site (hydrogen-related DB) is created at a certain time t after the first switching of the Si-H bond occurs. In figure 1, these processes are illustrated in simple cubic lattice sites, in which the Monte Carlo computer simulation of such a Si-H bond switching and diffusion processes is performed. The mutual distance of two types of DBs, i.e., a normal DB and a hydrogen-related DB, has been estimated to be about 13Å from the width of spin packet of the ESR line due to DBs [1].

In Section 2, the Monte-Carlo computer simulation as well as its results are presented. In Section 3, these issues are discussed. In Section 4, a conclusion is drawn.

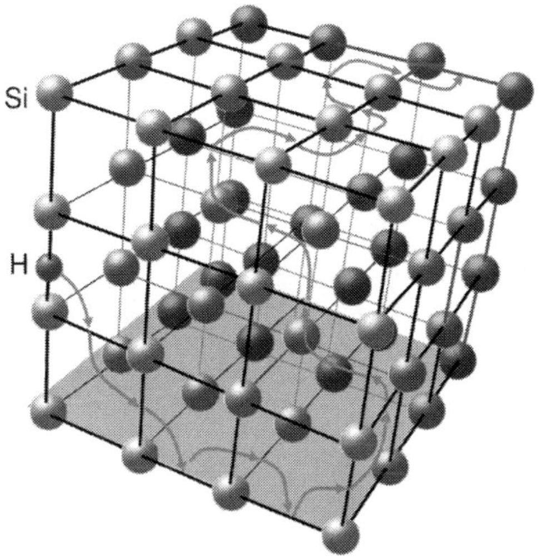

Figure 1. A schematic illustration of hydrogen movements in a simple cubic lattice.

2. Monte-Carlo computer simulation

Monte-Carlo computer simulation is performed in a simple cubic lattice, that is, different 10 lattices with 300×300×300 sites are attempted. The hydrogen-silicon bond energy distributes with a Gaussian shape of the density of states, as given by

$$D(E) = \frac{1}{2\pi\sigma}\exp\left[-\frac{(E-E_0)^2}{2\sigma^2}\right] \tag{1}$$

where σ and E_0 are the standard deviation and the central energy, respectively. We put

$$E_0 = k_B T_0 \text{ and } \sigma = k_B T_\sigma, \tag{2}$$

where k_B is the Boltzmann constant.

The calculated mean square hopping distance of hydrogen, $\langle r^2 \rangle$, is shown as a function of the number of steps, s, for various lattice temperature, T, in figure 2. The log $\langle r^2 \rangle$ vs. log s plot forms a slope in the range of large steps, as shown in figure 2.

$$\langle r^2 \rangle = a\, s^\alpha, \tag{3}$$

The value of α depends on T, as shown in figure 3.

As mentioned in Section 1, the distance between a normal DB and a hydrogen-related DB is r = 13Å. This distance corresponds to the hopping distance of hydrogen to create two type of DBs. We obtain

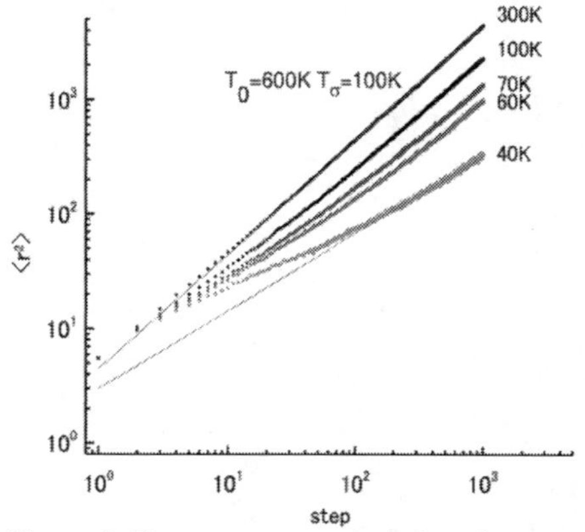

Figure 2. The mean square hopping distance of hydrogen versus the number of steps curves for various temperatures in the case of $T_0 = 600$ K and $T_\sigma = 100$ K. The line expresses (3), i.e., $<r^2> = a\, s^{\alpha}$.

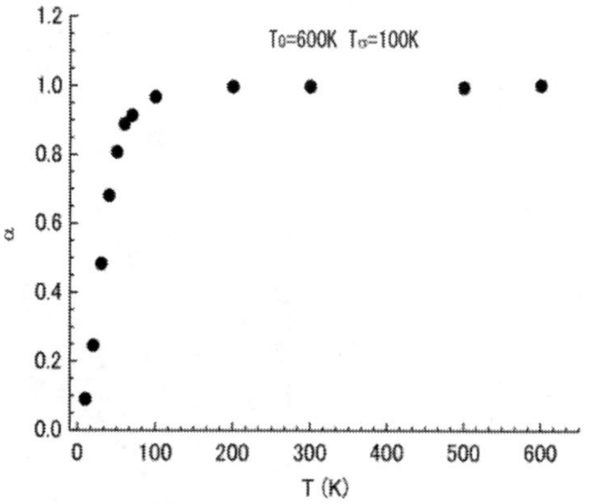

Figure 3. The exponent α in (3) versus temperature curve.

from figure 2 that the number of step corresponding to r = 13 Å is 37.4. Thus, the time necessary for creating two types of DBs is estimated from the Si-H bond switching time and the number of steps, i.e., is 37.4. The Si-H bond switching time is estimated to be the reversal of the Si-H bond switching frequency, that is given by

$$\nu = \nu_0 \exp\left[-\frac{\Delta E}{k_B T}\right], \qquad (4)$$

where ν_0 and ΔE are the pre-factor and the barrier height, respectively. We take $\gamma_0 = 1 \times 10^{12}$ s^{-1} and ΔE = 40 meV [2]. For T = 300 K, the time necessary for creating two types of DBs, Δt, is

$$\Delta t = \nu_0^{-1} \exp\left[\frac{\Delta E}{k_B T}\right] \times (\text{number of steps})$$

$$= 1.8 \times 10^{-10}\, s, \qquad (5)$$

In the following, we consider how many two types of DBs are created through the Si-H bond switching and movement during 1 hr illumination from xenon-arc lamp with infrared cut at 1.3 Wcm^{-2} with G (generation rate of free carriers) = 1.9×10^{22} $cm^{-3}s^{-1}$ [3]. The light-induced DB creation begins with self-trapping of a hole at a specific weak Si – Si bond. After then, the Si – H bond switching and movement are repeated to create two types of DBs separated by 13 Å, as shown in figure 4. Now, we take the case of 1 hr illumination, during which the above light-induced DB creation process is repeated by 2.0×10^{13} (= 3600 s /1.8×10^{-10} s). It is assumed that eight Si – Si bonds are broken by hole trapping during one period, i.e., the time necessary for hydrogen reaching on the surface of radius 13 Å. Here, the efficiency of bond breaking, η, is taken to be 0.1 [4]. The ratio of the self-trapped hole density in specific weak Si – Si bonds to the total density of trapped holes in weak Si – Si bonds, ξ, is taken to be 10^{-3} at room temperature[4]. The hole density, p, is estimated from (5) in [5] to be 1.9×10^{14} cm^{-3}.

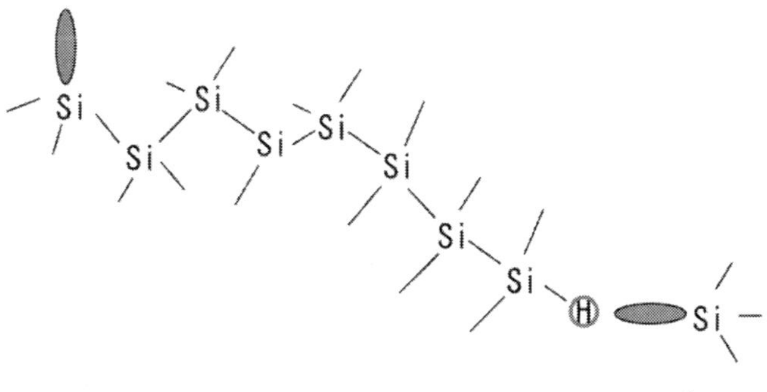

Figure 4. Atomic configuration of two types of DBs separated by 13Å.

Then, the light-induced DB density is given by

$$N_d = p\ \xi\eta^8 \times (2.0 \times 10^{13}), \tag{6}$$

The above values of p, ξ, and η are inserted into (6), then we obtain N_d =3.9 $\times 10^{16}$ cm^{-3}. This is in a good agreement with the observed value of 3.0 × 10^{16} cm^{-3} [3], as will be mentioned in Section 3.

3. Discussion

First, we discuss the light-induced DB creation considered in the Monte-Carlo computer simulation. The mutual distance between a normal DB and a hydrogen-related DB is estimated to be 13 Å, as mentioned in Section 1. This distance almost corresponds to the atomic configuration of two types of DBs, as shown in Fig. 4. This is of course a typical configuration corresponding to r = 13 Å. For simplicity, it is assumed that eight weak Si – Si bonds are broken when a hole is self-trapped at each weak bond, as mentioned in Section 2. The observed light-induced DB density has been measured by electron spin resonance measurements for various types of a-Si:H samples and for illumination times, e.g., 3 × 10^{16} cm^{-3}, for 1 hr illumination by xenon-arc lamp with infrared cut for device-quality a-Si:H samples prepared at substrate temperature of 250 °C [3].

Secondly, we discuss the temperature dependence of α in (3). As shown in figure 3, the value of α tends to 1 above 100 K that is equal to T_σ, corresponding to almost the full width of half-magnitude (FWHM) of the Gaussian shape in a unit of temperature. This is in contrast with the case of exponential density of states of hydrogen-site energy, in which α is proportional to temperature [5].

4. Conclusion

Two types of DBs, i.e., a normal DB and a hydrogen-related DB, are created through the Si – H bond switching and movement under illumination. The time necessary for separating two types of DBs with 13Å is estimated from the Monte-Carlo computer simulation in a simple cubic lattice. Using this time, light-induced DB density is estimated and is compared with the observed density in a good agreement.

References

[1] Morigaki K and Hikita H 2007 *Phys. Rev.* B**76** 085201
[2] Morigaki K 1988 *Jpn. J. Appl. Phys.* **27** 163
[3] Morigaki K, Takeda K, Hikita H and Roca i Cabarrocas 2007 *Solid State Commun.* **142** 232
[4] Morigaki K1990 *Jpn. J. Appl. Phys.* **29** L1582
[5] Morigaki K and Hikita H 2011 *Phys. Status Solidi* C**8** 2564

Journal of Physics: Conference Series **619** (2015) 012014 doi:10.1088/1742-6596/619/1/012014

Investigation of isochronal annealing on the optical properties of HWCVD amorphous silicon nitride deposited at low temperatures and low gas flow rates

T F G Muller[1], S Jacobs[1], F R Cummings[1], C J Oliphant[2], G F Malgas[1] and C J Arendse[1]

[1]Department of Physics, University of the Western Cape, Private Bag X17, Bellville 7535, South Africa

[2]National Metrology Institute of South Africa, Private Bag X34, Lynwood Ridge, Pretoria 0040, South Africa

E-mail: tmuller@uwc.ac.za

Abstract. Hydrogenated amorphous silicon nitride (a-SiN_x:H) is used as anti-reflection coatings in commercial solar cells. A final firing step in the production of micro-crystalline silicon solar cells allows hydrogen effusion from the a-SiN_x:H into the solar cell, and contributes to bulk passivation of the grain boundaries. In this study a-SiN_x:H deposited in a hot-wire chemical vapour deposition (HWCVD) chamber with reduced gas flow rates and filament temperature compared to traditional deposition regimes, were annealed isochronally. The UV-visible reflection spectra of the annealed material were subjected to the Bruggeman Effective Medium Approximation (BEMA) treatment, in which a theoretical amorphous semiconductor was combined with particle inclusions due to the structural complexities of the material. The extraction of the optical functions and ensuing Wemple-DeDomenici analysis of the wavelength-dependent refractive index allowed for the correlation of the macroscopic optical properties with the changes in the local atomic bonding configuration, involving silicon, nitrogen and hydrogen.

1. Introduction

The dielectric function, and therefore refractive index of hydrogenated amorphous silicon nitride (a-SiN_x:H) can be tailored according to the requirements for its intended application. This is achieved by varying the deposition conditions, thus ultimately altering the structural properties, and hence the optical properties of the material. In this study thin films were deposited using HWCVD from $SiH_4/NH_3/H_2$ gas mixtures with gas flow rates and wire temperature deliberately chosen lower than typically seen in literature [1-2] to reduce gas, thermal and power budgets. Changes in structural and optical properties in the annealed state were investigated and related to bonding configurations in a-SiN_x:H. It was found that successful modelling of UV-vis reflectance spectra of a-SiN_x:H necessitates the application of complex virtual models for both as-deposited and annealed films.

2. Experimental Procedure

Silicon rich a-SiN_x:H thin films of 710 nm nominal thickness and N/Si = x = 0.26 were simultaneously deposited in 10^{-7} mbar vacuum on 7 cm × 4 cm Corning®7059 glass and crystalline silicon (100) substrates in a MVSystems® HWCVD chamber as described elsewhere [3]. The low Ta filament temperature of 1490°C and substrate heater temperature of 240°C were combined with low gas flow rates of 5 sccm, 7 sccm and 20 sccm respectively for SiH_4, NH_3 and H_2. The overall gas pressure in the deposition chamber was maintained at 100 μbar for 30 minutes. After deposition the glass and Si substrates were sectioned into 1 cm × 1 cm squares and isochronally annealed for 30 minutes in 1.3×10^{-6} mbar vacuum. At maximum annealing temperature of 1000°C only the Si substrate sample was annealed due to choice of temperature resistant transparent substrate. Fourier transform infrared (FTIR) spectra were collected in transmission mode on the c-Si substrate samples and thereafter corrected for coherent and incoherent reflections [4-5]. The Si-N stretching and the Si-H stretching mode centred around 850 cm^{-1} and 2100 cm^{-1} [6] were investigated for bonding configurations as N-H bending and stretching modes at 1200 cm^{-1} and 3330 cm^{-1} respectively were absent from the as-deposited samples. Elastic Recoil Detection (ERD) data from a 3 MeV $^4He^+$ ion beam experiment was used to probe the total hydrogen content. UV-Vis spectroscopy was performed in reflection mode on a spectral range of 200-1000 nm. The spectra were analysed with SCOUT ® simulation software [7], employing an iterative function with fitting parameters according to the O'Leary, Johnson and Lim (OJL model) for an amorphous semiconducting material [8], combined with Bruggeman Effective Media Approximations (BEMA) [7]. X-ray diffraction (XRD) and Raman spectroscopy probed the long range order information of the network. Atomic Force Microscopy (AFM) and High Resolution Scanning Microscopy (HRSEM) were used to study the surface features of the films.

3. Results and Discussion

XRD and Raman spectroscopy results (not shown) confirmed the absence of long range bonding order in the silicon matrix due to the absence of crystalline peaks, for annealing temperatures up to 700°C. HRSEM and AFM surface studies on the as-deposited films revealed distinct areas on the surface of the film rich in columnar structures, while some areas appear smooth due to the growth process. The low-temperature anneals (300°C - 500°C) exhibited smooth features with average rms roughness of 2.36 ± 0.53 nm. At annealing temperatures ≥ 600°C two regions of roughness were discerned: (i) blistered regions where we postulate that trapped gas escaped and broke the surface of the film and (ii) a relatively smooth surface where a decrease in rms roughness is detected. As shown in figure 1, at 1000°C the blistered region (4.99 ± 0.45) showed a pitted surface with holes extending into the film in this region. The smoother region still retained a small rms roughness.

Figure 1. Topographical AFM graphs of (a) smooth region (b) pitted region at 1000°C, and HRSEM viewgraphs of (c) smooth region (2µm scale) (d) pitted region (100 µm scale) at 1000°C.

Overall, a decrease in roughness is observed for the annealing series, as the columnar growth sizes decreased with an increase in annealing temperature. The total hydrogen content in the films is seen to decrease as the annealing temperature is increased, as shown in figure 2. This signifies effusion of hydrogen from the material as the annealing temperature is increased. At 1000°C only surface hydrogen is detected. Complimentary information is given by FTIR difference spectra for the main nitrogen and hydrogen containing modes, calculated as the difference between each sample in its annealed and as-deposited state, as shown in figures 3 and 4. A positive result implies an increase in vibrational intensities for a specific bonding configuration, and a negative result indicates a loss in intensity. The difference spectra can also detect increase of another bonding type due to scavenging of hydrogen or nitrogen atoms from another vibrational mode. There is good agreement between the decrease in the total hydrogen content in the films at high annealing temperatures from ERD analysis and the 2100 cm^{-1} mode in the FTIR difference spectra. Figure 4 further suggests that after an initial decrease in the Si-N bond density no clear pattern for this mode can be discerned. However, at 1000°C anneal temperature a large increase in Si-N mode density is observed. A nitrogen-rich interface between the SiN_x and the underlying c-Si substrate act as possible source of the extra nitrogen available for bonding was discussed in previous work [2]. Increasingly complicated optical models were implemented and discarded to improve the optical fitting result, since the basic OJL construction proved ineffectual, as did a BEMA consisting of an OJL semiconductor containing voids (particles of air). A film-surface interface, refractive index variation along the depth of the thin film, and a substrate-interface [2] in a generic heterogeneous optical model was then developed which successfully modelled all as-deposited and annealed states. It consisted of three virtual layers constituting one thin film. Each virtual layer contained a BEMA matrix which was built up from (OJL amorphous semiconductor + small amounts of Si crystallites for ordering of the network + dielectric functions of generic Si_3N_4 from the database to steer the fitting parameters). Air was inserted as particle material in the overall matrix. The OJL parameters accounted for variations of disorder from near-stoichiometric material types. A BEMA of SiONO-SiO_2 was inserted on top of the stack and a software defined formula to account for scattering effects was implemented. It was found that the high temperature anneal produce material with silicon characteristics from the silicon-rich as-deposited state, which determined the starting value refractive index. We also postulate that effusion of hydrogen and nitrogen bonding type rearrangements also contribute to the optical properties at the respective annealing temperatures. This was concluded from single oscillator Wemple-DiDomenico (WD) analysis [9] performed on the wavelength dispersed refractive index data, and absorption coefficient curves (not shown).The WD results are displayed in figure 5. The single oscillator energy E_0, suggestive of an 'average band gap' of the material provides quantitative information on the total band structure whereas the so-called Tauc gap probes the optical properties near the fundamental band gap of the material.

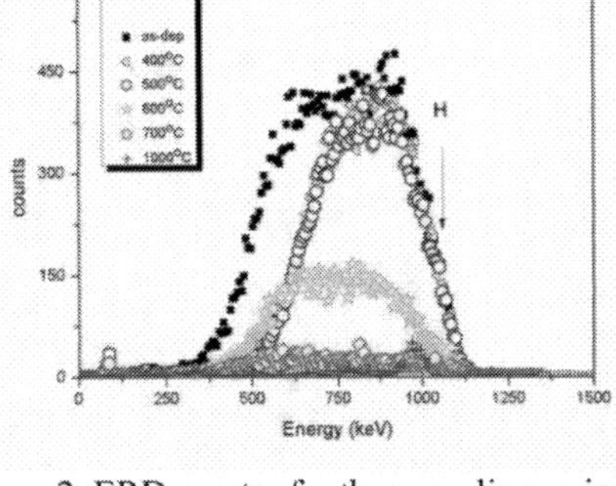

Figure 2. ERD spectra for the annealing series.

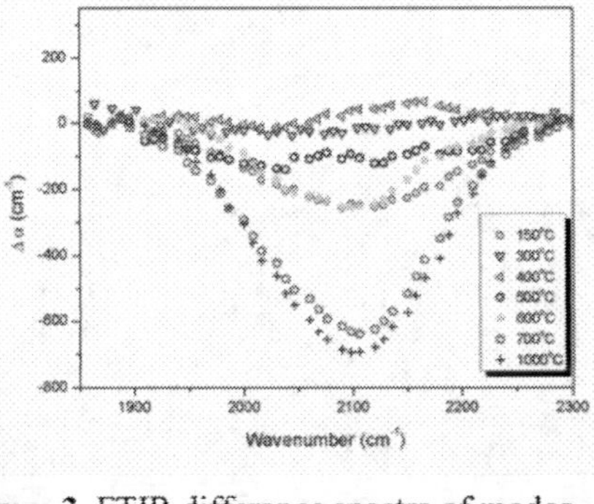

Figure 3. FTIR difference spectra of modes centred around 2100 cm^{-1}.

The oscillator energy E_0 can be empirically related to the optical gap. The dispersion energy E_d, which is a measure of the strength of the interband optical transitions, and the static refractive index n_0, are displayed for the anneal series together with E_0. It is observed for the series that both n_0 and E_d start to increase at 600°C, while E_0 decreases; this is also the first anneal temperature where the greatest loss in hydrogen bonded in Si-H mode occurs, which would influence the density of the thin film.

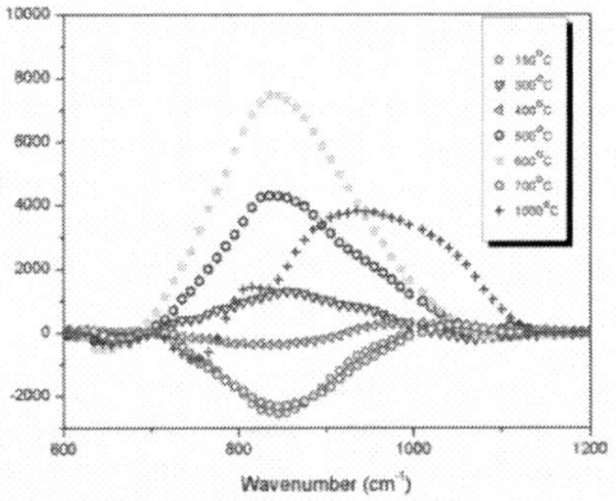

Figure 4. FTIR difference spectra of modes centred around 850 cm^{-1}.

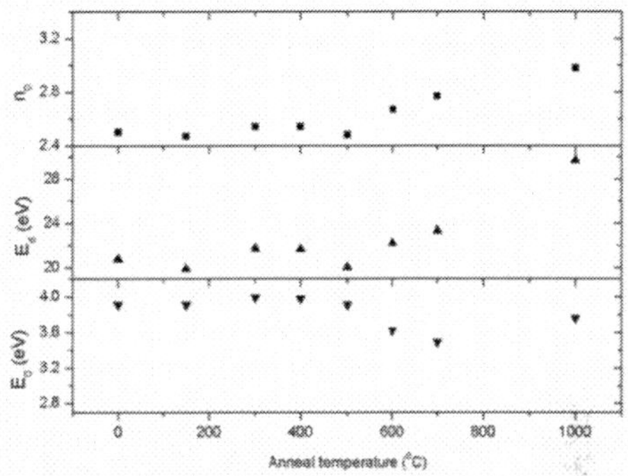

Figure 5. Optical data generated from WD analysis for anneal series.

The variation in the values of the optical parameters would therefore depend on x for the starting material, which determines the microstructure of the material and the manner in which the local bonding environments change at the different annealing temperatures. The analysis therefore makes it possible to undertake an in-depth study of annealing experiments on films with different x values in our deposition series' on low temperatures and gas rates.

4. Conclusion

It was established that hydrogen effuses from the thin films as the annealing temperature increased, with bond rearrangements occurring at the different annealing temperatures. A complex model normally reserved for spectroscopic ellipsometry (SE), was based on the physical and structural characteristics of the film-substrate system and then applied on UV-vis reflectance data, which is a more readily available measurement technique. Wemple-DeDominici analysis proved a valuable tool to relate the microstructure of the material to the optical properties.

Acknowledgements
The authors thank the National Research Foundation of South Africa for financial support.

References

[1] Liu F, Ward S, Gedvilas L, Keyes B, To B, Wang Q, Sanchez E and Wang S 2004 *J. Appl. Phys.* **96(5)** 2973
[2] Oliphant C J, Arendse C J, Muller T F G and Knoesen D 2013 *Appl. Surf. Sci.* **285P** 440
[3] Adams A 2013 *Hot-Wire Chemical Vapour Deposition of Silicon-Nitride Thin Films* (MSc. Thesis, University of the Western Cape, Bellville, South Africa)
[4] Brodsky M H, Cardona M and Cuomo J 1977 *Physical Review* B **16** 16
[5] Maley N 1992 *Phys. Rev.* B **46** 46
[6] Verlaan V, van der Werf C H M, Arnoldbik W M, Goldbach H D and Schropp R E I 2006 *Phys. Rev. B* **73** 195333
[7] Theiss W 2006, Hard and Software for optical spectroscopy, http://www.wtheiss.com
[8] O'Leary S K, Johnson S R and Lim P K 1997 *J. Appl. Phys.* **82** 3334
[9] Wemple S H and DiDomenico M 1971 *Phys. Rev. B* **3** 1338

Polariton-condensation effects on photoluminescence properties in a CuBr microcavity

Masaaki Nakayama, Katsuya Murakami, Yoshiaki Furukawa and DaeGwi Kim

Department of Applied Physics, Graduate School of Engineering, Osaka City University, Sugimoto, Sumiyoshi-ku, Osaka 558-8585, Japan

nakayama@a-phys.eng.osaka-cu.ac.jp

Abstract. We have investigated the photoluminescence (PL) properties in a CuBr microcavity with HfO_2/SiO_2 distributed Bragg reflectors at 10 K from the viewpoint of cavity-polariton condensation. From the excitation-power dependence of PL spectra detected at an in-plane wave vector of $k_{//}=0$, we found that the PL characteristics of the lower polariton branch markedly change with a threshold nature: a drastic increase (decrease) in the intensity (band width). In addition, the PL energy exhibits a large blueshift, ~10 meV, around the threshold excitation power, reflecting strong polariton-polariton interactions. The blueshifted PL energy is far below the bottom energy of the cavity photon. In addition, the estimated density of photogenerated electron-hole pairs at the threshold excitation power is two orders lower than the Mott-transition density for the formation of electron-hole plasma. The above results consistently demonstrate the occurrence of the cavity-polariton condensation.

1. Introduction

Exciton polaritons in semiconductor microcavities, the so-called cavity polaritons, are bosonic quasiparticles resulting from the strong coupling between excitons and cavity photons. The bosonic properties have attracted much attention in Bose-Einstein condensation [1-6] and polariton lasing from polariton condensates without population inversion [7-9]. Recently, wide-gap semiconductors such as GaN [4, 7, 8], ZnO [6, 9-12], and copper halides (CuCl, CuI, and CuBr) [13-15] have been adopted as the active layers of microcavities because the excitons have large binding energies leading to high thermal stability of the cavity polariton: This is advantageous in applications of microcavities. In our previous works on copper halide microcavities [13-15], we demonstrated giant Rabi splitting energies of the order of 100 meV and room-temperature stability of the cavity polaritons. Especially, in a CuBr microcavity, we found that there is negligible population of the cavity polaritons in the bottleneck region under a weak excitation condition [15]. This fact suggests that the relaxation process of the cavity polaritons in the lower polariton branch (LPB) is hardly affected by a bottleneck effect, which is favorable for the cavity-polariton condensation.

In the present work, we have focused on the polariton-condensation effects on PL properties in a CuBr microcavity with HfO_2/SiO_2 distributed Bragg reflectors (DBRs) at 10 K. With an increase in excitation power, the intensity (width) of the PL band of the LPB at $k_{//}=0$ abruptly increases (decreases) with a threshold nature accompanying with a large blueshift, ~10 meV, due to polariton-polariton interactions. These phenomena reflect the cavity-polariton condensation. We discuss the stability of the cavity-polariton condensation from aspects of the Mott transition for the formation of electron-hole plasma and a change from the strong coupling to the weak coupling due to the large blueshift of the LPB energy.

2. Experimental details

The sample of a CuBr microcavity with HfO_2/SiO_2 DBRs was prepared on a (0001) Al_2O_3 substrate. The thickness of the CuBr active layer was λ, which is given by $\lambda = \lambda_{EX} / \sqrt{\varepsilon_b}$, corresponding to an effective resonant wavelength of the lowest-lying exciton, where λ_{EX} is the resonant wavelength of the Z_f exciton, 418.5 nm [15], and ε_b is the background dielectric constant, 4.06 [16] in CuBr: λ=208 nm. The bottom and top DBRs consisted of 17.5 and 12.5 periods, respectively, and each DBR was terminated by the HfO_2 layer. The HfO_2 and SiO_2 layers were fabricated by rf magnetron sputtering. The CuBr active layer, which corresponds to the cavity in this case, the so-called bulk cavity, was grown by vacuum deposition in 2×10^{-6} Pa. The quality factor of the microcavity was estimated to be Q=~810. It was confirmed from X-ray diffraction patters that the crystalline CuBr layer is just oriented along the [111] direction. The details of the sample preparation were described in Ref. [15].

In the PL measurements, the excitation light source was a mode-locked Ti:sapphire laser with a pulse duration of 110 fs and a repetition rate of 76 MHz. The energy of the fundamental laser light was doubled for the excitation: 3.351 eV that is higher than the LPB energy at an in-plane wave vector $k_{//}=0$ by ~0.45 eV. The incident angle of the excitation light was normal to the sample surface, $\theta=0°$, and the spot diameter on the surface was ~30 μm. The detection angle of the PL was fixed to $\theta=0°$, which corresponds to detecting the PL at $k_{//}=0$. A spread of the detection angle was ±2°. The PL spectra were detected with a cooled charge couple device attached to a 32-cm single monochromator with a resolution of 0.15 nm. In addition, angle-resolved reflectance spectra were measured to characterize the cavity-polariton dispersions. All the optical measurements were performed at 10 K.

3. Results and discussion

At first, we describe the fundamental properties of the cavity polaritons. In a CuBr crystal, there exist three kinds of excitons labeled Z_f, $Z_{1,2}$, and Z_3 in order of energy. The Z_f exciton peculiar to CuBr corresponds to a triplet exciton, the oscillator strength of which is markedly enhanced by mixing between the triplet and singlet excitons [17]. The $Z_{1,2}$ (Z_3) exciton is attributed to the degenerate heavy-hole and light-hole excitons (the split-off-hole exciton). Thus, the strong coupling between the Z_f, $Z_{1,2}$, and Z_3 excitons and cavity photon produces four cavity-polariton branches: the LPB, middle polariton branch 1 (MPB1), MPB2, and upper polariton branch (UPB). Figure 1 shows the incident angle dependence of the energies of observed reflectance dips due to the LPB, MPB1, and MPB2, which are indicated by the solid circles, open circles, and solid squares, respectively. The solid curves indicate the results of the analysis of the cavity-polariton dispersions with use of a phenomenological Hamiltonian for the strong coupling between the three excitons and cavity photon: The details of the analysis are described in Ref. [15]. The dashed horizontal lines and dashed curve show the exciton energies [15] and calculated cavity-photon dispersion, respectively. The bottom energy of the cavity photon is E_0=2.977 eV, and the detuning from the Z_f-exciton energy is Δ=+13 meV. The Rabi splitting energies were estimated to be 33, 114, and 89 meV for the Z_f, $Z_{1,2}$ and Z_3 excitons, respectively.

Hereafter, we discuss the PL properties in the CuBr microcavity. Figure 2 shows the PL spectra at various excitation-power densities, where the reflectance spectrum at $\theta=0°$ is depicted on the top for reference. At the lowest excitation-power density of 0.4 μJ/cm^2, the PL-peak energy just agrees with the reflectance-dip energy of the LPB. This demonstrates that the PL band originates from the LPB at $k_{//}=0$. With an increase in excitation-power density, the PL-peak energy exhibits a remarkable blueshift due to polariton-polariton interactions [1-3, 6-9, 18]. At the excitation-power density of 6.0 μJ/cm^2, the intensity and width of the PL band suddenly increases and decreases, respectively, in comparison with those at 5.2 μJ/cm^2. The drastic changes of the PL properties are discussed below.

Figures 3(a) and 3(b) show the integrated intensity and full width at half maximum (FWHM) of the LPB-PL band, respectively, at $k_{//}=0$ as a function of excitation-power density. It is evident that the PL

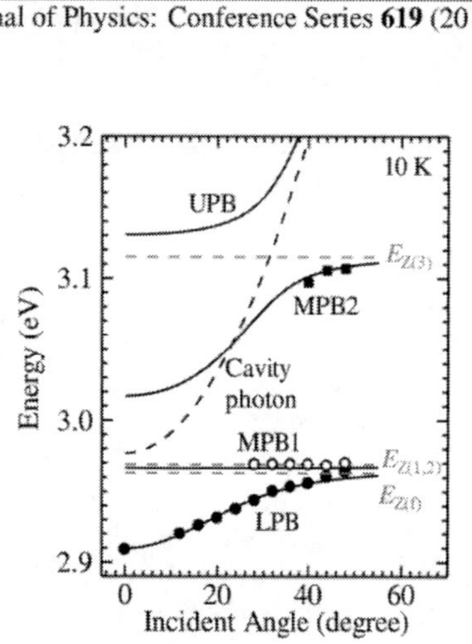

Figure 1. Incident angle dependence of the energies of observed reflectance dips (solid circles, open circles, and solid squares). The solid curves indicate the results of the analysis of the cavity-polariton dispersions.

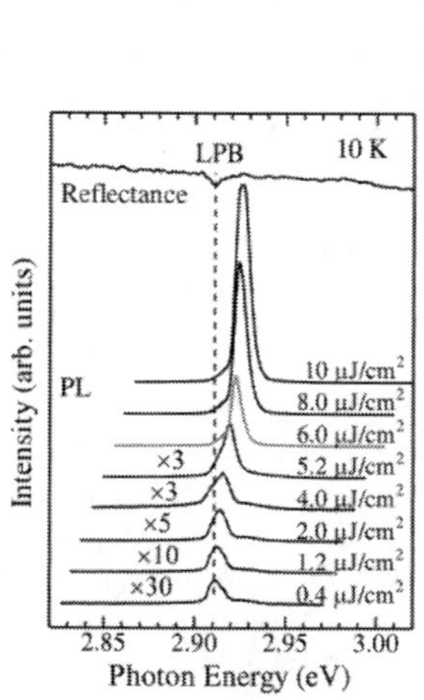

Figure 2. PL spectra at various excitation-power densities, where the reflectance spectrum at θ=0° is depicted on the top for reference.

intensity exhibits a threshold-like increase at 6.0 μJ/cm². The dashed line in Fig. 3(a) indicates the fitted result below the threshold, which exhibits almost linear excitation-power-density dependence of the PL intensity: $I_{\mathrm{PL}} \propto P^{1.2}$. The narrowing of the FWHM just correlates with the threshold-like increase of the PL intensity. These facts indicate a drastic increase of the cavity-polariton population in the LPB at $k_{//}$=0 and population shrinkage in momentum space; namely, the cavity-polariton condensation occurs [1-6]. The extent of PL-band narrowing is not so marked. This seems to be due to the relatively low quality factor and/or coexistence of condensates and non-condensates.

Figure 3(c) shows the peak energy of the LPB-PL band at $k_{//}$=0 as a function of excitation-power density. The blueshift is considerably larger than those in the previous reports, a few meV [1-3, 6-9]; namely, the polariton-polariton interaction in the CuBr microcavity is relatively strong. The highest PL-peak energy is 2.925 eV, which is far below the cavity-photon energy at $k_{//}$=0: E_0=2.977 eV. This fact indicates that the strong coupling is fully kept in spite of the large blueshift, which is owing to the giant Rabi splitting energy. According to Ref. [18], the blueshift of the PL energy is approximately proportional to the cavity-polariton density at $k_{//}$=0 in the LPB. It is evident from Fig. 3(c) that the blueshift remarkably increases around the threshold excitation-power density, 6.0 μJ/cm². This behavior reflects a drastic increase of the cavity-polariton density at $k_{//}$=0, which is consistent with the

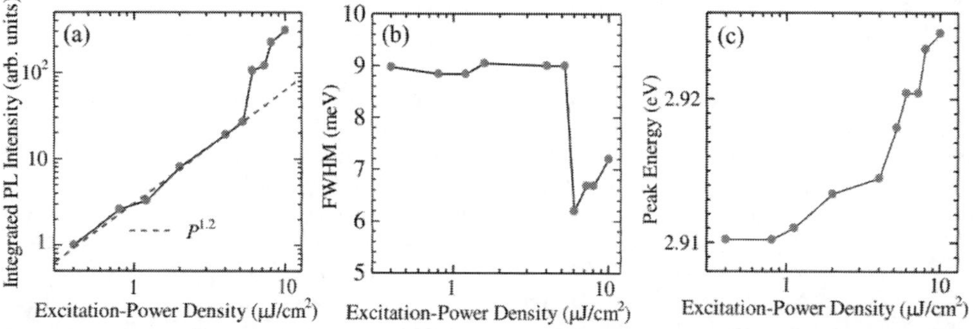

Figure 3. (a) Integrated intensity, (b) FWHM, and (c) peak energy of the LPB-PL band at $k_{//}$=0 as a function excitation-power density. The solid lines are guides for the eye. The dashed line in (a) indicates the fitted result below the threshold.

excitation-power-density dependence of the PL intensity shown in Fig. 3(a). We have to consider the Mott transition in a high density regime. According to Ref. [19], the Mott-transition density is given by $n_M=(2/9)3/(4\pi a_B^3)$ in the framework of the random phase approximation, where a_B is the exciton Bohr radius, 1.25 nm in CuBr [16]: $n_M=2.7\times10^{19}$ cm^{-3}. Taking account of an absorption coefficient of CuBr, 8.2×10^4 cm^{-1} [20] at the excitation energy, the density of photogenerated electron-hole pairs is estimated to be $\sim4.4\times10^{17}$ cm^{-3} at the threshold excitation-power density, which is two orders lower than the Mott-transition density. Thus, the Mott transition is fully negligible in the present case.

4. Conclusions

We have investigated the polariton-condensation effects on the PL properties in the CuBr microcavity. The excitation-power dependence of the intensity and width of the LPB-PL band at $k_{//}=0$ exhibits the threshold-like behavior: a drastic increase (decrease) in the intensity (width). A large blueshift of the PL-peak energy, ~10 meV, due to the polariton-polariton interaction correlates with the excitation-power dependence of the PL intensity. The blueshifted energy of the LPB PL at the threshold excitation power is ~50-meV lower than the bottom energy of the cavity photon. This indicates that the strong coupling is fully kept. The estimated density of photogenerated electron-hole pairs at the threshold excitation power is two orders smaller than the expected Mott-transition density. Consequently, the PL properties described above consistently demonstrate the occurrence of the cavity-polariton condensation in the CuBr microcavity.

References

[1] Richard M, Kasprzak J, André R, Romestain R, Dang Le Si, Malpuech G and Kavokin A 2005 *Phys. Rev. B* **72** 201301.
[2] Balili R, Hartwell V, Snoke D, Pfeiffer L and West K 2007 Science **316** 1007.
[3] Kasprzak J, Solnyshkov D D, André R, Dang Le Si and Malpuech G 2008 *Phys. Rev. Lett.* **101** 146404.
[4] Levrat J, Butté R, Feltin E, Carlin J F, Grandjean N, Solnyshkov D and Malpuech G 2010 *Phys. Rev. B* **81** 125305.
[5] Deng H, Haug H and Yamamoto Y 2010 *Rev. Mod. Phys.* **82** 1489.
[6] Li F, Orosz L, Kamoun O, Bouchoule S, Brimont C, Disseix P, Guillet T, Lafosse X, Leroux M, Leymarie J, Mexis M, Mihailovic M, Patriarche G, Réveret F, Solnyshkov D, Zuniga-Perez J and Malpuech G 2013 *Phys. Rev. Lett.* **110** 196406.
[7] Christopoulos S, Baldassarri Höger von Högersthal G, Grundy A J D, Lagoudakis P G, Kavokin A V, Baumberg J J, Christmann G, Butté R, Feltin E, Carlin J F and Grandjean N 2007 *Phys. Rev. Lett.* **98** 126405.
[8] Christmann G, Butté R, Feltin E, Carlin J F and Grandjean N 2008 *Appl. Phys. Lett.* 93, 051102.
[9] Guillet T, Mexis M, Levrat J, Rossbach G, Brimont C, Bretagnon T, Gil B, Butté R, Grandjean N, Orosz L, Réveret F, Leymarie J, Zúñiga-Pérez J, Leroux M, Semond F and Bouchoule S 2011 *Appl. Phys. Lett.* **99** 161104.
[10] Nakayama M, Komura S, Kawase T and Kim D 2008 *J. Phys. Soc. Jpn.* **77** 093705.
[11] Médard F, Zuniga-Perez J, Disseix P, Mihailovic M, Leymarie J, Vasson A, Semond F, Frayssinet E, Moreno J C, Leroux M, Faure S and Guillet T 2009 *Phys. Rev. B* **79** 125302.
[12] Kawase T, Kim D, Miyazaki K and Nakayama M 2011 *Phys. Stat. Sol. B* **248** 460.
[13] Nakayama M, Miyazaki K, Kawase T and Kim D 2011 *Phys. Rev. B* **83** 075318.
[14] Nakayama M, Kameda M, Kawase T and Kim D 2011 *Phys. Rev. B* **83** 235325.
[15] Nakayama M, Kanatani Y, Kawase T and Kim D 2012 *Phys. Rev. B* **85** 205320.
[16] Madelung O 2004 *Semiconductors: Data Handbook*, 3rd ed. (Springer, Berlin) p.254.
[17] Suga S, Cho K and Bettini M 1976 *Phys. Rev. B* **13** 943.
[18] Ciuti C, Schwendimann P, Deveaud B and Quattropani A 2000 *Phys. Rev. B* **62** R4825.
[19] Haug H and Schmitt-Rink S 1984 *Prog. Quant. Electr.* **9** 3.
[20] Goto T and Ueta M 1967 *J. Phys. Soc. Jpn.* **22** 488.

Defects in hydrogenated amorphous silicon created by intense pulsed illumination at low temperature and the decay of their density

C Ogihara[1], A Nakayama[1], K Yamaguchi[1] and K Morigaki[2*]

[1] Department of Applied Science, Yamaguchi University, Ube 755-8611, Japan.
[2] Department of Electrical-System Engineering, Hiroshima Institute of Technology, Miyake, Saeki-ku, Hiroshima 731-5193, Japan.
[*] Present address: C-305, 2-12 Wakabadai, Inagi, Tokyo 206-0824, Japan.

E-mail: ogihara@yamaguchi-u.ac.jp

Abstract. We present density of dangling bonds of silicon created in a-Si:H films by illumination of intense pulsed light at low temperature. The density of the photo-created dangling bonds decreases at room temperature with increasing time from illumination with a decay time estimated to be approximately 4.2×10^6 s (48 days). The results are compared with the intensity and lifetime of defect PL in a-Si:H films measured at various time from the pulsed illumination.

1. Introduction

Light-induced creation of defects in a-Si:H, which is known as the Staebler-Wronski effect [1], is an interesting phenomenon. Dominant defects in a-Si:H are dangling bonds of silicon (Si DBs). Si DBs in a-Si:H are detected by EPR measurements as a signal observed at $g = 2.0055$. The light induced creation of the Si DBs is observed as increase of the intensity of the EPR signal after illumination of intense light. However the mechanism of the defect creation has not been fully understood. Si DBs in a-Si:H act as non-radiative recombination centres as has been shown by optically detected magnetic resonance measurements [2] . The light-induced creation of the Si DB causes decrease of photoluminescence (PL) of the principal peak at 1.3 – 1.4 eV in intensity. On the other hand, there are defects which act as radiative recombination centres in a-Si:H, as indicated by observation of defect PL of the energy less than 0.9 eV depending on the defect density. The illumination causes the increase of the defect PL in intensity [3], indicating that the radiative defects are also created by illumination together with the Si DBs.

We have investigated the photo-induced phenomena in a-Si:H by monitoring PL. Pulsed illumination at low temperature causes significant enhancement of defect PL [4] and is suitable for the detailed study of the radiative defects responsible for the defect PL. However, the properties of the photo-created defects may be affected by the condition of illumination such as intensity of the light and the temperature of the a-Si:H film during the illumination. Further study of the properties of the photo-created defects is desired for the case of pulsed illumination at low temperature. In this paper we present our recent results of EPR measurements for a-Si:H after the pulsed illumination together with those of defect PL.

2. Experiment

The a-Si:H films used in this study were prepared in a capacitively coupled glow discharge reactor from a mixture of SiH_4 and H_2. The illumination was done at 8 – 10 K by a pulsed light of 2.48 eV from a YAG-OPO laser system for 0.5 h with a repetition frequency of 11 Hz . The intensity of the pulsed light was approximately 2×10^{-2} mJ/cm^2. The a-Si:H films were kept at room temperature after the illumination.

The EPR measurements were performed for the a-Si:H films prepared at the substrate temperature of $T_s = 120$ °C and those after illumination of the pulsed light. The frequency and power of the microwave were 9.45 GHz and 1.0 mW, respectively. The intensity of the signal from Si DBs at $g = 2.0055$ was plotted as a function of the time from the illumination. In order to plot the intensity of the signal from photo-created DBs, the contribution of native DBs, estimated from the measurements of the a-Si:H film as grown, was subtracted.

The defect PL of 0.83 eV was measured under excitation of continuous wave light of 1.95 eV (635nm) for the a-Si:H films prepared at $T_s = 250$ °C after illumination of the pulsed light of 2.48 eV. Non-exponential decay of the PL in a-Si:H is understood as a superposition of exponential decays with lifetime distribution [5]. The broad featureless lifetime distributions of the defect PL were observed by means of frequency resolved spectroscopy (FRS). Characteristic lifetimes, $\tau_{1/2}$, which correspond to the intermediate values of the lifetime distributions [6, 7], were also estimated from the results of the FRS measurements. The intensities, I, and $\tau_{1/2}$ of the defect PL were obtained at various temperatures in the range of 10 – 200 K.

3. Results and Discussion

First we present the results of the density of Si DB after the illumination obtained from EPR measurements. The a-Si:H film used for EPR measurements contains a number of native Si DBs since it was prepared at low T_s. The intensity of the EPR signal obtained for the a-Si:H film before illumination, which represents the number of the native Si DBs, N_{s0}, was subtracted from those obtained for the a-Si:H films after illumination, N_s. Thus the number of photo-created DBs, $N_s - N_{s0}$, is plotted as a function of the time from the illumination, t. Figure 1(a) shows decay of the number of the photo-created Si DBs in the a-Si:H film measured for several months after the illumination. The result indicates that Si DBs in the a-Si:H film created by the pulsed illumination at low temperature are partially annealed out at room temperature. We have also found from EPR measurements of the a-Si:H film for several years after the pulsed illumination that some photo-created Si DBs have a much longer decay time. The result in figure 1(a) is fitted by least squares method with an equation

$$N_s - N_{s0} = a \exp(-t/b) + c \quad , \qquad (1)$$

where a and b are constants characterizing an exponential decay and c is a constant approximately describing other components of much longer decay time. We obtained a curve with $b \simeq 4.2 \times 10^6$ s (48 days) and $c/a \simeq 0.23$. The curve is also shown by a broken line in figure 1(a).

Figure 1(b) shows the number of the photo-created Si DBs including those measured for several days after illumination. A significant decrease of the number of Si DBs at room temperature with increasing time after the illumination is seen. Due to the difference of the condition of the illumination, e.g. the area of the illuminated surface, the number of the photo-created DBs for the results in figure 1(b) is actually smaller than that for the results in figure 1(a) and the error in figure 1(b) seems to be larger than that in figure 1(a). However, the most of the results in figure 1(b) are likely described by (1) with the same b and c/a as above, which is also shown by a broken line in figure 1(b).

Next we present the intensity and characteristic lifetime of the defect PL measured at various temperatures in the range of 12– 200 K. The measurements were done for a high quality a-Si:H

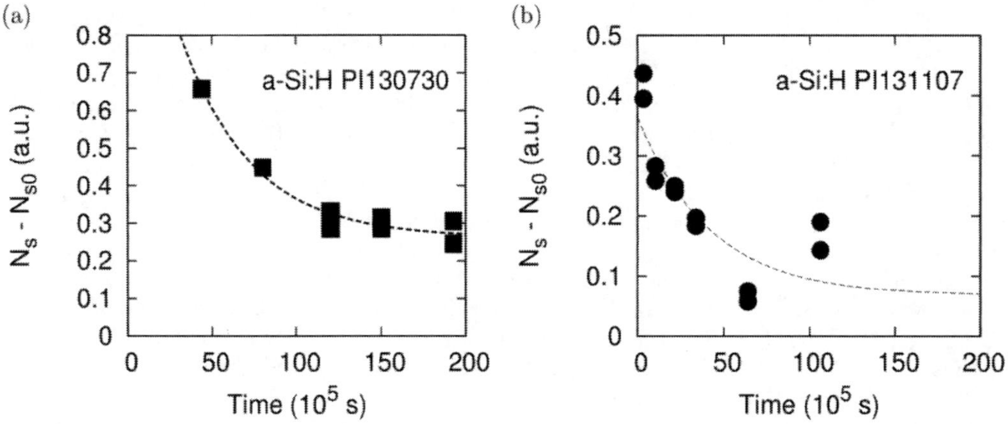

Figure 1. (a) Decay of the number of the photo-created Si DBs in the a-Si:H film obtained for several months after the illumination, and (b) that obtained for shorter time after illumination. The curve of (1) with $b \simeq 4.2 \times 10^6$ s (48 days) and $c/a \simeq 0.23$ is also shown by a broken line.

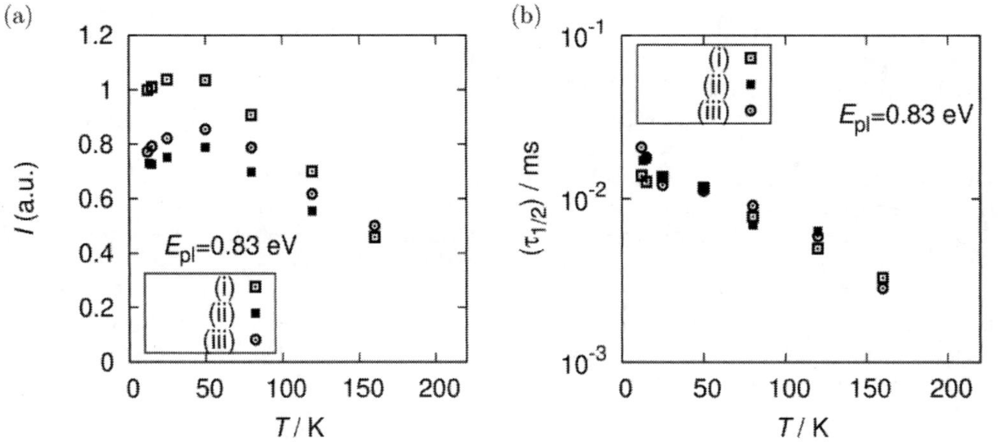

Figure 2. (a) Intensity and (b) characteristic lifetime of the defect PL of 0.83 eV in a-Si:H after illumination of the pulsed light. The results were measured at (i) 9×10^4 s, (ii) 3.5×10^5 s and (iii) 1.12×10^6 s after illumination.

film prepared at $T_s = 250$ °C. The defect PL from the a-Si:H film as grown is too weak to obtain the FRS results, because of the low defect density. Therefore the FRS signal from the a-Si:H film after illumination is due to photo-created radiative defects. Figures 2(a) and (b) show the intensity and the characteristic lifetime of the defect PL of 0.83 eV, respectively, measured at different time after illumination and plotted as functions of the temperature, T. The results do not strongly depend on the time after illumination, although the time after illumination is still shorter than b obtained from the above EPR results. The characteristic lifetime observed at low temperature is approximately 20 μs and much smaller than that we have previously reported for various a-Si:H films [8]. The short lifetime may be due to high density of Si DBs which act non-radiative recombination centres.

4. Conclusions

We have observed that Si DBs in the a-Si:H film created by the pulsed illumination at low temperature are partially annealed out at room temperature. The decay time of the photo-created DBs is estimated to be 4.2×10^6 s (48 days). The intensity and lifetime of defect PL due to radiative defects have also been obtained at various time after illumination. However, the results of the defect PL do not strongly depend on the time after illumination.

Acknowledgments

We would like to thank Prof. A. Kai for her help in the EPR measurements carried out by using a spectrometer in Organization of Research Initiatives, Yamaguchi University.

References

[1] Staebler D L and Wronski C R 1977 *Appl. Phys. Lett.* **31** 292
[2] Morigaki K 1999 *Physics of Amorphous Semiconductors* (Singapore: World Scientific)
[3] Pankove J I and Berkeyheiser J E 1980 *Appl. Phys. Lett.* **37** 705
[4] Ogihara C, Nomiyama T, Yamamoto H, Nakanishi K, Harada J, Yu X and Morigaki K 2006 *J. Non-cryst. Solids* **352** 1064
[5] Stachowitz R, Schubert M and Fuhs W 1994 *Phil. Mag.*B **70** 1219
[6] Ogihara C, Yu X, Nomiyama T, Yamamoto H and Morigaki K 2007 *J. Mater. Sci. Materials in Electronics* **18** S103
[7] Ogihara C, Yu X and Morigaki K 2008 *J. Non-Cryst. Solids* **354** 2121
[8] Ogihara C and Morigaki K 2012 *Phys. Stat. Sol.*C **9** 2574

Journal of Physics: Conference Series **619** (2015) 012017 doi:10.1088/1742-6596/619/1/012017

Changes in screen-printed ZnO/$CuInSe_2$ p-n junction before and after laser ablation

A. Ogurcovs, Vj. Gerbreders, E. Tamanis, A. Gerbreders
G. Libert's Innovative Microscopy Centre, Daugavpils University,
1 Parades Str., Daugavpils, LV-5401, LATVIA

E-mail: max.kovac@gmail.com

Abstract. In the experiments, ZnO and $CuInSe_2$ semiconductor powders were used (the latter produced by the high-temperature synthesis method). Microstructural properties of the powders were analyzed using X-Ray diffraction and SEM. The experimental ZnO/CuInSe2 samples were prepared by the screen printing technique. The V–I measurements of the samples indicated the presence of a p-n junction in the ZnO/$CuInSe_2$ contact zone. The samples were exposed to artificial solar radiation, during which an exponential decay of photo-emf was observed. It was shown that the starting conductive properties of the p-n junction restores in ~ 15 min after the exposition. Laser ablation technique was applied to ZnO/CuInSe2 p-n junction with purpose to improve its performance and quality. After several attempts, significant changes in electrical properties of samples was observed. Improvenent of p-n junction can be achieved only at single attempt at certain amount of delivered energy. Futher ablation attempt leads to degradation of p-n junction of samples.

1. Introduction

Chalcopyrite $CuInSe_2$ (CISe)-based materials are most promising for thin-film photovoltaic devices owing to their high absorption coefficient (α~10^5/cm), tunable bandgap (1.7-1 eV) material that is dependent on the composition and higher efficiency than other thin-film solar cells [1]. Various technical methods aimed at improving the performance and decreasing the production cost of CISe solar cell with single-crystal and thin-films CISe have been reported [2]. CISe absorber layers can be prepared using several procedures, such as sputtering [3], electro-plating deposition [4], evaporation [5], and printing techniques [6]. Despite the advantages offered by above mentioned methods, thin film semiconductors are not useful in devices without thermal annealing to form the desired semiconductor phase, to improve crystallinity, and to minimize structural and electronic defects [7]. The main purpose of this work is to verify if laser ablation (LA) is compatible with a damage-free processing of CISe compound and improvement of quality of p-n junction formed by ZnO/CuInSe2 contact. Laser ablation is quite different from conventional furnace annealing (FA) and or rapid thermal annealing (RTA). The main advantage that laser ablation offers is selective heating of the sample. Some works [8] reporting attempts in processing $CuInSe_2$ using laser ablation with little or no success, however more work needs to be done in order to understand laser ablation of $CuInSe_2$ [9].

2. Experiment

For the experiments, ZnO and $CuInSe_2$ powders were taken. Using a laboratory grinding machine Retsch PM100, the average size of ZnO particles was reduced to 300 nm, and $CuInSe_2$ particles – to 2.5 μm. After the milling process, ZnO powder was annealed in a muffle furnace at 500°C for 120 min. CuInSe2 was obtained by high-temperature synthesis. Initial components (Cu, In, Se) were put in the evacuated (air-less) quartz jar. The jar was slowly heated (20°C/hour) till 1000°C. The homogenous melt was obtained within 48 hours at 1000 °C, then the jar was slowly cooled (7°C/hour). The total time of $CuInSe_2$ synthesis took 240 hours. After a preliminary mechanical grinding, the average grain size was reduced to 7 microns. As a surface active agent, polyvinyl butyral $(C_8H_{14}O_2)_n$ was used. The paint for screen printing was prepared by mixing ZnO and $(C_8H_{14}O_2)_n$ in the proportion of 97:3; the same proportion was applied for the $CuInSe_2$ paint. The glass substrates with 180 nm DC magnetron sputtered ITO layer were coated with a 3 μm thin $(C_8H_{14}O_2)_n$ + ZnO layer using the screen printing method. After drying the first layer, a 10 μm layer of the same composition $(C8H_{14}O_2)_n$ + $CuInSe_2$ was covered onto the first layer. The prepared samples were dried for 20 min in muffle furnace at 70 °C. A 300 nm magnetron-sputtered cobalt layer was used as a back-contact. Each layer was coated with a 1mm shift relative to each other (in a sandwich structure) to prevent short-circuiting of the $CuInSe_2$ layer on the ITO layer. Laser ablation was performed by the scanning laser system "TruMARK 6000" in CW mode with wavelength λ = 1060 nm. Laser focus diameter was 50 μm, distance between each track – 50 μm, scanning speed – 450 mm/s. Laser radiation were directed through the glass and from the CuInSe2 side (Fig 1.). Chemical and surface analysis was performed using SEM "Vega Tescan LMU" with "INCA x-ray microanalysis" attachment. A solar simulator "Oriel Sol2A" with a 140W lamp and multimeter "Picotest M3500A" were used in photoelectric studies of the samples before and after laser ablation.

3. Results and Discussion

Samples were processed from CuInSe2 side with 0.5W of laser power (1% of maximum power); this led to melting of separate grains to larger areas (Figure 1). Irradiated surface is rather heterogeneous, containing multiple cracks and cavities. Total thickness of the CuInSe2 layer was ~ 15 μm, thickness of meltdown area is ~ 8 μm, which has almost no noticeable effect on photovoltaic properties of the sample. An increase of laser power didn't affected thickness of the melting layer and led to the destruction of working properties of CuInSe2 compound.

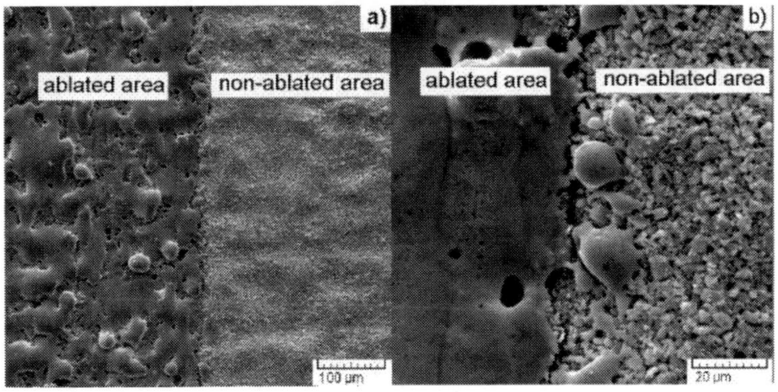

Figure 1. SEM images of the ablated and non-ablated CuInSe2 areas.

Melting of Culnse2 grains without significant change in stoichiometry (Table 1), can be achieved by applying a certain amount of the laser radiation dose, going beyond these limits leads to irreversible changes in the structure of CuInSe2, and loss of functional properties of the sample. Chemical analysis of an irradiated CuInSe2 area indicates reduced level of Se, most likely, this is due to the evaporation of this substance as a result of the intense heat.

Table 1. Chemical analysis of the sample before and after irradiation

	Average atomic ratio %		
	Cu	**Se**	**In**
non-ablated area	26.644	48.651	24.705
ablated area	27.591	42.775	29.663

From above mentioned experiments, it can be concluded that irradiation from CuInse2 side is non-effective, therefore, an attempt of ablation of sample from ZnO side was made. During an irradiation of p-n junction from through ZnO layer, laser power was increased to 1W, due to insufficient transparency of ZnO layer. Sample treatment was carried out in one attempt with scanning laser beam on 2 $cm^{\wedge 2}$ large surface area. Chemical analysis revealed no changes in the chemical composition and structure of ZnO, indicating the absence of absorption of laser radiation. Measurement results of the dark current-voltage characteristics of the sample before and after processing (through ZnO layer) are shown in Figure 2.

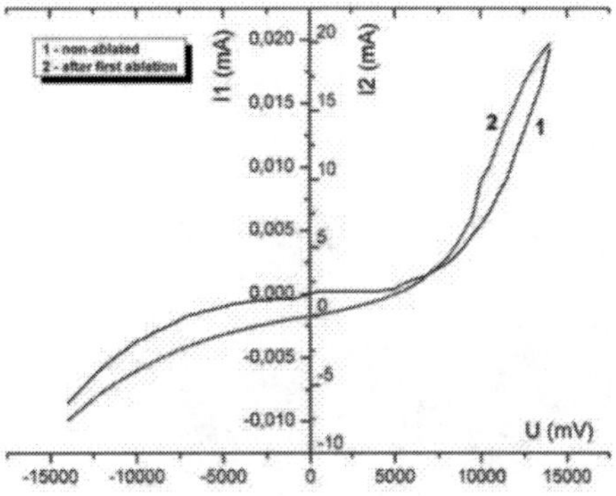

Figure 2. Dark voltage current curves of ZnO/CuInSe2 p-n junction before and after ablation.

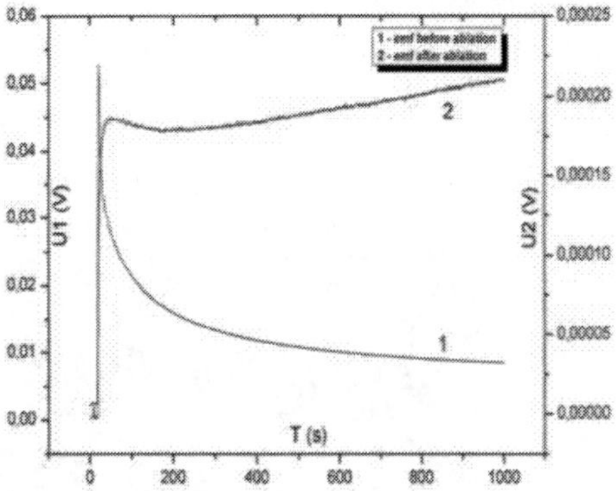

Figure 3. Time base of the photo-emf level of the sample before ablation (curve 1) and after (curve 2).

After ablation, the quality of the current-voltage curve has improved significantly, but most importantly, the dark resistance of the p-n junction decreased by 2 orders of magnitude. This fact is apparently associated with improvement of contact between the layers of ZnO and CuInSe2. Subsequent treatment of the same sample leads to irreversible degradation of the properties of the p-n junction. Destruction of p-n junction, presumably related to the formation of various phases based on the elements constituting CuInSe2, such as SuSe2, InSe2, as well as Cu and In pure form, that shunts the p-n junction in the contact zone with ZnO. Comparative analysis of the photovoltaic properties of the sample before and after irradiation revealed significant changes in the dynamics of the photo-emf. Sample before ablation indicated an abrupt rise of photo-emf level to $V_{ph}(max)$= 5.5E-3 V, with followed exponential decay to 330E-6 V within ~ 15 minutes (Fig.3. Curve 1). Completely different picture was observed after ablation of the sample. After sample illumination of an artificial solar spectrum, immediate rise of photo-emf level to V_{ph}=200E-6 and subsequent slow rise to V_{ph}=330E-6 was observed (Fig.3. Curve 2). Such dynamics of photo-emf before irradiation, presumably related to the large number of defects in the contact zone of ZnO/CuInSe2, which negates the charge separation in the potential barrier. An interesting fact is that the final value of the photo-emf reaches the same value ~ 330E-6 V. The observed decay of V_{ph} might be a result of a decrease in the height of the potential barrier under the influence of solar radiation or formation of the charge carrier recombination centers. The potential barrier's height can decrease with Fermi's level in $CuInSe_2$ changing due to generation of the electrically active point defects, with the donor and acceptor levels created. If the generated defects are the charge carrier recombination centers, this will lead to decrease in the photo-emf level because of a reduced (under direct sunlight during V_{ph} measurements) stationary stream of the excited charge carriers. In this case, most likely, changes in the properties of the potential barrier are caused by generation of defects in the area adjacent to $ZnO/CuInSe_2$ contact [10]. In turn, such generation might be due to flaws in the $CuInSe_2$ synthesis process, or in the mechanical milling of starting materials followed by residual stress in the crystalline lattice of experimental materials.

4. Conclusion

Properly chosen laser parameters can significantly improve properties of the p-n junction in the contact ZnO/CuInSe2 zone. Quality improvement of p-n junction is possible only at a certain limits of laser radiation, irradiation beyond these limits leads to irreversible degradation of the p-n junction. Recrystallization of CuInse2 causes annealing of the large amount of defects that positively affects the photovoltaic properties of samples. This method has prospects of use in the production because of its cheapness and simplicity.

5. Acknowledgements

This study was supported by ESF project: Nr.2013/0029/1DP/1.1.1.2/13/APIA/VIAA/029.

6. References

[1] Wyatt K Metzger Ingrid L Repins & Miguel A Contreras 2008 Long lifetimes in high efficiency Cu(In,Ga)Se_2 solar cells *Appl. Phys. Lett.* **93** 022110

[2] Tatsuo Fukano Shin Tajima and Tadayoshi Ito 2013 Enhancement of Conversion Efficiency of Cu_2ZnSnS_4 Thin Film Solar Cells by Improvement of Sulfurization Conditions *Appl. Phys. Express* **6** 062301

[3] J A Thornton T C Lommasson H Talieh B H Tseng 1988 Reactive sputtered CuInSe2 *Solar Cells* **24** P 1-9

[4] J F Guillenmoles A Lusson P Cowache S Massaccesi J Vedel D Lincot 1994 Recrystallization of electrodeposited copper indium diselenide thin films in an atmosphere of elemental selenium *Adv. Mater.* **6** 376

[5] A M Gabor J R Tuttle D S Albin M A Contreras R Noufi A M Hermann 1994 High-efficiency $CuIn_xGa_{1-x}Se_2$ solar cells made from $(In_x,Ga_{1-x})_2Se_3$ precursor films *Appl. Phys. Lett.* **65** 198

[6] S A AL Kuhaimi S Bahammam 1990 All Screen-Printed CdS/$CuInSe_2$ Solar Cells *Jpn. J. Appl. Phys.* **29** 1499

[7] Ashish Bhatia Phillip Dale Matt Nowell Michael Scarpulla 2010 Pulsed Laser Processing of Electrodeposited $CuInSe_2$ Photovoltaic Absorber Thin Films *MRS Proceedings* **1268**

[8] Joliet M C Antoniadis C Andrew R and Laude L D 1985 Laser induced synthesis of thin CuInSe2 films *Appl. Phys. Lett.* **46**(3) 266-267

[9] Ashish Bhatia Helen Meadows Alexandre Crossay Phillip J Dale Michael A Scarpulla 2012 Pulsed and continuous wave solid phase laser annealing of electrodeposited CuInSe2 thin films *SPIE Proceedings* **8473**

[10] D Seghier, H P Gislason. Opto-electrical characterization of electrically active point defects in Al_xGa_{1-x}, 2008 *J. Phys. D: Appl. Phys.* **41** 095109.

Phase transitions and Raman scattering spectra of $TlGaSe_2$

R. Paucar[1], H. Itsuwa[1], K. Wakita[1], Y. Shim[2], O. Alekperov[3], N. Mamedov[3]

[1]Chiba Institute of Technology, 2-17-1 Tsudanuma, Narashino, Chiba, Japan
[2]Osaka Prefecture University, 1-1 Gakuen-cho, Naka-ku, Sakai, Osaka, Japan
[3]Institute of Physics, Azerbaijan National Academy of Sciences, H. Javid ave. 33, Baku AZ-1143, Azerbaijan

E-mail: kazuki.wakita@it-chiba.ac.jp

Abstract. In this work we examined the phase transitions in the ternary thallium chalcogenide $TlGaSe_2$ by studying the temperature dependence of the Raman spectra with the aid of confocal microscopy. The unpolarized Raman scattering spectra of $TlGaSe_2$ single crystals were measured over the temperature range 78– 300 K (which includes the range of the successive phase transitions) in the frequency region of 50 – 300 cm^{-1}. The Raman spectra exhibited 12 lines at 300 K, but the number of lines rose to 17 at 78 K. In the temperature interval between 107 K and 120 K, where the phase transitions take place, the temperature dependence of the phonon frequencies showed discontinuities for several of the Raman lines.

1. Introduction

Ternary thallium chalcogenides with chemical formula $TlMeX_2$ (where Me = In, Ga and X = Se, S, Te) belong to a family of semiconductors with a chain or layered structure. These compounds exhibit phase transitions at low temperature. Detailed information about the phase transitions and other properties of these compounds has been collected in a topical review by Panich [1]. In one way or another, the properties are related to successive phase transitions, the nature of which is not yet completely understood.

Many previous studies of $TlGaSe_2$ which belongs to $TlMeX_2$ family have been reported [2-5]. It is commonly accepted that the temperature-dependent structural changes in $TlGaSe_2$ are manifestations of a phase transition from a normal (N) to a commensurate (C) state via an intermediate incommensurate (I) phase. For these phase transitions Ti (N-I transition) and Tc (I-C transition) are around 107 K and 120 K, respectively. As a result of this transition, the unit cell in the ferroelectric C-phase is quadrupled in the crystallographic direction c. However, a manifestation of this quadrupling in the Raman spectra of $TlGaSe_2$ has not been observed.

Although the Raman spectra of $TlGaSe_2$ have been studied comprehensively, the details of the line shapes and their temperature dependence are poorly understood. We decided to study the phase transition in$TlGaSe_2$ by measuring the temperature dependence of the Raman spectra using a confocal spectroscopy system.

2. Experimental details

Single crystals of layered $TlGaSe_2$ were grown by the Bridgman-Stockbarger method. The unpolarized Raman spectra were recorded in the frequency range 50– 300 cm^{-1} in a backscattering geometry such that the electrical field vectors of the incident and scattered light were nearly perpendicular to the normal to the layer plane and c– axis. The sample was mounted in a special vibration-free cryostat and its temperature was varied between 78– 300 K. (This includes the temperature range where the series of phase transitions is believed to occur.) A 532 nm diode-pumped solid-state (DPSS) laser was used as the light source. The spectral resolution was better than $\pm$ 0.8 cm^{-1}.

3. Results and discussion

The Raman spectra of a $TlGaSe_2$ crystal at 78 and 300 K are shown in Figure 1. In the frequency region of 50– 300 cm^{-1} the spectra exhibit at least 12 Raman modes at 300 K and the number rises to 17 at 80 K. Table 1 summarizes the numerical values of the modes frequencies shown in Figure 1. This table also includes Raman peaks reported by other researches [2 - 4] in the frequency range of 50– 300 cm^{-1}.

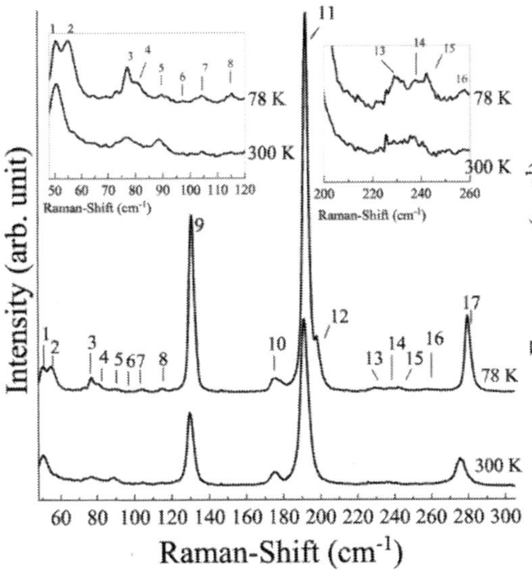

Figure 1. Raman spectra of $TlGaSe_2$ at 78 K and 300 K. (Numbers correspond to the list in Table 1.)

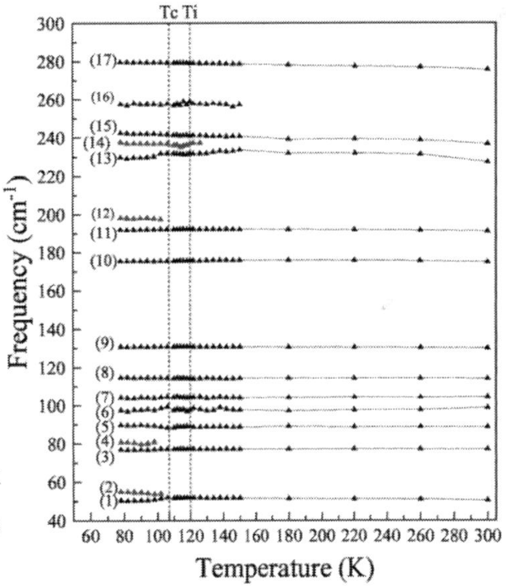

Figure 2. Temperature dependence of the mode frequencies of the Raman spectra of $TlGaSe_2$. The dashed lines represent guides for the eyes. (Number correspond to the list in Table 1.)

Each mode in the measurement Raman spectra was deconvoluted into Lorentzian peaks, using a least squares procedure. The temperature dependence of the modes is shown in Figure 2. As the temperature decrease, additional Raman lines suddenly appear at frequencies of about 56, 82, 199 and 237 cm^{-1}, this is an indication of a structural phase transition at low temperatures [2]. More detailed information about the temperature dependences of the frequency and width (FWHM) of the lines is given in Figure 3. We can clearly observe that in the temperature interval between Tc = 107 K and Ti = 120 K, where the phase transition takes place, the temperature dependence of the phonon frequency and the width of the peak show discontinuities for the highest intensity Raman modes (1, 9, 11 and 17). These results partly agree with those reported earlier [5], where pressure dependence of phonon frequencies in $TlGaSe_2$ also showed discontinuities for modes 1, 11 and 17. Furthermore, from Figure 3 we can also observe also that the mode width (FWHM) exhibits broadening with the temperature increases. This fact explains why Raman modes are clearly resolved at low temperatures, while some closely spaced modes are not resolved at high temperatures.

Table 1. Raman modes frequencies (ν_i) in $TlGaSe_2$ at low and room temperature.

Number of the mode	Ref. [2] ν_i (cm^{-1})		Ref. [3] ν_i (cm^{-1})	Ref. [4] ν_i (cm^{-1})	This work ν_i (cm^{-1})	
	85 K	300 K	300 K	300 K	80 K	300 K
1	53	54	54	53	50.4	50.5
2	56	-	-	-	55.1	-
-	74	-	-	-	-	-
3	79	-	-	-	78.9	77.1
4	83	82	82	80	80.8	-
5	87	88	88	-	89.9	88.8
-	92	-	-	91	-	-
6	-	-	-	-	97.4	98.8
7	106	105	105	105	104.1	104.6
8	118	-	-	-	114.9	114.1
9	133	133	133	132	130.8	129.9
-	-	-	163	-	-	-
10	-	-	-	177	175.5	174.9
11	194	194	194	193	192.0	191.2
12	199	198	198	-	198.2	-
-	204	-	-	-	-	-
13	231	230	230	229	229.8	227.2
14	240	240	240	-	237.6	-
15	244	-	-	-	242.5	236.8
-	250	250	250	249	-	-
16	260	-	-	-	257.6	-
-	-	268	268	-	-	-
17	280	278	278	278	279.7	275.7
-	285	-	-	-	-	-

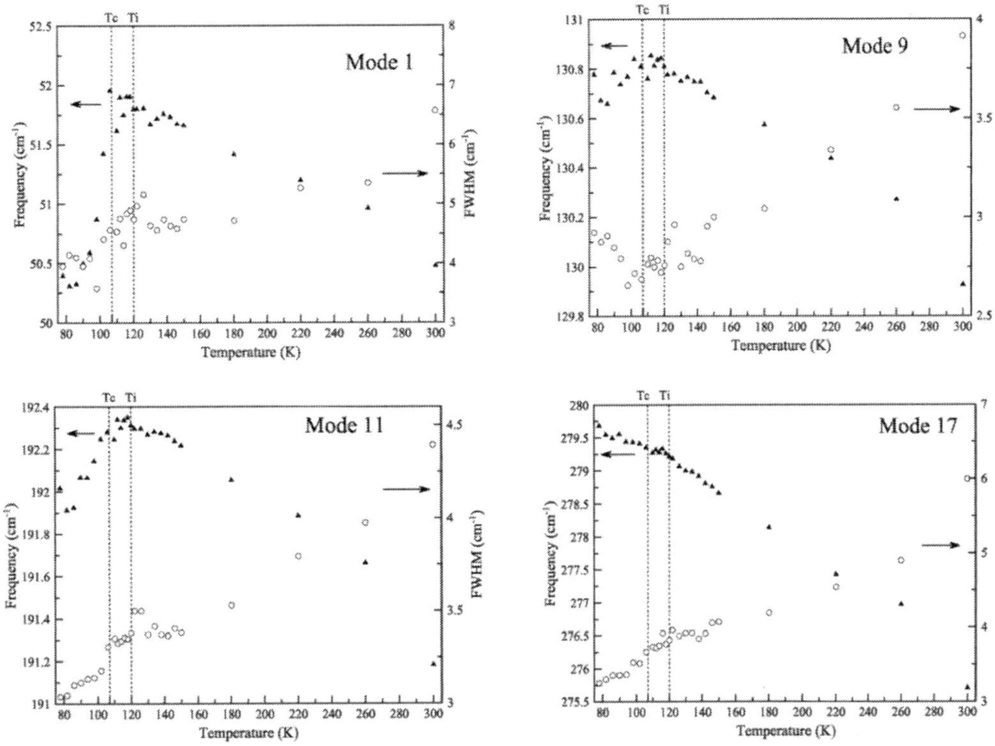

Figure 3. Temperature dependence of frequency and width (FWHM) of Raman modes. (Numbers correspond to the list in Table 1.)

4. Conclusions

We have studied the temperature dependence of the frequency and peak-width of the Raman lines in $TlGaSe_2$ including the temperature range of the reported ferroelectric phase transition through an intermediate incommensurate phase. Our results indicate a complicated picture that cannot be fully explained within the theoretical framework of phase transitions previous reported for $TlGaSe_2$. More studies of the broadening and peak position us a function of the temperature are necessary to fully understand the results obtained in this study. A re-examination of the Raman scattering spectra over a wider range of frequencies and temperatures as well as polarized Raman measurements are under way.

Acknowledgements

This work was supported in part by Japan Ministry of Education, Culture, Sport, Science and Technology under the grant-in-aid No 23560371.

References

[1] Panich, A. M. J. Phys.: Condens Matter 2008, 20, 293202.
[2] K. Allakhverdiev, R. Sardarly, F. Wondre, and I. F. Ryan, Phys. Status Solidi B 88, K5 (1978).
[3] B. N. Marvrin, et.al. Fiz. Tverd. Tela (Leningrad) 19, 2960-2963 (October 1977).
[4] W. Henkel et al., Phys. Rev. B 26 (1982) 3211.
[5] E. A. Vinagrodov, et al. Phys. Stat. (b) 95, 383 (1979).

An analysis of electrical and optical characteristics of a type II superlattice for optical switching applications

M Rahman, A S M S Rouf and F M Mohammedy
Department of Electrical and Electronic Engineering, *Bangladesh University of Engineering and Technology*, Dhaka -1000, Bangladesh.

E-mail: muktadir@ieee.org

Abstract. The present work demonstrates an analysis of electronic and optical characteristics of InAs/GaSb Type II superlattice based photodetectors. The electronic characteristics are analyzed by developing a model of the Type II superlattice. The 8 band k.p method is implemented to deduce the wavefunctions. The effects of temperature on zero-bias resistance-area product (R_0A) are also included in the model. The newly proposed M-Structure design method is also implemented for our model. Electrical and optical properties of the material such as dark current density and absorption coefficient are calculated. At 50mV reverse bias, the dark current density is found equal to 1.5×10^{-4} A/cm^2. These calculations are done based on the approximation based models of reflectivity. These results were used to demonstrate the variation of optical properties with applied bias voltage to be used as a novel optical switching technique in WDM based communication networks.

1. Introduction

In recent times III-V semiconductor materials have achieved growing interest for potential new device applications. This interest originates from their excellent optical and electronic properties. One particular example is the InAs/GaSb broken gap Type II superlattice (T2SL) which is finding ever increasing applications in lasing operations and as IR photodetectors for larger effective mass and tunability of the energy gap.

Novel engineering techniques have contributed in the significant reduction of Auger Recombination as well. But the greatest advantage is undoubtedly the tunability of cutoff wavelength between 3μm to 20μm, covering almost the entire EM wave spectrum. Currently the development of these materials has entered a phase where cost effectiveness and mass production have become important concerns. As a result, numerous research activities are currently being conducted to counter the problems involved.

In this paper, such a device was modeled with experimentally verified parameters using the 8 band k.p method. These techniques have been employed only for the LWIR and MWIR range until now. Our results were matched with experimental results and found to be consistent. We demonstrated a correlation between the bias voltage and the cutoff wavelength of the absorption coefficient, which has the potential to be used as a novel optical switching technique within the LWIR range.

2. Modeling of the type II superlattice

The wave functions of the superlattice structure were determined using the 8 band k.p method. The appropriate boundary conditions were applied at the interfaces between layers and the Schrodinger's equation was solved. For this purpose, the interactive education tool nextnano™ was used. The miniband approximation for the T2SL based structure was successfully included in the simulation scheme.

The M-structure is constructed by inserting a thin barrier of AlSb in the middle of the GaSb layer of the superlattice[1,3,4] to enhance the carrier effective mass and provide better control over the band edges. The wavefunctions of the M-Structure was determined through the same method.

3. Characteristics of the T2SL Structure

The diffusion current due to thermally generated minority carriers diffusing into the depletion region is given by

$$J_{diff} = n_i^2 \sqrt{qkT}\left(\frac{D_p}{N_d L_p} + \frac{D}{N_a L_n}\right)\left(e^{qV/kT} - 1\right) \tag{1}$$

The current due to generation and recombination of electron-hole pairs in the depletion region is approximately

$$J_{GR} = \frac{q n_i d(V)\sinh(-qV/2kT)}{\tau_{GR}\, q(V_{bi} - V)/2kT} f(b),$$
$$f(b) = \int_0^{\infty} \frac{du}{u^2 + 2bu + 1}, \text{ where } b = \exp\left(\frac{-qV}{2kT}\right) \tag{2}$$

The Zener tunneling current is given by

$$J_T = \frac{q^3 F(V) V}{4\pi^2 \hbar^2}\sqrt{\frac{2m_T}{E_g}} \exp\left(-\frac{4\sqrt{2m_T}E_g^{3/2}}{3q\hbar F(V)}\right) \tag{3}$$

The Trap-assisted Tunneling Current is given by

$$J_{trap} = \frac{q^2 m_T V M^2 N_t}{8\pi\hbar^3} \exp\left(-\frac{4\sqrt{2m_T}(E_g - E_t)^{3/2}}{3q\hbar F(V)}\right) \tag{4}$$

If the total leakage current is consisted of diffusion, generation- recombination and surface leakage current, it can expressed as,

$$J_T = J_D + J_{gr} + \left(\frac{p}{A}\right) J_S \tag{5}$$

Where p/A is perimeter-to-area ratio of the diode. The surface leakage current J_S can be disregarded due to its very low value.
The curve of Current vs. Reverse Bias for the M-Structure has been displayed in the following figure. The device area was taken to be 300μm×300μm [2].

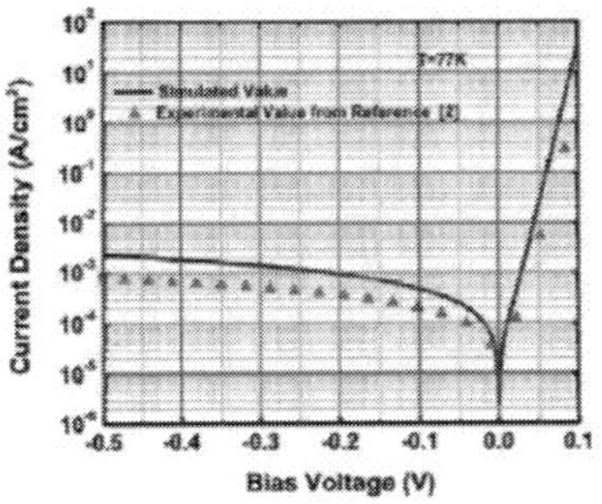

Figure 1. Current Density vs. Bias Voltage Characteristics of M-Structure Design.

5. Novel optical tuning applications in WDM

The cutoff wavelength obtained from the absorption coefficient was plotted against variations of applied bias voltage. As evident from the figure, a negative bias of 0.1 volt displays a variation of almost 50nm. The channel spacing for WDM is 0.8nm [6]. This clearly implies that the T2SL based photodiode has a distinct advantage of tunability over a vast range which may be very effective in WDM networks.

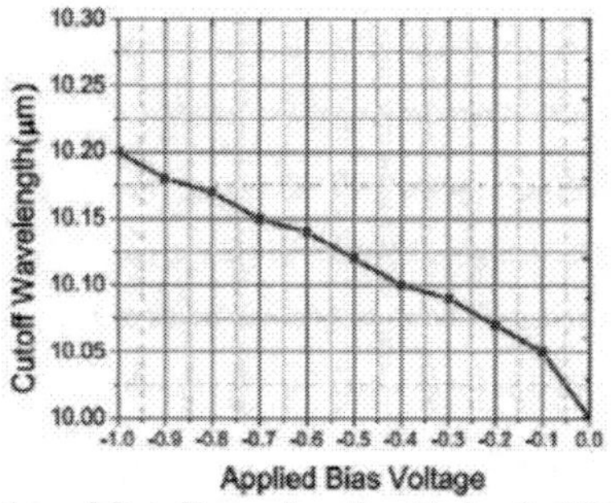

Figure 2. A variation of Cutoff wavelength with Applied Bias Voltage.

Temperature is a very important consideration in any photo detection scheme. A variation of the cutoff wavelength of the basic structure against the temperature was also displayed. This plot also displayed a significant variation in the cutoff wavelength.

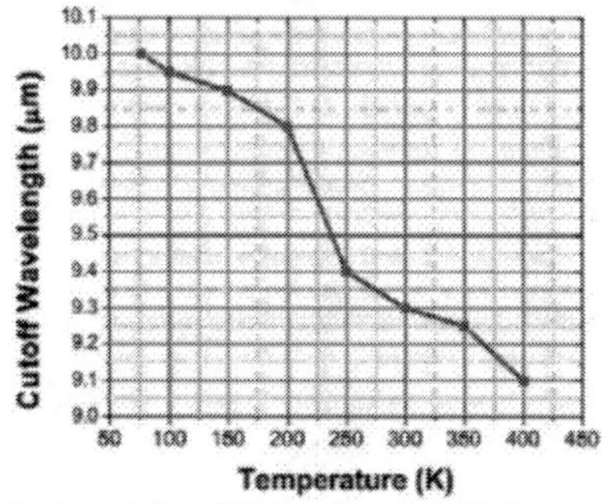

Figure 3. A variation of Cutoff Wavelength with Temperature.

This variation of the cutoff wavelength indicates that the T2SL based InAs/GaSb photodetector may be used in a wide range of temperature scale. As seen from fig. 4, a variation from 400K to 77K displays a change of almost 1 μm in cutoff wavelength.

This discrepancy, that can be seen in some of the simulated curves, can be attributed to the fact that, we did not put in the substrates between the superlattice structure, as a result, the wavefunctions are transmitted instead of reflecting inside the superlattice. This caused a minor disparity in the results of the current density and detectivity simulations when compared with experimental curves.

6. Conclusion

This analysis of the electrical and optical properties effectively demonstrates that the Type-II superlattice is a promising material system, viable for the cost reduction and size expansion of infrared imagers in the third generation. The dark current of 1.5×10^{-4} A/cm^{2}at 50mV reverse bias also displays satisfactory match with experimental values. The M-structure model also displayed promising results. The value of dark current density at 50 mV reverse bias was found to be 5×10^{-4} A/cm^2. A change in temperature also displays change of almost 1μm in cutoff wavelength. The superlattice based photodetector can become the material of choice for the IR camera market. Moreover, WDM promises to be the central technology in the all-optical networks of the future and our research provides the scope for a control parameter of the WDM based network. The modeling and analysis carried out in this thesis work can undoubtedly contribute to the performance enhancement of such photodiodes for the detection of wavelength in the near to mid infrared range.

References

[1] Nguyen B M, Chen G, Hoang M and Razeghi M 2011 *IEEE Journal of Quantum Electronics* **7945** 79451O.

[2] Mou S 2009 Theory and Experiment of Antimony-based Type-II Superlattices Infrared Photodetectors (Illinois: University of Illinois at Urbana Champaign)

[3] Nguyen B M, Razeghi M, Nathan V and Brown G J 2007 Proc. SPIE **6479** 64790S-1-64790S-10.

[4] Razeghi M, Nguyen B M, Delaunay P Y, Huang E K, Pour S A, Manukar P and Bogdanov S 2010 Proc. Of SPIE **7467** 74670T-1.

[5] Chuang S I 1995 *Physics of Optoelectronic Devices* (Wilzy-Interscience publication).

[6] Agarwal G P 2012 *Fiber-Optic Communication Systems* (New Jersey: John Wiley & Sons).

[7] Davies J H 1998 The Physics of Low-Dimensional Semiconductor Devices (Cambridge University Press).

[8] Hoang A M, Chen G, Haddadi A, Pour S A and Razeghi M 2012 *Applied Physics Letters* **100** 211101.

Photonic crystal thin-film micro-pressure sensors

J Sabarinathan, A Bakhtazad, B Poulsen, M Zylstra

Department of Electrical and Computer Engineering, Western University, London, Ontario, N6A 5B9, Canada

Email: jsabarin@uwo.ca

Abstract. We have engineered both the optical and mechanical behaviour of a suspended line defect photonic crystal waveguide to demonstrate an improved sensor response. Utilizing thinner than the conventional photonic crystal membrane increases the mechanical sensitivity and the dynamic range of the sensor. The mechanical sensitivity is also increased by optimizing the membrane pad connected to the core photonic crystal waveguide. The pad optimization also ensures maximum pad flatness under uniform pressure. The overall sensitivity obtained is as high as 6 % optical transmission change per Pa of pressure.

1. Introduction

Photonic crystal structures are increasingly being used for a variety of micro-sensor applications. An important application is the development of highly sensitive microscale pressure sensors for use in medical catheters and in lab-on-chip applications [1-2]. While capacitive-based devices have been widely used for pressure sensors, the frequency range of such devices makes them highly susceptible to electromagnetic interference in the MHz to DC spectral range. In contrast, optical microscale sensors overcome this limitation by being sensitive only to dielectric effects in the optical domain, providing significant performance advantages.

2D planar photonic crystals have been reported to produce micro-pressure sensors with high sensitivity and very small footprint area compared to optical fibre based sensors [3-7]. Point defects in 2D photonic crystal slabs can be used for resonators and sensors by utilizing the very sensitive shift of resonance wavelength by refractive index variation of GaAs [3] or the deformation of the silicon photonic crystal [4] under applied pressure. Realization of such a sensor needs delicate fabrication equipment and high resolution spectrum analyser. Line defects slab photonic crystal can be used similarly and its parameters can easily be adjusted for wider transmission bandwidth and dynamic range and we have previously reported high sensitivity waveguide based PC membrane pressure sensors [5-7].

In this paper, we present significant improvements to the optical and mechanical modelling of the sensor design to create a more robust and stable sensor while simultaneously increasing the overall sensitivity. A thinner than conventional silicon slab photonic crystal has been chosen to increase the dynamic range of the sensor. The engineering of the membrane and its optimization toward maximizing the sensitivity are explained.

This paper is organized as follows. In the next section, we present the basic principle of the sensor. In section three, we present the optical design of the sensor beginning with of SOI wafer dimensions and photonic crystal geometrical parameters needed to achieve good transmission in the telecom range of spectrum 1.3–1.6 μm while improving sensitivity. 3D Finite Difference Time Domain (FDTD) analysis is used and we show how the optical response can be improved by switching to a thinner than conventional slab photonic crystal. In section four we describe the mechanical modelling using finite element method and the optimisation of the geometry to achieve higher sensitivity followed by conclusion in section five.

2. Basic Principal of operation

The micro-pressure sensor is essentially an air-bridge silicon line-defect photonic crystal slab waveguide. In Figure 1(a) the schematic of the sensor is shown after fabrication. As can be seen, the photonic crystal is etched into the top silicon layer of a silicon on insulator (SOI) wafer and then the air-bridge is formed by removal of the buried oxide layer under the PC waveguide.

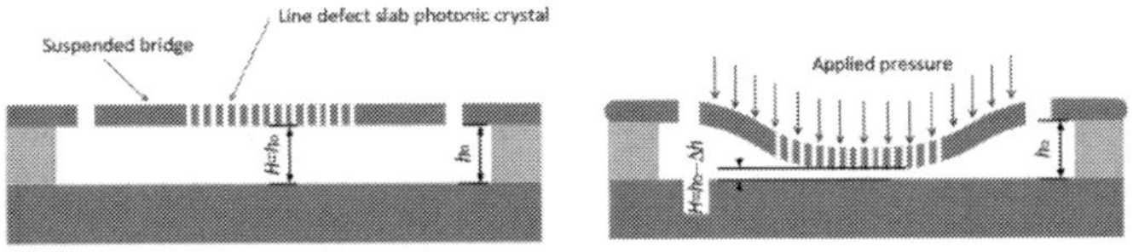

Figure 1 The SOI photonic crystal sensor under (a) no pressure and b) under uniform pressure over the top surface.

Under a uniform applied pressure, the air-bridge membrane will deform towards the substrate as shown in Figure 1(b). The principle of operation of the device is based on the sensitivity of the PC waveguide transmission to the surrounding material in its proximity. When the bridge is brought closer to the substrate, the evanescent field of the line defect waveguide interacts with the substrate and leaks optical power to the substrate. The overall transmission sensitivity of this PC waveguide based micro-pressure sensor can be expressed as a multiplication of the independent optical and mechanical sensitivities respectively given by

$$\underbrace{S_t}_{\text{Total Sensitivity}} \equiv \frac{\partial T}{\partial P} = \underbrace{\frac{\partial T}{\partial h}}_{\text{Optical Sensitivity}} \times \underbrace{\frac{\partial h}{\partial P}}_{\text{Mechanical Sensitivity}} \tag{1}$$

where T is the optical transmittance, P is the applied pressure and h is the photonic crystal height over the substrate.

3. Optical design

In order to improve the optical sensitivity of the sensor we recognized that we had to deviate our design from conventional PC device structure with 220nm device layer to an SOI wafer with top silicon device layer with a thickness of 145nm thickness. The benefit of moving toward to a thinner membrane is two fold. As device layer gets thinner, the line defect photonic crystal made out of it confines increasingly less light in the vertical direction which makes it more susceptible to the surrounding environment. In other words, as the membrane approaches the substrate (on applying pressure), we expect the line defect photonic crystal waveguide mode to see the substrate sooner. As we will show later, it causes the optical sensitivity curve to broaden, and its peak shifts toward higher membrane/substrate gap height. Furthermore, a thinner membrane is also much more mechanically sensitive. The closed form sensitivity equation for thin circular membrane indicates that the sensitivity increases by cubic power of membrane thickness [8]. Typically, the host photonic crystal for a line defect waveguide is a hexagonal lattice with TM light polarization (electric field in the plane of the slab). The lattice geometry has to be chosen so the band gap covers the wavelength of interest. The band diagram is obtained using a 3D plane wave expansion method. For this a lattice constant (a = 550nm), the hole diameter (D = 320nm) are chosen so the band gap extends from 1415- 1720 nm.

A line defect PC waveguide is made by removing a row of holes in ΓK direction of the host photonic crystal [9]. In this design, the line defect length was selected as 25 periods. The width of the input/output channel waveguides to the PC line defect waveguide ($w = \sqrt{3}a \approx 950\,\text{nm}$) was chosen to ensure good coupling of the light into and out of the structure. 3D FDTD simulations were used to obtain the transmission spectrum of the line defect PC waveguide as designed for different heights of the bridge above the substrate as shown in Figure 2a. The transmission decreases as the bridge height decreases and the optical power couples into the substrate as predicted. In Figure 2b the transmission at a target wavelength of $\lambda_c = 1550\,\text{nm}$ and the transmission sensitivity versus bridge height are plotted. The best optical sensitivity is found to be 0.4%/nm, and it is centered on the bridge height of 280nm. Our investigation shows that the sensitivity curve width (the sensor's dynamic range) is wider for thinner bridge; however sensitivity is a bit lower. More importantly, the sensitivity peak shifts toward higher bridge gap heights as the bridge thickness reduces. Table 1 compares these designs.

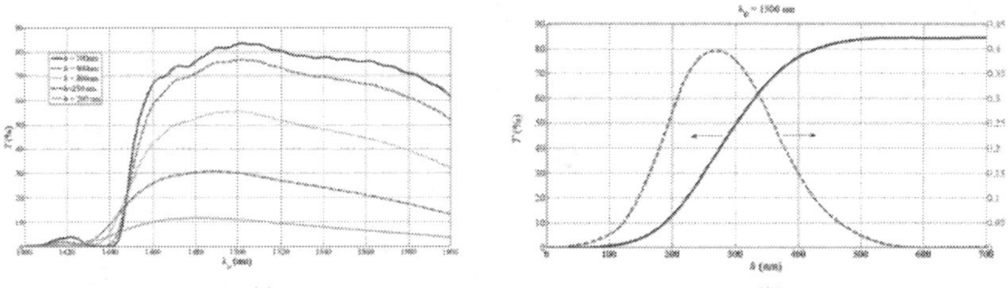

(a) (b)

Figure 2(a) Transmission spectrum of the PC line defect waveguide at different bridge heights. (b)Transmission at wavelength of 1500nm and sensitivity versus bridge substrate gap height

Table 1: Comparison of the previous [5] and current design

Slab height(nm)	Period(nm)	Hole size(nm)	Max. sensitivity (%/nm)	Max. sensitivity depth (nm)	Sensitivity FWHM (nm)
380	450	300	0.5	160	150
145	550	320	0.4	280	200

4. Mechanical design

The second parameter to consider is the mechanical design of the suspended bridge which will be important for the sensitivity as described previously and also the dynamic response of the sensor to applied pressure. The overall sensor sensitivity can be tailored by keeping the optical design (i.e. the PC waveguide configuration) of the bridge intact while altering the mechanical shape of the bridge. As a benchmark, we started with a simple rectangular platform of dimension 14.30μm × 14.29 μm supported by input/output waveguides of length 22.85 μm on both sides (so the total bridge length is 60 μm). To evaluate the behavior of the membrane under uniform pressure, we simulated the structure using a finite element model with appropriate boundary conditions. When the structure is under 1 Pa uniform pressure, the displacement at the center of the rectangular membrane was found to be 3.37 nm. Using Eq.(1), the best total sensitivity obtained is calculated to be $S_t = 3.37 \times 0.4 = 1.35\ (\%/\text{Pa})$. Although the maximum displacement occurs at the edges of the rectangular membrane, however the membrane flatness under uniform pressure as defined by the following equation:

Flatness(%) = 100 X displacement at centre/maximum displacement (2)

is as large as 99.22%.

To improve the mechanical sensitivity, one could increase the length of the supporting input/output waveguide, which would effectively reduce the stiffness of the central platform. However a better approach is to increase the applied force by increasing the area exposed to pressure while keeping the sensor footage constant. Based on this idea, a novel mechanical structure was designed around the photonic crystal waveguide. The new membrane has a 2D PC in the center with the same optical PC line defect waveguide as described previously, but with a larger suspended area. The proposed mechanical design is shown in Figure 3a where the total size of the membrane is kept at 60 μm. The support thickness was optimized by varying the angle φ as defined in Figure 3a to maximize the pad flatness. Figure 3b shows the sensitivity and flatness curves versus φ. As can be seen, sensitivity increases as φ increases, but flatness is maximized at $\varphi = 70.5^0$ where the sensitivity is 16.54nm/Pa and flatness is 95.7%.

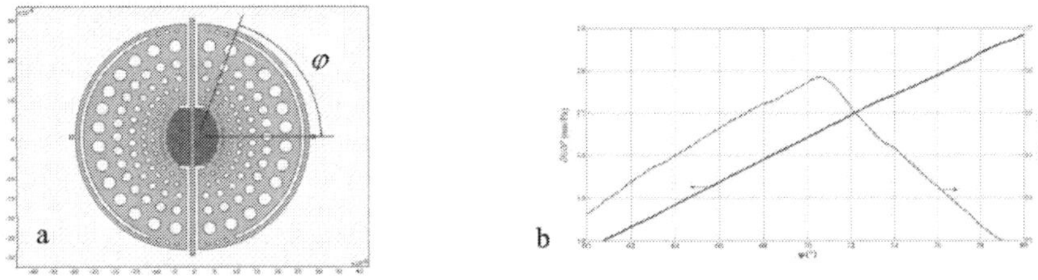

Figure 3 (a)The new membrane design (b) Sensitivity curve and flatness versus support angle φ as defined in (a)

The central platform displacement was found to be 16.54nm/Pa and the overall bridge surface displacement is very flat. Compared to the simple rectangular platform used as a benchmark, the sensitivity is improved by about 5.5 times. Using Eq.(1), one can deduce that the overall sensitivity will increase by 4.9 times ($S_t = 16.54 \times 0.4 = 6.62\ (\%/\text{Pa})$), however the new membrane is slightly less flat compared to the conventional one. Table 2 summarizes the performance of the conventional and the new optimized pads for a couple of sensor sizes. As can be seen the sensitivity of the pad sensor increases by about 5 times as the size increases from 45μm to 60μm.

Table 2, Sensor performance comparison (new/conventional membrane design)

Sensor area(μm)	Mechanical Sensitivity(nm/Pa) (new/conventional)	Flatness, % /nm (new/conventional)
45	(3.21/1.11)	(96.6/98.1)
60	(16.54/3.37)	(95.7/99.2)

5. Conclusion

We have shown that how thin line defect photonic crystal waveguide augmented by engineered petal membrane shaped flaps can exhibit very high sensitivity to applied pressure. The dynamic range of the sensor shows improvement and best operation point shifts toward higher slab/substrate gap width. The mechanical shape of the membrane was optimized to obtain maximum flatness under uniform applied pressure.

References

[1] Lu Z, et al, 2008 "Extracellular superoxide dismutase deficiency exacerbates pressure overload induced left ventricular hypertrophy and dysfunction," Hypertension, vol. 51, no. 1, pp. 19-25.

[2] Miller Inc. 2010 "Cardiovascular Ultra-Miniature Single Sensor Mikro-Tip Pressure Transducer Catheters." http://www.millarinstruments.com/pdf/005-0950B.pdf.

[3] Biallo D, et al, 2007 "High sensitivity photonic crystal pressure sensor," Journal of the European Optical Society-Rapid Publications, vol. 2.

[4] Tung B T, et al, 2011 "Investigation of strain sensing effect in modified single-defect photonic crystal nanocavity," Optics Express, vol. 19, no. 9, pp. 8821-8829.

[5] Bakhtazad A, Sabarinathan J, Hutter J, 2010 "Mechanical Sensitivity Enhancement of Silicon Based Photonic Crystal Micro-Pressure Sensor." International Symposium on Optomechatronic Technologies (ISOT), Toronto.

[6] Sabarinathan J, Bakhtazad A, Huo X, Hutter J, 2010 "Photonic crystal Pressure Sensors", US Patent, filed May, 2010.

[7] Wang Y, Bakhtazad A, Sabarinathan J, 2011 "Reflection mode 2-dimensional photonic crystal slab waveguide based micro-pressure sensor," Proceedings of the SPIE, vol. 8007, 800711.

[8] Di Giovanni M, 1982 Flat and Corrugated Diaphragm Design Handbook Marcel Dekker, Inc.

[9] Chutinan A, Noda S, 2000, "Waveguides and waveguide bends in two-dimensional photonic crystal slabs," Physical Review B, vol. 62, no. 7, pp. 4488-4492.

ICOOPMA2014 IOP Publishing
Journal of Physics: Conference Series **619** (2015) 012021 doi:10.1088/1742-6596/619/1/012021

Terahertz emission from asymmetric, doped quantum wells under resonant pumping

Nathan Shammah[1] [a], **Chris C. Phillips**[2], **and Simone De Liberato**[1]

[1]School of Physics and Astronomy, University of Southampton, Southampton, SO17 1BJ, UK
[2]Physics Department, Imperial College London, London SW7 2AZ, UK

E-mail: [a]nathan.shammah@soton.ac.uk

Abstract. When the intersubband transitions of a doped quantum well are strongly driven by a coherent optical pump, the bare energy levels are split into dressed doublets, a phenomenon known as the ac Stark (or Autler-Townes) effect. In time domain this can be understood as Rabi oscillations of the confined electrons between two subbands. Here we show how, by employing asymmetric quantum wells, the selection rule preventing emission from an upper to a lower state of the same doublet can be lifted. The fluorescence spectrum of such strongly-driven system is then marked by an additional peak that would appear at the Rabi frequency, in the terahertz portion of the electromagnetic spectrum.

1. Introduction

In the terahertz THz region of the electromagnetic spectrum the increasing demand for sources of radiation is challenged by a lack of compact, tunable and efficient devices– the so-called "THz gap"– while potential applications span several fields, from medical imaging to gas detection and product-quality control[1]. Here we illustrate how THz light with tunable frequency could be obtained by strongly driving with a coherent optical pump the intersubband transitions of a doped quantum well (QW)[2]. Asymmetry in the emitters wavefunction, has recently been explored also in other schemes[3, 4, 5, 6, 7, 8].

2. Driven Dynamics

In a 2DEG confined in the x-y plane, the bulk electronic bands are split into parabolic subbands, as shown in Figure (1) (a). Doping of the QW can be used to raise the Fermi energy, $\hbar\omega(k_F)$, where k_F is the Fermi wavevector, up to partially filling the first conduction subband. The ground state of the electronic system is then given by $|G\rangle = \bigotimes_{|\mathbf{k}|<k_F} c^\dagger_{1,\mathbf{k}} |0_\mathbf{k}\rangle$, where $c^{(\dagger)}_{j,\mathbf{k}}$ destroys (creates) an electron in the j-th subband, $j = \{1, 2\}$, with in-plane momentum $\mathbf{k}$, the standard fermionic commutation relations holding, $\{c_\alpha, c^\dagger_\beta\} = \delta_{\alpha,\beta}$, and where $|0_\mathbf{k}\rangle$ is the vacuum state. The Hamiltonian of the free system is given by

$$H_0 = \hbar \sum_\mathbf{k} \omega_{1,\mathbf{k}} c^\dagger_{1,\mathbf{k}} c_{1,\mathbf{k}} + \omega_{2,\mathbf{k}} c^\dagger_{2,\mathbf{k}} c_{2,\mathbf{k}} \tag{1}$$

and the optically active transition is the intersubband transition (ISBT) between the first and the second conduction subband occurring at the frequency $\omega_{12} = \omega_{2,\mathbf{k}} - \omega_{1,\mathbf{k}}$. If a coherent

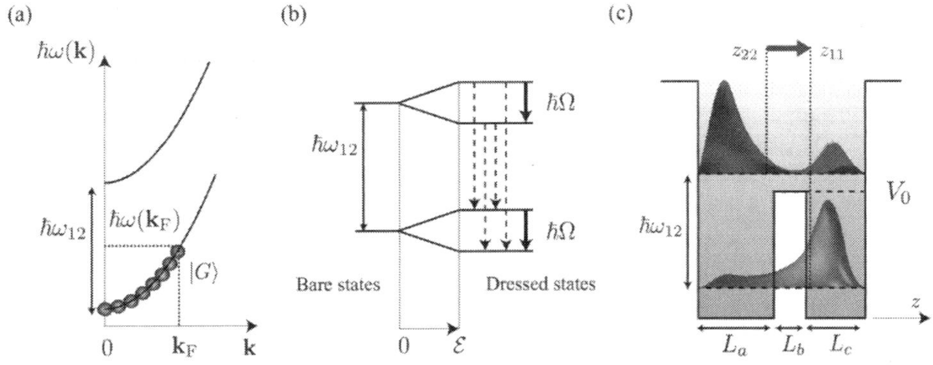

Figure 1. (a) Dispersion in k-space of the two conduction subbands of the doped QW. (b) Dipole transitions at the Rabi frequency Ω are allowed in a broken-symmetry two-level system (thick arrows). (c) Asymmetric QW structure and wavefunctions, with the asymmetric dipole $e(z_{22} - z_{11})$ highlighted; $L_a + L_b + L_c = L_{QW}$, V_0 is the central barrier potential energy.

optical pump of frequency ω_L shines on the system, it can drive the ISBTs. The Hamiltonian in the rotating wave approximation thus reads

$$H = H_0 + \frac{\hbar\Omega}{2}\sum_{\mathbf{k}} e^{-i\omega_L t} c^{\dagger}_{2,\mathbf{k}+\mathbf{q}'} c_{1,\mathbf{k}} + e^{i\omega_L t} c^{\dagger}_{1,\mathbf{k}} c_{2,\mathbf{k}+\mathbf{q}'}, \tag{2}$$

where Ω, known as Rabi frequency, is the pump-ISBT coupling strength, that depends linearly upon the pump amplitude; the transitions are tilted by the in-plane photonic wavevector contribution $\mathbf{q}'$, which is fixed by the pump. Given a state $|\varphi\rangle$ subject to the action of H, we can define $|\varphi'\rangle \to U|\varphi\rangle$, where U is a unitary transformation, that satisfies the Schrödinger equation with respect to the new Hamiltonian H', $H' = UHU^{\dagger} + i\hbar U^{\dagger}\dot{U}$. By choosing $U = e^{iH_0 t/\hbar}$, the Hamiltonian in Eq. (2) at resonance $\omega_L = \omega_{12}$ becomes

$$H' = \frac{\hbar\Omega}{2}\sum_{k}(c^{\dagger}_{2,\mathbf{k}+\mathbf{q}'} c_{1,\mathbf{k}} + c^{\dagger}_{1,\mathbf{k}} c_{2,\mathbf{k}+\mathbf{q}'}). \tag{3}$$

In this rotating frame H' has become completely time independent and, as all transitions now occur between resonant states, the free energies have disappeared. Indeed, we approximated $\omega_{1,\mathbf{k}+\mathbf{q}} \simeq \omega_{1,\mathbf{k}}$, which is a reasonable assumption since the photonic momentum is much smaller than the electronic one. Under strong pumping $\Omega > \gamma$, with γ being the subband line width, the system becomes dressed, the bare energy levels of the two subbands being split by $\hbar\Omega$ into two doublets[9, 10], as shown in Fig.(1) (b).

3. Interaction Hamiltonian

Incoherent photon emission can scatter electrons between states belonging to different Rabi doublets, as shown in Fig.(1) (b) by the dashed arrows. Under strong pumping, this effect gives rise to the "Mollow triplet" in the frequency fluorescence spectrum[11]. The effect can be described considering a two-level system, $|\varphi\rangle = \{|\varphi_1\rangle, |\varphi_2\rangle\}$, interacting with the vacuum electric field

$$V = -\mathbf{d}\cdot\mathbf{E}, \tag{4}$$

where **d** is the dipole moment and **E** is the vacuum electric field. In symmetrical systems, only the off-diagonal terms $V_{ij} = \langle \varphi_i | V | \varphi_j \rangle$ with $i \neq j$, are non-zero. Recently however, it has been pointed out that the development in controlling the design of quantum emitters can be used to engineer asymmetric systems, where the parity of the eigenfunction is not well defined[4]. Then the optical transitions between the upper and lower state of each doublet are no more forbidden $\langle \varphi_i | V | \varphi_i \rangle \neq 0$, as shown in Figure (1) (b) (thick arrows). The dipole operator $\mathbf{d} = e\mathbf{r}$, where e is the electron charge, and $\mathbf{r} = (\mathbf{x}, z)$, can be written in second quantisation by projecting it on the electron states in the two subbands

$$\mathbf{r} = \sum_{\substack{\mathbf{k},\mathbf{k}' \\ j,j'=\{1,2\}}} \hat{\mathbf{r}}_{j,\mathbf{k}}^{j',\mathbf{k}'} c_{j,\mathbf{k}}^{\dagger} c_{j',\mathbf{k}'}. \tag{5}$$

The dipole matrix elements are determined by the in-plane symmetry of the system. The spatial wavefunction of an electron in the j-th subband is $\varphi_j(\mathbf{r}) = e^{i\mathbf{k}\cdot\mathbf{x}} \psi_j(z)$ and thus

$$\hat{\mathbf{x}}_{j,\mathbf{k}}^{j',\mathbf{k}'} = \hat{\mathbf{x}}_{\mathbf{k}}^{\mathbf{k}'} \delta_{j,j'} \qquad \hat{z}_{j,\mathbf{k}}^{j',\mathbf{k}'} = \hat{z}_{jj'} \delta(\mathbf{k} - \mathbf{k}') \tag{6}$$

with

$$\mathbf{x}_{\mathbf{k}}^{\mathbf{k}'} = \int e^{i(\mathbf{k}-\mathbf{k}')\cdot\mathbf{x}} \mathbf{x} d\mathbf{x} \qquad z_{ij} = \int \psi_i^*(z) z \psi_j(z) dz. \tag{7}$$

From Eq. (6) we can see that only the z component of the field can see the asymmetry, and as such, as usual for ISBTs, we can consider just a transverse magnetic (TM) polarised field, whose z component is given by

$$\mathbf{E}_z = \sum_{\mathbf{q},q_z} E_0(q, q_z) e^{i\mathbf{q}\cdot\mathbf{r}} (a_{q_z,-\mathbf{q}}^{\dagger} + a_{q_z,\mathbf{q}}) \sin\theta \mathbf{e_z}, \tag{8}$$

where $\mathbf{e_z}$ is a unit vector, θ is the angle with the z-axis, $E_0 = (\hbar\omega_{q,q_z}/2\epsilon_0\epsilon_r\mathcal{V})^{1/2}$ is the field amplitude, $\omega_{q,q_z} = c\sqrt{q^2 + q_z^2}$, where c is the speed of light, $\mathcal{V}$ is the quantisation volume, and ϵ_0 and ϵ_r are the absolute and relative permittitivities. The interaction of Eq. (4), in the same rotating frame used for Eq. (3) is then rewritten using Eq. (5), Eq. (8), and using the z component in Eq. (7) as

$$V' = -\sum_{\mathbf{k},\mathbf{q},q_z} (a_{q_z,-\mathbf{q}}^{\dagger} + a_{q_z,\mathbf{q}}) \{ \lambda_{12,q,q_z} (e^{i\omega_{12}t} c_{2,\mathbf{k}+\mathbf{q}}^{\dagger} c_{1,\mathbf{k}} + e^{-i\omega_{12}t} c_{1,\mathbf{k}}^{\dagger} c_{2,\mathbf{k}-\mathbf{q}}) + \sum_{j=\{1,2\}} \lambda_{jj,q,q_z} c_{j,\mathbf{k}+\mathbf{q}}^{\dagger} c_{j,\mathbf{k}} \} \tag{9}$$

where $[a_\alpha, a_\beta^\dagger] = \delta_{\alpha,\beta}$ and $\lambda_{ij,q,q_z} = e z_{ij} E_0(q, q_z) \sin\theta$.

4. THz emission rates

The rate of photon emission per unit time, up to first order in perturbation, is given by Fermi's golden rule, which depends upon the matrix element of V' in Eq. (9). The diagonalisation of Eq. (3) and the derivation of the different scattering processes leading to THz emission can be found in Ref. [2]. Here we just give a qualitative analysis: in Eq. (9) the terms in λ_{12,q,q_z} correspond to intersubband transitions that will give rise to Mollow-like peaks in the fluorescence frequency spectrum, similarly to the case of a symmetrical two-level system. As the properties of the spectrum in many-body systems can be richer than in a single two-level system

[12], they are subject to current investigation[13]. On the other hand, the terms in λ_{11,q,q_z} and λ_{22,q,q_z} correspond to intrasubband transitions and are peculiar of asymmetric QWs in which $\Delta z = z_{11} - z_{22} \neq 0$[2], as the one shown in Fig.(1) (c).

We obtain that the rate of emission per electron due to the latter (asymmetric) part of the interaction in Eq. (9) is

$$\Gamma_{Asym} = \gamma_0 |\Delta z|^2 \Omega^3, \tag{10}$$

where $\gamma_0 = \frac{e^2\sqrt{\epsilon_r}}{12\epsilon_0 \pi \hbar c^3}$. This THz emission line, centered at the Rabi frequency Ω, can be Purcell-enhanced by employing a subwavelength microcavity[5] to tailor the photonic density of modes[14]. For example with a planar microcavity of length L_{cav},

$$\Gamma^{cav}_{Asym} = \frac{3\pi c \gamma_0 |\Delta z|^2 \Omega^2}{8\sqrt{\epsilon_r} L_{cav}}. \tag{11}$$

Since $\Omega = e z_{12} \mathcal{E}/\hbar$, where $\mathcal{E}$ is the optical pump amplitude, with achievable values of the order of $\mathcal{E} \simeq 10^6 - 10^7$ V m^{-1}, and z_{12} that is of the order of the QW length L_{QW}, $L_{QW} \simeq 10$ nm, the emission can be tuned to span the THz range.

5. Conclusions

We have considered the Rabi oscillations occurring between the two conduction subbands of a 2DEG confined in a doped QW under resonant, coherent optical pumping. When the QW confining structure is not symmetrical, the interaction with the vacuum electromagnetic field leads to the emission of light centered at the Rabi frequency[2], which is tuned by the field strength of the optical pump. This nonlinear effect could thus empower solid-state THz emitters of tunable frequency for optoelectronic applications.

6. Acknowledgments

S.D.L. acknowledges financial support as a Royal Society Research Fellow. S.D.L. and N.S. acknowledge support from the Engineering and Physical Sciences Research Council (EPSRC), research Grant No. EP/L020335/1.

References

[1] Tonouchi M 2007 *Nature Phot.* **1** 97
[2] Shammah N, Phillips C and De Liberato S 2014 *Phys. Rev. B* **89** 235309
[3] De Liberato S, Ciuti C and Phillips C C 2013 *Phys. Rev. B* **87** 241304
[4] Kibis O V, Slepyan G Y, Maksimenko S A and Hoffmann A 2009 *Phys. Rev. Lett.* **102** 023601
[5] Kavokin K V, Kaliteevski M A, Abram R A, Kavokin A V, Sharkova S and Shelykh I A 2010 *Appl. Phys. Lett.* **97** 201111
[6] Christmann G, Askitopoulos A, Deligeorgis G, Hatzopoulos Z, Tsintzos S I, Savvidis P G and Baumberg J J 2011 *Appl. Phys. Lett.* **98** 081111
[7] Ridolfo A, Leib M, Savasta S and Hartmann M J 2013 *Phys. Rev. Lett.* **109** 193602
[8] Savenko I G, Kibis O V and Shelykh I A 2012 *Phys. Rev. A* **85** 053818
[9] Autler S H and Townes C H 1955 *Phys. Rev.* **100** 703
[10] Dynes J F, Frogley M D, Beck M, Faist J and Phillips C C 2005 *Phys. Rev. Lett.* **94** 157403
[11] Mollow B R 1969 *Phys. Rev.* **188** 1969
[12] C. Sánchez Munõz C, del Valle E, González Tudela A, Müller K, Lichtmannecker S, Kaniber M, Tejedor C, Finley J J and Laussy F P 2014 *Nature Phot.* doi:10.1038/nphoton.2014.114
[13] Shammah N and De Liberato S *in preparation*
[14] Todorov Y, Sagnes I, Abram I and Minot C 2007 *Phys. Rev. Lett.* **99** 223603

Compact, diode-pumped, solid-state lasers for next generation defence and security sensors

M. Silver, S.T. Lee, A. Borthwick, I. McRae, D. Jackson and W. Alexander

Thales UK, 1 Linthouse Road, Glasgow, G51 4BZ, UK

mark.silver@uk.thalesgroup.com

Abstract. Low-cost semiconductor laser diode pump sources have made a dramatic impact in sectors such as advanced manufacturing. They are now disrupting other sectors, such as defence and security (D&S), where Thales UK is a manufacturer of sensor systems for application on land, sea, air and man portable. In this talk, we will first give an overview of the market trends and challenges in the D&S sector. Then we will illustrate how low cost pump diodes are enabling new directions in D&S sensors, by describing two diode pumped, solid-state laser products currently under development at Thales UK. The first is a new generation of Laser Target Designators (LTD) that are used to identify targets for the secure guiding of munitions. Current systems are bulky, expensive and require large battery packs to operate. The advent of low cost diode technology, merged with our novel solid-state laser design, has created a designator that will be the smallest, lowest cost, STANAG compatible laser designator on the market. The LTD delivers greater that 50mJ per pulse up to 20Hz, and has compact dimensions of 125x70x55mm. Secondly, we describe an ultra-compact, eye-safe, solid-state laser rangefinder (LRF) with reduced size, weight and power consumption compared to existing products. The LRF measures 100x55x34mm, weighs 200g, and can range to greater than 10km with a single laser shot and at a rep rate of 1Hz. This also leverages off advances in laser pump diodes, but also utilises low cost, high reliability, packaging technology commonly found in the telecoms sector. As is common in the D&S sector, the products are designed to work in extreme environments, such as wide temperature range (-40 to +71^{0}C) and high levels of shock and vibration. These disruptive products enable next-generation laser sensors such as rangefinders, target designators and active illuminated imagers.

1. Introduction

Solid-state lasers are well suited for long range sensor applications in the defence and securities (D&S) sector. Q-switched sources can generate very high peak power (100's MW), short duration (typically ns) pulses with good beam quality. This allows the signals to propagate and be detected over large distances. In this paper, we will describe developments in solid-state lasers used in two key sensor functions, namely, laser target designation (LTD) and laser range finding (LRF). LTDs are used for laser guiding of munitions. According to a recent report [1], by 2020, 90% of all munitions will be laser guided. The shift to laser guided munitions will lead to unprecedented target accuracy and hence decrease in collateral damage, with the majority of munitions landing within 2m of their intended target. LRFs are used in sensors heads and target locators and use compact eye-safe laser transmitters to measure distances of 10's of km to an accuracy of better than 1m [2].

Next generation systems will require a step-reduction in size, weight and power (SWaP) for application in un-manned autonomous vehicles (UAVs) and ultra-light, compact, man-portable systems, examples of which are shown in Figure 1. In addition to low SWaP, sensor products in the D&S sector are required to work in the harshest environments in the world. Typically, products need to operate over a very wide temperature range and withstand high levels of shock and vibration. Furthermore, the unit production volumes for the next generation sensors are likely to be much higher than traditionally encountered in the sector due to a big increase in the number of sensor platforms. As a result, cost is becoming a strong driver in these markets.

Figure 1: Reductions in Size, Weight and Power (SWaP) are key drivers for the next generation of sensors, such as hand-held target locators (left) or gimbals for UAVs (right).

We describe the development of new LTDs and LRFs with improved SWaP and lower cost. The key enabler for the new designs is the recent availability of low cost, high performance, high reliability, semiconductor laser diodes, driven by demand in market sectors such as advanced manufacturing and communications [3]. The new pump sources give solid-state lasers, typically, a 10-fold improvement in efficiency compared to flash-lamp systems, allowing big gains in compactness and reliability. In addition, we have produced designs that are compatible with low cost, high reliability, high volume, package technologies, that are commonly found in the telecoms sector. Technologies such as hermetically sealing, low out-gassing materials, and high precision component alignment, along with the possibility of automation, will bring huge benefits.

In the next two sections, we describe the design of the products and present experimental results.

2. Laser Target Designator

2.1. Laser Design

Q-switched, Nd:YAG lasers were adopted near the beginning of target designation in the 1960's due to their ability to deliver high output energies with high efficiency. This led to the seekers on the laser-guided munitions being tuned to 1064nm, which remains the chosen wavelength in the Standardised Agreement for designators (STANAG 3733).

Developing diode pumped Nd:YAG solid state lasers for application as target designators poses several design challenges. Firstly, operating a Nd:YAG laser over a broad ambient temperature range raises the issue of non-uniform pump absorption. Around the ideal pump wavelength of 808nm, the laser diodes tune at approximately 0.3nm/°C. Thus, a system specified to operate over an ambient temperature range of –46°C to +71°C can result in the pump wavelength drifting by 32.8nm. Unfortunately, the optical absorption of Nd:YAG varies dramatically over this wavelength range [4]. Despite the absorption profile being smoothed out somewhat by the 4nm laser pump line-width, the resulting absorption length required to achieve 90% absorption of the pump ranges from 6 to 50mm over temperature, as shown in Figure 2. Stabilising the pump absorption by reducing the pump wavelength shift using either temperature or wavelength control [5] are not practical due to system electrical power consumption requirements and their higher cost.

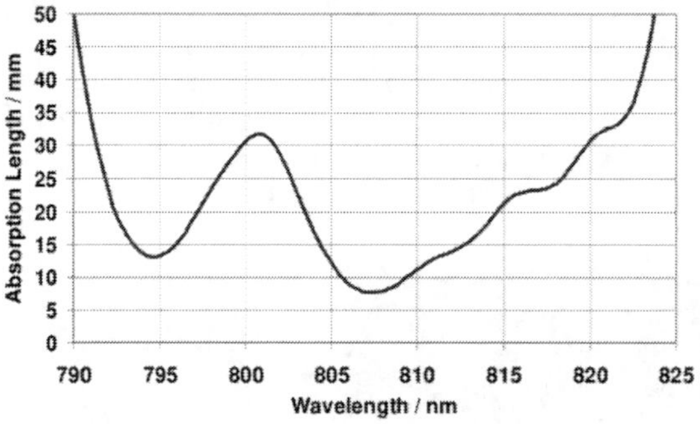

Figure 2: 90% absorption length of 1 atomic % Nd^{3+}:YAG.

A solution is to adopt an end-pumped geometry to allow for the large differences in absorption length over the temperature range[4]. We end-pump a 50mm laser rod with a diode pump source consisting of a 10- bar, 808nm, QCW stack, with each bar fitted with a cylindrical fast axis collimation lens. The output of the entire array is focused and collimated into the end of the rod and results in constant laser gain over temperature. Crucially, we have also eliminated the parasitic lasing effects that can limit the output energy in end-pump designs.

The second key system challenge is the need for a relatively long laser pulse width. Although not specifically stated in the STANAG requirement, a pulse width of between 10-20ns is favoured by the target seekers, but is also required to avoid issues such as laser induced coating damage due to the greater than 50mJ required for designation. To achieve a pulse width in this range, the cavity length has to be of the order of 20cm long. Hence, in order to benefit from the compact diode pumps, the laser cavity needs to be folded several times to reduce size.

The straight resonator is first folded into a U-shaped geometry, made up of a corner cube at one end and a plane mirror substrate at the other. The plane mirror substrate is coated in two regions to act as both the high-reflector mirror and the output coupler. The action of the corner cube, coupled with this configuration of end mirrors, results in a resonator that is significantly less sensitive to misalignment. The U-shaped resonator is further reduced in size with a second fold using two extra mirrors. Each of these fold mirrors reflects both arms of the laser resonator and hence does not adversely affect misalignment insensitivity. This was confirmed by modeling the resonator misalignments in Zemax via a Monte-Carlo analysis, as shown in Figure 3. The double folded cavity, coupled with the compact end-pumped geometry, allows the designator to be fitted into dimensions of (LxWxH) 125x70x55mm.

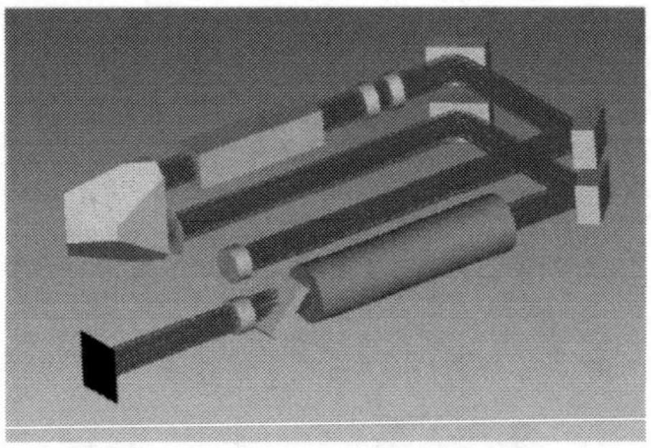

Figure 3: Folded laser resonator model used for misalignment sensitivity calculations.

Journal of Physics: Conference Series **619** (2015) 012022 doi:10.1088/1742-6596/619/1/012022

Q-switching is achieved using a $LiNbO_3$ Pockels cell, in conjunction with a polariser and waveplate. The waveplate has the dual function of zeroing the retardation due to the corner cube and setting the remaining retardation to a quarter wave. In a double pass, the combination of corner cube and waveplate act as a half wave plate and, in conjunction with the polariser, this creates the low-Q state of the resonator. The Pockels cell is used at its quarter-wave voltage to switch to the high-Q state.

2.2. *Experimental Results*

The laser has been tested over a range of temperatures with no active cooling. The output pulse energy with time at 20Hz is shown in Figure 4 for –35°C, room temperature and +55°C operation. The output pulse energy of the laser is selectable by changing the pump pulse duration. This allows for the capability of using the laser in a low-energy mode for short-range operation.

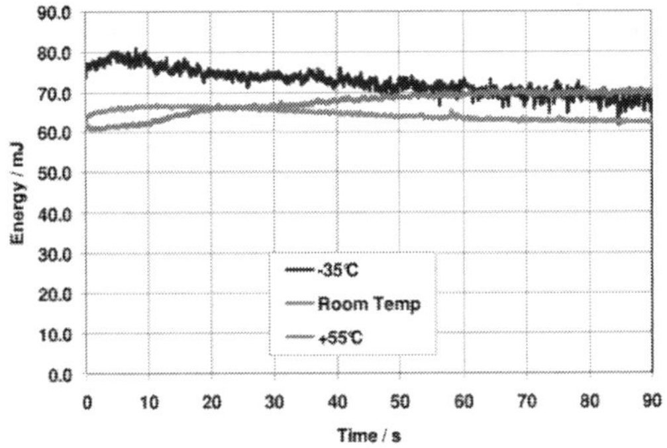

Figure 4: Output pulse energy at 20Hz

3. Eye-safe, Diode pumped, Solid-state, Laser Range Finder

The second laser system that benefits from low cost, compact, laser diode arrays is for application in laser range finders (LRFs). LRFs need to be eye-safe for most applications, and hence the 1500-1600nm emission wavelength range is most attractive. Furthermore, diode pump solid-state sources are ideal for long-range, high accuracy time-of-flight LRFs. They have very high peak powers to allow propagation over 10's of km, and narrow pulse widths (nanoseconds) to allow accurate time-stamping of the returned signal.

3.1. *Laser Design*

Our laser is based on a diode-pumped, Erbium:Ytterbium (Er:Yb) Glass (1535nm), passively Q-switch solid-state laser. Figure 5 shows a schematic of the resonator design. The laser gain block is formed from an Er:Yb glass laser rod pumped by 2x1cm, 940nm pump diodes. In order to maximize efficiency, the components size should be matched closely to the optical mode. The laser rod is less than 1.5mm in cross-section in order to maximize the gain efficiency from the 1mm diameter laser spot.

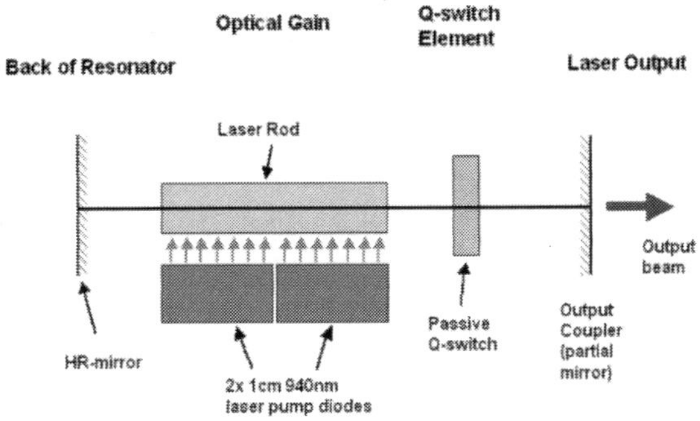

Figure 5: Exploded view of laser resonator configuration

Positioning and handling of such small components lends itself to utilising many of the technologies already developed for low cost, compact, high reliability telecoms style packaging. As a result, our laser has an efficient pump arrangement that need only be passively aligned (to within +/- 50um) to ensure efficient pumping of the gain region. Furthermore, the laser rod is mounted to produce good thermal contact to the transmitter package. The laser cavity is formed by a highly reflecting (HR) back reflector deposited on the laser rod, and a separate, partially reflecting output coupler (OC) at the front. Q-switching is achieved with a Cobalt (Co^{2+}) doped Spinel ($MgAl_2O_4$). The output coupler is actively aligned to achieve the required energy and beam profile. Finally, there is an integrated photodetector to allow the laser driver to shut-down the pump current when a Q-switch pulse is emitted. The whole resonator is hermetically sealed within a package that measures (LxWxH) 40x14x10mm, as shown in Figure 6.

The design and assembly technique ensures a compact, low cost, high reliability product, which can be easily integrated into a host system. We have developed a LRF with the transmitter, which measures 100x55x34mm and weighs less than 200g. Furthermore, the small size of the components means that the design is suitable for assembly automation to significantly reduce cost.

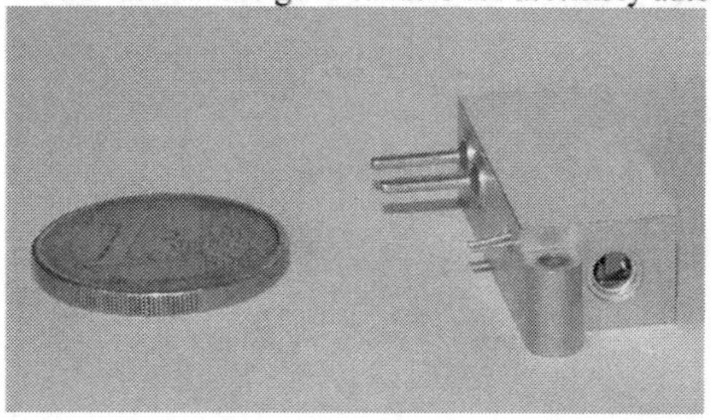

Figure 6: Image of transmitter package

3.2. Experimental Results

Devices have been tested over a range of temperatures (-38 to +71^{0}C) and at repetition rates up to of 1Hz. The device was mounted onto an aluminium base to act as a heat sink and passively cooled. The laser fired from the first shot at all temperatures. Figure 7(a) shows that the pulse energy over temperature was around 1mJ, and shows good uniformity. There was no evidence of thermal runaway as illustrated by a 500 shot run (8.2mins) shown in Figure 7(b). The pump time varied from 1.7ms at low temperature, up to 2.1ms at high temperature, as shown in Figure 8(a). The laser output is a typical Q-switch pulse, as shown in Figure 8(b) and varies between 6 and 8ns over temperature. The beam profile is a single TEM_{00} gaussian, as shown in Figure 8(c) with a temperature variation of the beam divergence shown in Figure 8(d). The electrical power drawn at 1Hz is around 0.8W. As is typical for Q-switch solid-state lasers, the electrical to optical efficiency is quite low (<1%).

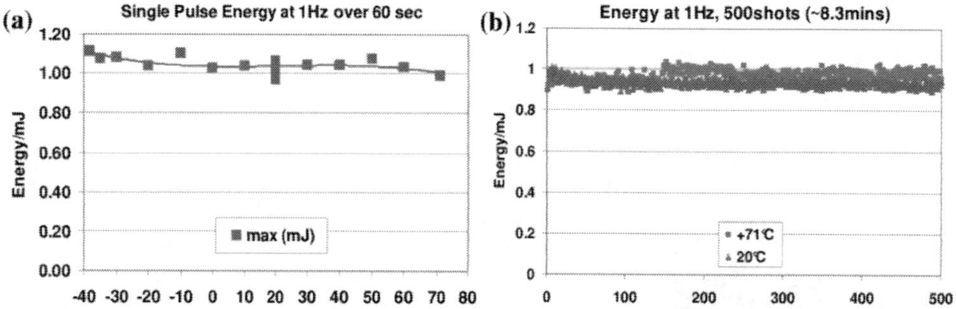

Figure 7 (a) max pulse energy at 1Hz over 60 seconds, (b) extended firing over 500 shots at 1Hz

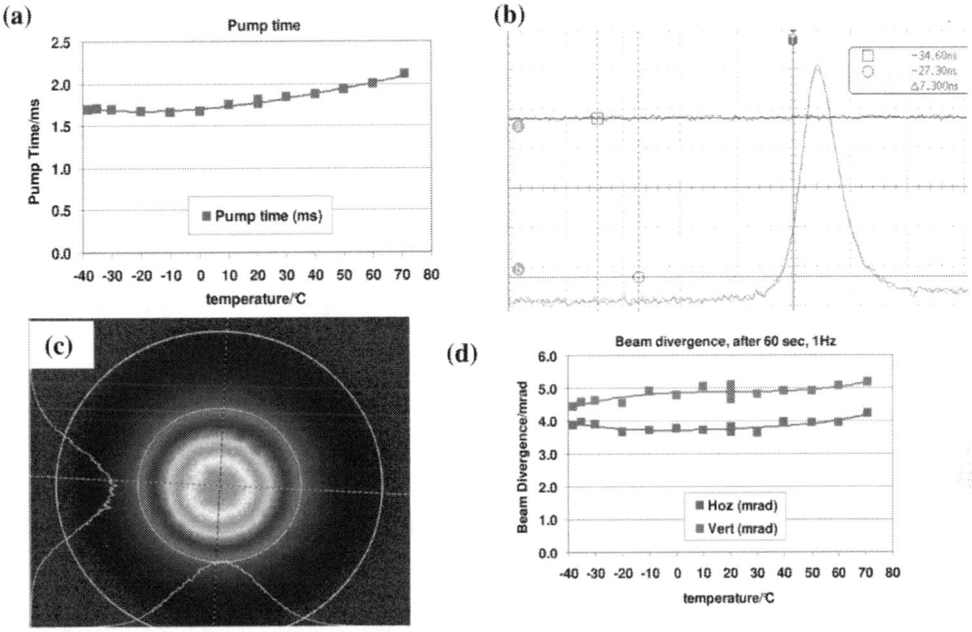

Figure 8: Performance over temperature (a) pump time, (b) typical Q-switch pulse shape, (c) laser beam far-field at 25°C, (d) beam divergence ($1/e^2$) over temperature.

4. Conclusion

The two applications described above demonstrate the potential of compact diode pumps and telecoms-style manufacturing techniques applied to defence and security applications. We would expect further improvements in SWaP over the next few years, driven by the demand for higher performance pump heads for the advanced manufacturing sector. Furthermore, costs will decrease, leading to the sensors becoming more commoditised than is typical in the D&S sector. Under these conditions, we would expect to see, for example, high performance rangefinders and active imagers appearing in the low cost civilian UAVs that look set to emerge over in the coming years.

5. Acknowledgments

We would like to acknowledge my colleagues, S.M.W. Wills, M.B. Darlow, B. Horsfall. We would also like to thank our packaging partner, Gooch & Housego UK Ltd and in particular, M.C. Cox, I.W. Wilson, T.J. Cook, and A.M. Norman. This work is part funded by the TSB project LightMILES (101236).

References

[1] Work R O and Brimley S 2014 20YY Preparing for War in the Robotic Age *Center for New American Security*, www.cnas.org

[2] Forrester P A, Hulme K F 1981 Review: Laser Rangefinders, *Optical and Quantum Electronics* ***13**, 259-293.*

[3] Martinsen R 2007 Industrial markets beckon for high-power diode lasers *Optics & Laser Europe.*

[4] Schilling B W, *et al.* 2006 End-pumped 1.5um monoblock laser for broad temperature operation *Applied Optics* ***45** No. 25, 6607-6615.*

[5] McCarthy J C *et al* Athermal, Lightweight, Diode-Pumped, 1 micron Transmitter *Proceedings of SPIE Vol. 5707, 237-242.*

Dual-gate photo thin-film transistor: a "smart" pixel for high-resolution and low-dose X-ray imaging

Kai Wang [1,2], Hai Ou[3], and Jun Chen[3]

[1]Sun Yat-sen University-Carnegie Mellon University Joint Institute of Engineering, Sun Yat-sen University, Guangzhou 510275, People's Republic of China
[2]Sun Yat-sen University-Carnegie Mellon University Shunde International Joint Research Institute, Foshan 528300, People's Republic of China
[3]State Key Laboratory of Optoelectronic Materials and Technologies, Guangdong Province Key Laboratory of Display Material and Technology, and School of Physics and Engineering, Sun Yat-sen University, Guangzhou 510275, People's Republic of China
Emails: wangkai@andrew.cmu.edu(KW); stscjun@mail.sysu.edu.cn(JC)

Abstract. Since its emergence a decade ago, amorphous silicon flat panel X-ray detector has established itself as a ubiquitous platform for an array of digital radiography modalities. The fundamental building block of a flat panel detector is called a pixel. In all current pixel architectures, sensing, storage, and readout are unanimously kept separate, inevitably compromising resolution by increasing pixel size. To address this issue, we hereby propose a "smart" pixel architecture where the aforementioned three components are combined in a single dual-gate photo thin-film transistor (TFT). In other words, the dual-gate photo TFT itself functions as a sensor, a storage capacitor, and a switch concurrently. Additionally, by harnessing the amplification effect of such a thin-film transistor, we for the first time created a single-transistor active pixel sensor. The proof-of-concept device had a W/L ratio of 250μm/20μm and was fabricated using a simple five-mask photolithography process, where a 130nm transparent ITO was used as the top photo gate, and a 200nm amorphous silicon as the absorbing channel layer. The preliminary results demonstrated that the photocurrent had been increased by four orders of magnitude due to light-induced threshold voltage shift in the sub-threshold region. The device sensitivity could be simply tuned by photo gate bias to specifically target low-level light detection. The dependence of threshold voltage on light illumination indicated that a dynamic range of at least 80dB could be achieved. The "smart" pixel technology holds tremendous promise for developing high-resolution and low-dose X-ray imaging and may potentially lower the cancer risk imposed by radiation, especially among paediatric patients.

1. Introduction

Over the last two decades, flat panel X-ray detector (FPD) has become commercially available for various medical imaging modalities including chest radiography and mammography[1]. Silicon thin-film transistor (TFT) technology which has gained increasing popularity in large area manufacture is widely utilized in making FPDs. An FPD is essentially an amorphous silicon (a-Si:H) imaging sensor array containing millions of pixels. As a fundamental building block, a pixel typically consists of three components: sensor, storage, and switch to meet the needs of signal generation and collection, storage and readout. In terms of pixel architecture, a passive pixel sensor (PPS) is simple and includes a switching TFT and storage capacitor while an active pixel sensor (APS) differs by embedding an additional in-pixel amplifier TFT and thereby, leading to a higher signal to noise ratio (SNR) and reduced exposure dose requirement[2][3]. Nevertheless, in all current pixel architectures, either passive or active, the abovementioned three components are unanimously kept separate with a major disadvantage being trade off spatial resolution for a decent SNR.

In this work, we proposed and studied a "smart" pixel architecture that combines sensor, storage, and readout into one single dual-gate photo TFT, which apparently outperforms all conventional structures, not only for being much more space-efficient, especially for high-resolution X-ray imaging, but also for making a higher SNR possible upon TFT amplification.

Most previous reports on dual-gate TFTs were mainly focused on switching performance and applications in circuits, biosensors, and displays[4]. Such a device was primarily designed to allow for better control and reliability in threshold voltage operation. Recently, Kim et al. demonstrated a dual-gate HIZO/IZO Photo TFT intended for a transparent imager[5]. To our best knowledge, this was one of the few works to address photosensing applications of dual-gate TFTs. However, it still works as a passive pixel sensor and was only sensitive to short-wavelength light given the wide band gap of HIZO/IZO semiconductors.

Compared with existing works, our dual-gate TFT was set apart by its ease of production and the possibility of forming a compact single transistor APS so that in-pixel amplification can be readily achieved. This is of particular interest to low-dose X-ray detection such as pediatric imaging where patients are more susceptible to radiation dose.

2. Pixel Architecture

Fig. 1 depicts a schematic representation of the cross-sectional structure of an X-ray imaging pixel using an amorphous silicon dual-gate photo TFT. At the top of the pixel, there lies an X-ray scintillator (e.g. CsI) that converts X-ray to visible light. Underneath the scintillator, it is a four-terminal a-Si:H dual-gate photo TFT composed of a photo gate (Photo G-PG), a dark gate (Dark G-DG), a source electrode (S) and a drain electrode (D) to complete the whole signal collection, storage, and readout operation. In order to enhance photo absorption, a thick a-Si:H layer is generally used.

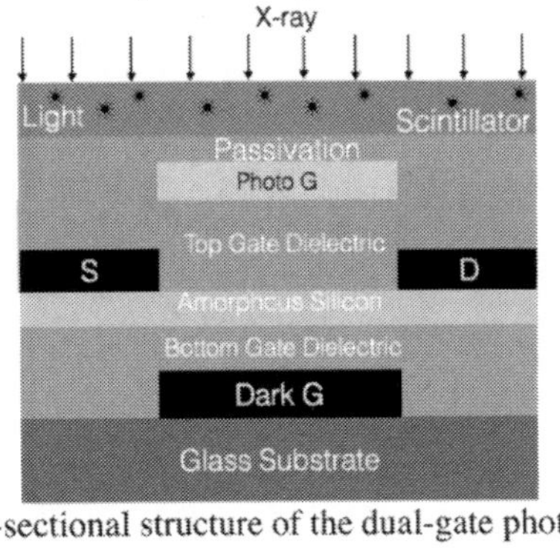

Fig. 1 The schematic cross-sectional structure of the dual-gate photo TFT in an X-ray imaging pixel. Dimensions are magnified to give a clear view.

3. Results and Discussion

3.1. Device Fabrication

We set out to design and fabricate a dual-gate photo TFT using a five-mask process: first of all, a 120 nm Cr film was sputtered and patterned on Corning 1737 glass substrate to form gate electrodes (Mask No. 1); a 220 nm SiNx/240nm a-Si:H/60nm n^+-a-Si:H triple-layer was then deposited consecutively at 200°C by using conventional PECVD consecutively without breaking vacuum and this was followed by the deposition of a 130 nm Cr film by sputtering; thirdly, Cr and n^+-a-Si:H layers were patterned by wet etching and dry etching to form source and drain electrodes (Mask No. 2); subsequently, a TFT island was created through etching the a-Si:H (Mask No. 3) layer; followed by a 220nm SiNx deposition atop to form a top gate dielectric through PECVD at 200°C; a 130nm ITO was then sputtered and patterned to form the top gate electrodes (Mask No. 4); in the end, a 200nm SiNx was deposited by PECVD to passivate the entire device and was patterned for via opening on the source, drain, and gate electrodes through dry etching of multiple layers (Mask No. 5).

The resulted device was examined under an optical microscope and layer thickness and interface information were obtained using scanning electron microscopy (SEM) and a cross-sectional interface was obtained through focused ion beam milling. Fig. 2 is a cross-sectional image, clearly demonstrating each layer and interfaces there between.

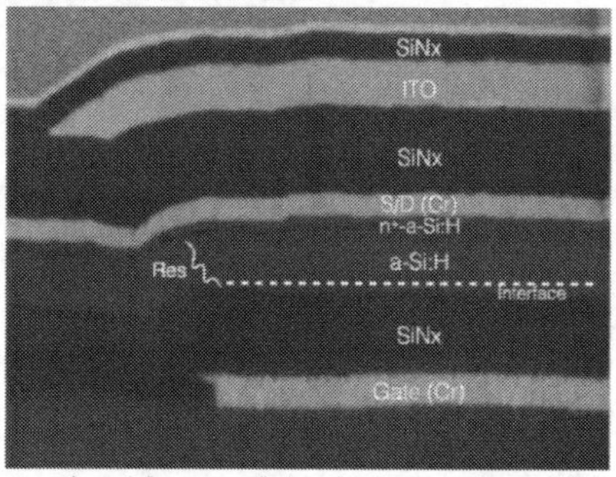

Fig. 2 The cross-sectional image of the dual-gate photo TFT by FIB-SEM

3.2. Device Characterization

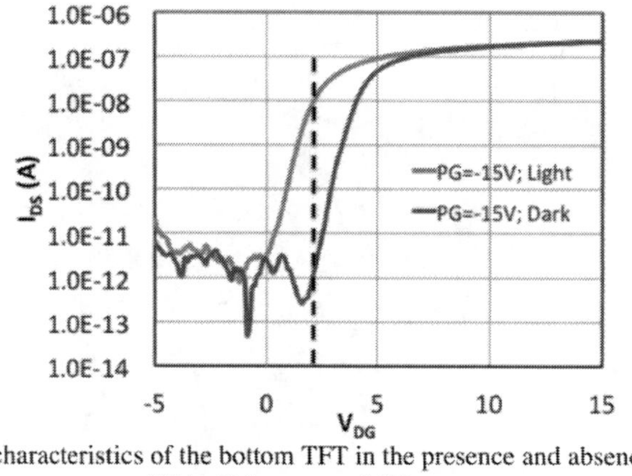

Fig. 3 The transfer characteristics of the bottom TFT in the presence and absence of light exposure. White light source: 3.8 Lux.

Upon light exposure, the transfer characteristics were modulated as shown in Fig. 3. The IV curve shifted towards negative gate voltage, indicating that the device could be easily turned ON under illumination. In another word, the threshold voltage was lowered as a result of light illumination. This served as experimental evidence of amplification effect. The photo-induced threshold voltage decrease conferred the device a signal gain when it was operated in the subthreshold voltage region. For instance, at a dark gate bias of ~2V, the output photocurrent I_{DS} can be enhanced by four orders of magnitude from ~10^{-12}A to ~10^{-8}A. Such signal gain mainly stems from the decrease in threshold voltage which is a function of the light intensity, the V_{PG}, and the top transistor capacitance.

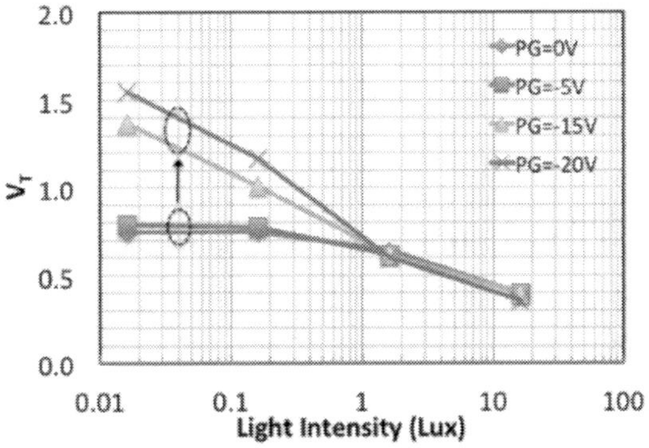

Fig. 4 Threshold voltage of the bottom TFT as a function of light illumination intensity. White light source; W/L=250/10um

To explore its potential application in low-dose light detection, the device was studied particularly under low-level light exposure. Fig. 4 illustrates the dependence of threshold voltage on light intensity. At a lower V_{PG} and lower light intensity, the threshold voltage change was not distinguishable. However, as the V_{PG} increased, the device became more sensitive since the threshold voltage also increased and accordingly, the threshold voltage change became more distinguishable (Fig. 4). The essential role that the V_{PG} played was to enhance the low-dose light sensitivity through increasing the original threshold voltage. The preliminary result showed a ~ 80dB dynamic range at low light levels. The device is therefore promising to work in the low-dose light exposure for low-dose indirect X-ray imaging.

4. Conclusion

We have successfully designed and fabricated a dual-gate photo TFT based on amorphous silicon TFT-LCD processes. With the sensor, storage, and readout functions combing in one device, the dual-gate photo TFT have demonstrated desirable features such as high resolution, high sensitivity, and simplicity. The device was sensitive to low-level light exposure as well and therefore could potentially work as a single transistor active pixel sensor for low-dose X-ray imaging especially for pediatric patients.

References

[1] M. J. Yaffe and J. A. Rowlands, "X-ray detectors for digital radiography.," *Phys. Med. Biol.*, vol. 42, no. 1, pp. 1–39, Jan. 1997.

[2] L. E. Antonuk, Q. Zhao, Y. El-Mohri, H. Du, Y. Wang, R. A. Street, J. Ho, R. Weisfield, and W. Yao, "An investigation of signal performance enhancements achieved through innovative pixel design across several generations of indirect detection, active matrix, flat-panel arrays.," *Med. Phys.*, vol. 36, pp. 3322–3339, 2009.

[3] K. S. Karim, A. Nathan, and J. A. Rowlands, "Amorphous silicon active pixel sensor readout circuit for digital imaging," *IEEE Trans. Electron Devices*, vol. 50, 2003.

[4] M.-J. Spijkman, K. Myny, E. C. P. Smits, P. Heremans, P. W. M. Blom, and D. M. de Leeuw, "Dual-gate thin-film transistors, integrated circuits and sensors.," *Adv. Mater.*, vol. 23, pp. 3231–3242, 2011.

[5] Sanghun Jeon, Seung-Eon Ahn, Ihun Song, Yongwoo Jeon, Young Kim, Sangwook Kim, Hyunsik Choi, Hojung Kim, Eunha Lee, Sungsik Lee, A. Nathan, J. Robertson, Changjung Kim, U-In Chung, Inkyung Yoo, and Kinam Kim, "Dual gate photo-thin film transistor with high photoconductive gain for high reliability, and low noise flat panel transparent imager," *2011 Int. Electron Devices Meet.*, pp. 14.3.1–14.3.4, 2011.

Journal of Physics: Conference Series **619** (2015) 012024 doi:10.1088/1742-6596/619/1/012024

Effects of surface modification on photoluminescence properties of self-assembled monolayer of CdSe quantum dots

Taichi Watanabe, Hiroki Yokota, Masaaki Nakayama, and DaeGwi Kim*

Department of Applied Physics, Graduate School of Engineering, Osaka City University, Sugimoto, Sumiyoshi-ku, Osaka 558-8585, Japan

*tegi@a-phys.eng.osaka-cu.ac.jp

Abstract. We have investigated surface-modification effects on photoluminescence (PL) properties of the self-assemble monolayer of CdSe quantum dots from the viewpoint of the temperature dependence of PL properties. The band-edge PL band is strongly activated by the surface modification and is observed as the main PL band, contrary to the fact that the defect-related PL band is dominant in the as-grown sample. From the analysis of PL decay dynamics, it is demonstrated that the contribution of a bound exciton state to PL processes would suppress a trapping process to defects.

1. Introduction

Semiconductor quantum dots (QDs) have been intensively investigated from the viewpoint of the fundamental physics and from the interest in the application to functional materials [1, 2]. CdSe QDs have been a model material for QD studies since the synthesis of size-controlled QDs with high photoluminescence (PL) yield has been well established [3, 4]. It is important to extract QDs from the solution and to disperse them into polymer films from the view point of application for optoelectronic devices. A typical method for preparing film samples is spin-coating and/or drop-casting of QD solutions. Compared to these popular methods, the formation of a self-assembled monolayer (SAM) of QDs is a simple and convenient method for preparing highly-homogeneous QD films [5, 6]. Although the thiol group, which is a constituent of the reagent for formation of the SAM, degrades the PL properties during the SAM formation process, we demonstrated that the band-edge PL intensity is increased remarkably with an additional dipping treatment in $Cd(ClO_4)_2$ aqueous solution with pH=10 [7]. The improvement of PL properties is due to surface modification with a $Cd(OH)_2$ layer.

In the present work, we have investigated the surface-modification effects on PL properties of the SAM of CdSe QDs from the viewpoint of the temperature dependence of PL dynamics. Although a defect-related PL band is dominant at low temperatures in the as-grown sample, the band-edge PL is observed as a main PL band after the surface modification of the SAM. The temperature dependence of the PL-decay time is explained using a four-state model which introduces a bound-exciton state into a conventional three-state model consisting of a ground state and two excited states; a lower lying dark-exciton state and a higher-lying bright-exciton state.

2. Experimental details

CdSe QDs with 3.3 nm in diameter were purchased from Sigma-Aldrich. The substrates of quartz were cleaned by immersion in fresh piranha solution [1/3 (v/v) mixture of 30% H_2O_2 and 98% H_2SO_4]

for 20 min. Then the substrates were rinsed with water. SAM of (3-mercaptopropyl) trimethoxysilane (MPTMS) molecule was deposited to link CdSe QDs to SiO_2 substrates. Next, the substrate was rinsed thoroughly with toluene. The SAM of the QDs linked by thiol group of a MPTMS was prepared in a QD/toluene solution. The samples were thoroughly rinsed with toluene and dried with a stream of nitrogen gas. An additional dipping treatment of the SAM of the CdSe QDs was performed in a 0.6 mM $Cd(ClO_4)_2$ aqueous solution with pH 10 according to the method in ref. [7] in order to improve PL properties. This process corresponds to the surface-modification treatment of the CdSe QDs with a $Cd(OH)_2$ layer.

The absorption spectra were measured with a double-beam spectrometer with a resolution of 0.2 nm. For PL measurements, the 325-nm line of a He-Cd laser was used as the excitation-light source, and the emitted PL was analyzed with a single monochromator with a spectral resolution of 0.5 nm. For measurements of PL-decay profiles, third-harmonic-generation light (355 nm) of a laser-diode pumped yttrium aluminum garnet laser with a repetition of 10 kHz was used as the excitation light. The pump fluence was 3 μJ/cm^2. The PL-decay profiles were obtained by a time-correlated single-photon counting method. The sample temperature was controlled using a closed-cycle helium-gas cryostat.

3. Results and discussion

Figure 1a shows absorption and PL spectra at 10 K of the SAM of the CdSe QDs. It is obvious that a defect-related PL band with a large Stokes-shift of ~0.5 eV is dominant at 10 K. This is because that the thiol group degrades the PL properties of the SAM of the CdSe QDs [8, 9]. Figure 1b shows absorption and PL spectra at 10 K of the SAM after the surface-modification treatment of the CdSe QDs with a $Cd(OH)_2$ layer. The PL spectrum is drastically changed by the surface modification. The band-edge PL band is strongly activated and is observed as the main PL band, contrary to the fact that the defect-related PL band is dominant in the as-grown sample. In order to understand the surface-modification effects on PL properties of the SAM of the CdSe QDs, we measured the temperature dependence of the absorption and PL spectra, and PL-decay profiles.

Figure 2 shows the temperature dependence of the absorption and PL spectra of the SAM after the

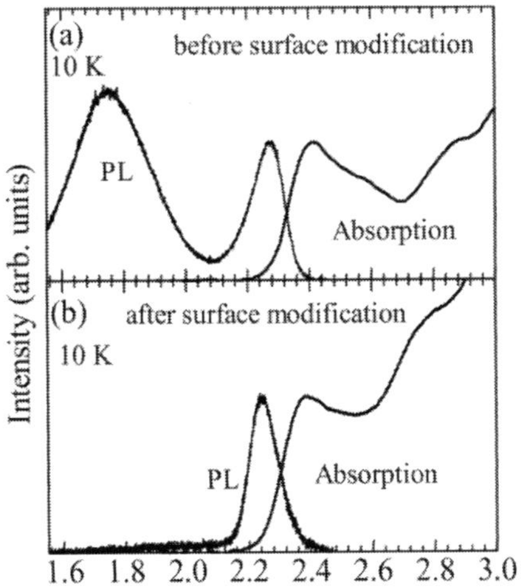

Figure 1 Absorption and PL spectra at 10 K the SAM of the CdSe QDs before ((a)) and after ((b)) the surface modification.

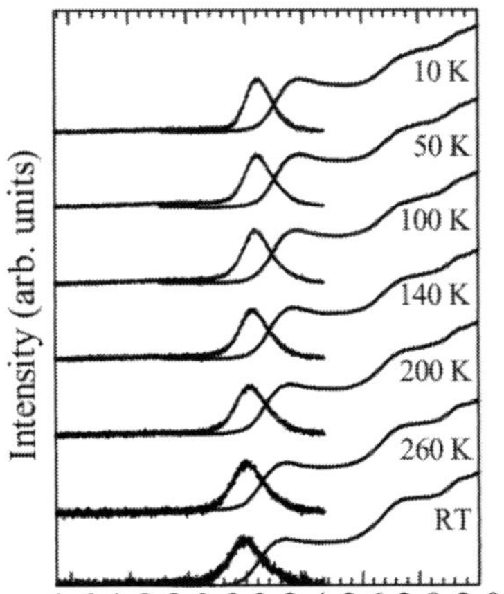

Figure 2 Temperature dependence of absorption and PL spectra of the SAM after the surface modification.

surface-modification treatment. The absorption- and PL-peak energies shift to the lower energy side with an increase in temperature. The intensity of the band-edge PL band at room temperature is almost 10 % of that at 10 K. Figure 3(a) shows the temperature dependence of the lowest absorption- and PL-peak energies. It is well known that the temperature dependence of the exciton energy in direct-gap semiconductors can be described by Varshni's law [10];

$$E(T) = E(0) - \alpha T^2 / (T + \beta), \qquad (1)$$

where $E(0)$ is the exciton energy at T=0 K, α is the temperature coefficient, and β is a parameter related to the Debye temperature of the crystal. The solid curve indicates the calculated result for the temperature dependence of the absorption and PL energies, where the parameter values of α=3.7×10^{-4} eV/K and β=150 K in a CdSe bulk crystal were used. The calculated result using the parameters in bulk crystal quantitatively explains the temperature dependence of the absorption energy, which indicates that the observed temperature dependence is an intrinsic optical property of the CdSe crystal. For the PL energy, the calculated result quantitatively explains the experimental one at the higher temperature than 140 K. In the low temperature region, the PL energy becomes lower than the calculated result. Thus, the Stokes-shift becomes larger in the lower temperatures. There is a possibility that "*bound-exciton state*" caused by the $Cd(OH)_2$ layer is formed on the low energy side of the free-exciton state as discussed in ref. [11]. The bound-exciton state would suppress a trapping process to defects, and the band-edge PL is observed as a main PL band. Thus, the temperature dependence of the PL dynamics was measured in order to clarify the contribution of the bound-exciton state to PL processes.

Figure 4 shows the temperature dependence of the decay time of the band-edge PL. In the temperature range lower than 180 K, the decay time becomes larger with an increase in temperature. In the usual case, as the temperature is increased, the nonradiative-decay rate related to the thermal quenching becomes larger, so that the PL intensity and the PL decay time are decreased. Although the band-edge PL intensity gradually decreases with an increase in temperature, the decay time becomes larger. The characteristic temperature dependence of the decay time in the SAM of the CdSe QDs was explained by a four-state model consisting of a ground state, a bound-exciton state, and two free-exciton states of a lower-lying dark-exciton state and a higher-lying bright-exciton state [11]. The decay time based on the four-state model is given as [11]

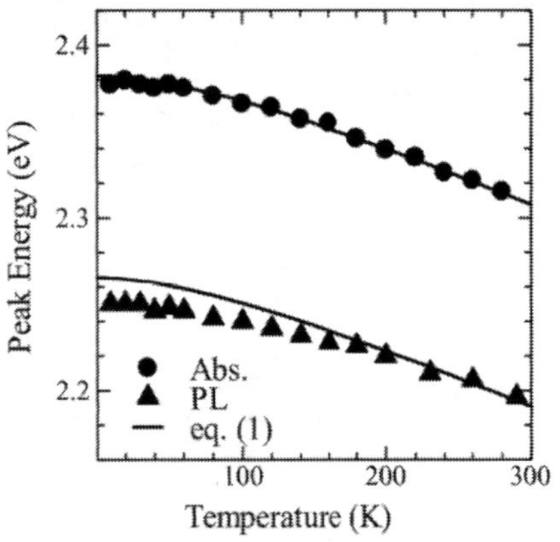

Figure 3 Temperature dependence absorption- and PL-peak energies. The solid curve indicates the calculated result using Varshni's law.

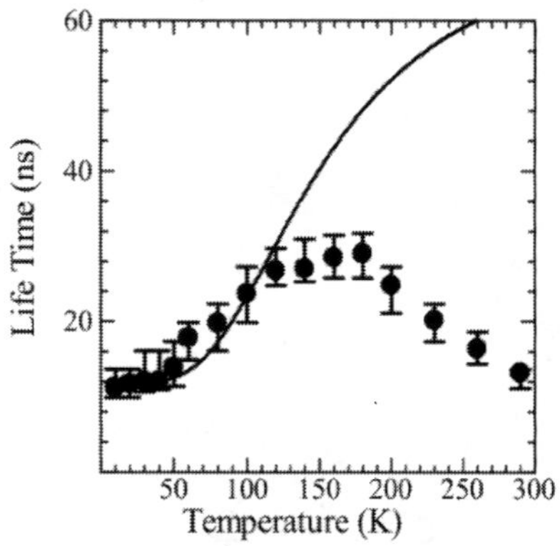

Figure 4 Temperature dependence of the decay time of the band-edge PL (●). The solid curve indicates the calculated result using eq. (2).

$$\frac{1}{\tau}=\frac{\dfrac{1}{\tau_{\mathrm{Bx}}}+d\dfrac{1}{\tau_{\mathrm{Dx}}}\exp\left(-\dfrac{\Delta E}{k_B T}\right)+d\dfrac{1}{\tau_{\mathrm{Brx}}}\exp\left(-\dfrac{\Delta E+\Delta_{ST}}{k_B T}\right)}{1+d\exp\left(-\dfrac{\Delta E}{k_B T}\right)+d\exp\left(-\dfrac{\Delta E+\Delta_{ST}}{k_B T}\right)}. \quad (2)$$

Here, τ_i denotes the radiative-decay time of bound-exciton (|Bx>) and "*free*"-exciton states (|Brx> and |Dx>), respectively. d represents the ratio of the density of state of |Bx> and "*free*"-exciton states (|Brx> and |Dx>). Δ_{ST} is the splitting energy between the bright- and dark-exciton states, and the bound-exciton state |Bx> lies below the dark-exciton state by an energy spacing of ΔE. The solid curve in Fig. 4 indicates the calculated result for the temperature dependence of τ using eq. (2) with τ_{Brx}=42 ns, τ_{Dx}=1000 ns, τ_{Bx} =12 ns, Δ_{ST}=14 meV, ΔE=26 meV, and d=18. The calculated result quantitatively explains the experimental one below 140 K. At higher temperatures, the calculated result still increases gradually, while the observed decay time begins to decrease. The discrepancy between the calculated and the experimental results can be attributed to the influence of the nonradiative recombination process because eq. (2) represents only the "*radiative*" decay rate in the four-state model. The magnitude of the energy difference between |Dx> and |Bx>, ΔE, in the as-grown SAM is ~17 meV [11]. Thus, ΔE becomes larger by the surface modification of the SAM. It is considered that an increase in the localization energy of the bound-exciton state would suppress a trapping process to defects and result in the activation of the band-edge PL.

4. Conclusions

The surface modification effects on PL properties of the SAM of CdSe QDs were investigated by measuring the temperature dependence of absorption and PL spectra, and PL dynamics. After the surface modification of the CdSe QDs with the $Cd(OH)_2$ layer, the band-edge PL is observed as a main PL band at every temperature between 10 K and RT. The decay time of the band-edge PL becomes larger slightly with an increase in temperature. The temperature dependence of the PL-decay time is explained using a four-state model including the bound-exciton state. The contribution of the bound exciton state to PL processes suppresses a trapping process to defects, and the band-edge PL is observed as a main PL band.

Acknowledgement

This work was supported by Grant-in-Aid for Scientific Research from the Japan Society for the Promotion of Science.

References

[1] Woggon U 1996 "Optical Properties of Semiconductor Quantum Dots (Springer)".
[2] Klimov V I 2004 "Semiconductor and Metal Nanocrystals (Marcel Dekker Inc)".
[3] Efros A L, Rosen M, Kuno M, Nirmal M, Norris D J and Bawendi M 1996 *Phys. Rev. B* 54, 4843.
[4] Donegá C M, Bode M and Meijerink A 2006 *Phys. Rev. B* 74, 085320.
[5] Bakkers E P A M, Roest A L, Marsman A W, Jenneskens L W, Steensel L I J, Kelly J J and Vanmaekelbergh D 2000 *J. Phys. Chem. B* 104, 7266.
[6] Marx E, Ginger D S, Walzer K, Stokbro K and Greenham N C 2002 *Nano Lett.* 2, 911.
[7] Yokota H, Okazaki K, Shimura K, Nakayama M and Kim D G, 2012 *J. Phys. Chem. C* 116, 5456.
[8] Aldana J, Wang Y A and Peng X 2001 *J. Am. Chem. Soc.* 123, 8844.
[9] Pradhan N, Battaglia D M, Liu Y and Peng X 2007 *Nano Lett.* 7, 312.
[10] Varshni Y 1967 *Physica* 34, 149.
[11] Kim D G, Yokota H, Shimura K and Nakayama M 2013 *Phys. Chem. Chem. Phys.* 15, 21051.

Indirect excitation of Eu^{3+} in GaN codoped with Mg and Eu

M Yamaga[1], H Watanabe[1], M Kurahashi[1], K P O'Donnell[2], K Lorenz[3] and M Boćkowski[4]

[1] Department of Mathematical and Design Engineering, Gifu University, Gifu 501-1193, Japan
[2] SUPA Department of Physics, Strathclyde University, Glasgow G4 0NG, UK
[3] CTN Instituto Superior Tecnico, Bobadela LRS 2686-953, Portugal
[4] IHPP Unipress, Polish Academy of Sciences, Warsaw 01-142, Poland

E-mail : yamaga@gifu-u.ac.jp

Abstract. Temperature-dependent Eu^{3+} luminescence spectra in GaN(Mg):Eu can be assigned to, at least, two distinct Eu^{3+} centres, denoted by Eu0 and Eu1. The splitting energy levels of the ^7F_J (J=1,2) multiplets for the Eu0 and Eu1 centres have been calculated using the equivalent operator Hamiltonian for C_{3v} crystal field with the addition of an odd parity distortion.

1. Introduction

Europium doped GaN is a candidate material of red light-emitting diodes and laser diodes [1-3]. Quantum efficiency at room temperature of the Eu^{3+} injection luminescence was less than 0.6 % [2], much lower than that of commercial blue InGaN devices (>30%); however, there are no commercial red InGaN diodes. In order to improve the quantum efficiency, that is, obtain sufficiently bright Eu^{3+} luminescence, it is very important to examine and optimize the energy transfer from the optical band-to-band excitation to the Eu^{3+} excited state [4,5]. The band-to-band excitation below 360 nm at low temperatures in GaN(Mg):Eu produces fairly strong $^5D_0 \rightarrow {}^7F_2$ luminescence lines of Eu^{3+} around 620 nm, suggesting that there are, at least, two different relaxation processes to Eu^{3+} as described in previous literature; one involves prompt relaxation from excitons produced by the band-to-band excitation immediately into the Eu^{3+} excited states [6-9]; the other involves slow relaxation to Eu^{3+} via the Auger process following donor-acceptor recombination, where V_N or Mg ions play the role of donors or acceptors, respectively [10,11].

Recent optical spectroscopy of Eu^{3+} in GaN doped with Mg (GaN(Mg):Eu) has revealed temperature-dependent hysteresis of the Eu^{3+} luminescence [5]. In this paper, we propose a structural model of Eu^{3+} luminescence centres in GaN(Mg):Eu, which are associated with Mg ions as acceptors..

2. Experimental procedure

GaN films, grown by hydride vapor phase epitaxy (HVPE) on (0001) sapphire substrates and doped with Mg (p-type) were implanted with 300 keV Eu ions [5]. The concentration of Mg was 6.4×10^{17} cm^{-3}. GaN has the Wurtzite crystal structure with the space group of $P6_3mc$. Eu and Mg ions substitute

Journal of Physics: Conference Series **619** (2015) 012025 doi:10.1088/1742-6596/619/1/012025

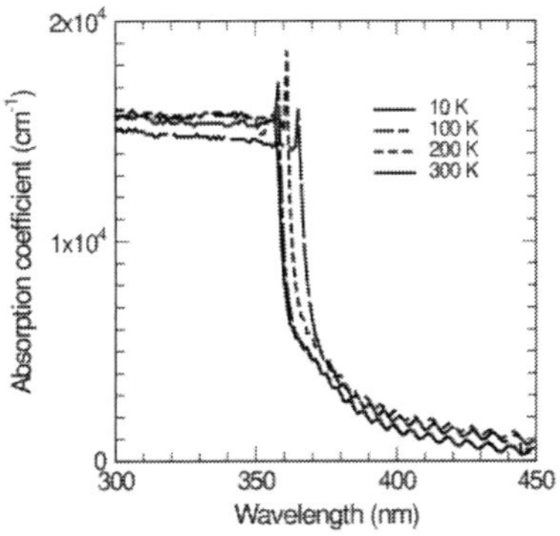

Figure 1. Temperature dependence of the absorption spectra for the GaN(Mg):Eu film.

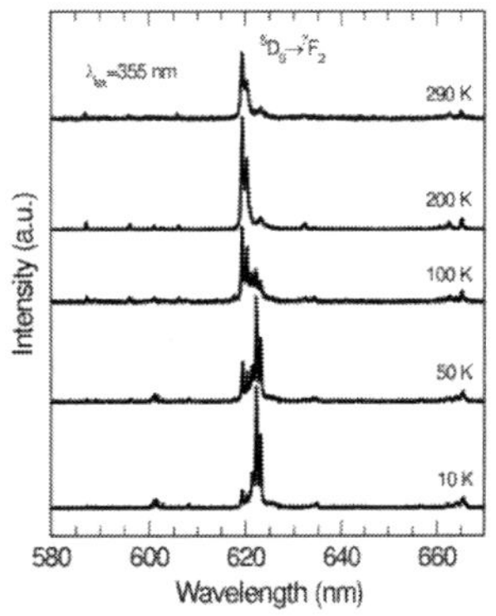

Figure 2. Eu^{3+} luminescence spectra for GaN (Mg) :Eu in the temperature range from 10 up to 290 K.

for Ga ions where a Mg ion has a possibility to be located at one of six first nearest neighbour Ga sites of a central Eu ion.

Optical absorption, excitation and emission spectra in the ultraviolet and visible regions were measured using the BL3B beam of the UVSOR facility in the Institute for Molecular Science at Okazaki in Japan. A sample was attached to a cold copper finger in a liquid helium cryostat with a temperature control system. Emission spectra in the temperature range of 10-300 K were measured using an Acton SpectraPro-300i monochromator combined with a Roper Spec-10 Si-CCD detector cooled by liquid nitrogen.

3. Experimental results

Figure 1 shows the absorption spectra of the GaN(Mg):Eu film in a temperature range from 10 to 300 K. The total film thickness is estimated to be ~4.2 μm from the interference fringes in the absorption spectrum at 300 K, while the implanted layer is only ~150 nm thick. The step near 360 nm corresponds to the band edge of GaN, whereas the sharp line around 360 nm is ascribed to excitons created in GaN. The absorption coefficient of such band-to-band transition is fairly large, >10^4 cm^{-1}. With increasing temperatures from 10 up to 300 K, the band gap shifts toward lower energy. Absorption lines due to Eu^{3+} could not be observed in these runs because of the weakness of the *4f-4f* transitions of Eu^{3+} and the low total Eu^{3+} content of the film.

Figure 2 shows Eu^{3+} luminescence spectra under 355 nm excitation for GaN(Mg):Eu in a temperature range from 10 up to 290 K. The luminescence spectrum is normalized to the maximum intensity of the $^5D_0 \rightarrow {}^7F_2$ line and the baseline of each spectrum is shifted for clarity. The luminescence spectra change markedly around 100 K on warming. The 622 nm lines (Eu1), dominant below 100 K, are replaced by 619 nm lines (Eu0) above 100 K. The Eu0 and Eu1 lines were first identified by O'Donnell [5]. The four groups around ~588, 596-608, 618-634 and 657-666 nm are assigned to the $^5D_0 \rightarrow {}^7F_J$ (J=0,1,2,3) intra-$4f^6$ transitions of Eu^{3+} [3,7,8].

In order to assign the luminescence lines to Eu0 and Eu1, the spectra around 600 and 620 nm in the regions of the $^5D_0 \rightarrow {}^7F_1$ and $^5D_0 \rightarrow {}^7F_2$ transitions were observed at high resolution (not shown). The two distinct Eu^{3+} luminescence lines with notation of Eu0 and Eu1 were best observed at 200 K and 10 K, respectively. The $^5D_0 \rightarrow {}^7F_0$ 'sensitive' line of Eu0 was visible, but that of Eu1 was too weak to observe. The $^5D_0 \rightarrow {}^7F_1$ transition of Eu0 is clearly split into three, while Eu1 lines comprise a doublet and a singlet. The $^5D_0 \rightarrow {}^7F_2$ lines of both Eu0 and Eu1 clearly split into five components. Line splittings for Eu0 are larger than for Eu1. This result suggests that the symmetry of the Eu0 centre is lower than that of the Eu1 centre, in agreement with the non-observation of the Eu1 $^5D_0 \rightarrow {}^7F_0$ transition.

4. Discussion

It is reasonable to assume that substitution of Eu^{3+} on a Ga^{3+} site provides a C_{3v} symmetry electrostatic potential and that substitution of Mg on a Ga site (Mg_{Ga}) is a perturbation upon Eu^{3+}. Such an odd-parity crystal field mixes opposite parity wavefunctions into the $^5D_0$ and ^7F_J (J=0-6) multiplets of the less symmetric Eu0 centre, resulting in breakdown of the parity selection rule. We consider the energy splitting of the ^7F_J (J=1,2) multiplets of both Eu0 and Eu1 centres. The equivalent operator Hamiltonian for C_{3v} crystal field [12,13] is given by

$$H_{c.f.} = B_2^0 O_2^0 + B_4^0 O_4^0 + B_4^3 (O_4^{-3} - O_4^3) + B_6^0 O_6^0 + B_6^3 (O_6^{-3} - O_6^3) + B_6^3 (O_6^{-6} + O_6^6) . \qquad (1)$$

The crystal field operator induced by the substitution of Mg is included in equation (1) as a perturbation term in a form of,

$$H_1 = B_2^2 O_2^2 . \qquad (2)$$

When the energy splitting of only ^7F_J (J=1,2) multiplets for the Eu0 and Eu1 centres is calculated, the sixth order terms in equation (1) can be omitted. The splitting energy levels of the $^7F_1$ and $^7F_2$ multiplets estimated from figure 2 are recalculated with the centre of gravity set equal to zero and summarized in Table 1.

Table 1. Experimental and calculated energy levels for $^7F_1$ and $^7F_2$ multiplets and optimized crystal field parameters in equations (1) and (2) for Eu0 and Eu1 with the energy unit (cm^{-1}).

Centre	Eu0		Eu1	
	(exp.)	(calc.)	(exp.)	(calc.)
$^7F_1$	137	138	122	122
	0	0	-53	-52
	-138	-138	-70	-70
B_2^0		-68		-62
$\lvert B_2^2 \rvert$		68		8
$^7F_2$	216	236	261	261
	169	162	-34	-38
	-89	-93	-57	-53
	-135	-143	-79	-80
	-161	-165	-90	-90
B_2^0		-34		-30
B_4^0		0.1		1.13
$\lvert B_4^3 \rvert$		39		33
$\lvert B_2^2 \rvert$		34		0.8

The observed energy splitting of the $^7F_1$ multiplet for Eu0 and Eu1 can directly estimate B_2^0 and B_2^2, since the energy separation between the doublet and the singlet is equal to 3 B_2^0 and the splitting energy of the doublet is equal to $2B_2^2$. The values of (B_2^0, B_2^2) and the calculated energy levels of the $^7F_1$ multiplet for the Eu0 and Eu1 centres are summarized in Table 1.

The splitting energy levels of the $^7F_2$ multiplet are obtained to solve the 5×5 determinant of equation (1) including equation (2). The splitting energy levels calculated using the values of B_2^0 and B_2^2 estimated from the $^7F_1$ multiplets of Eu0 and Eu1 are too large to fit the observed energy levels. The calculated energy levels of the $^7F_2$ multiplets of Eu0 and Eu1 and their optimized fitting parameters of the crystal field are summarized in Table 1. However, the values of B_2^0 and B_2^2 estimated from the $^7F_2$ multiplets are roughly half of those for the $^7F_1$ multiplet. At present, the origin of this difference is not clear.

The values of B_2^0, and B_4^3 for the Eu0 centre are a little larger than those for the Eu1 centre, but the B_2^2 value for Eu0 is much larger than that for Eu1. This calculation suggests that the Eu0 centre is strongly perturbed by the Mg substitution, whereas the Eu1 centre has close to C_{3v} symmetry and is assigned to a Eu^{3+} ion with Mg_{Ga} and/or V_N at a relatively long distance.

5. Conclusions

The luminescence of GaN(Mg):Eu features two main centres, Eu0 and Eu1, linked by photochromic hysteresis. Crystal field analysis suggests that Eu0 is less symmetric than Eu1, which is approximately C_{3v}.

Acknowledgement

This work was supported by the Use-of-UVSOR Facility Program (25-515) of the Institute for Molecular Science. M. Yamaga is indebted to the Ogawa Science and Technology Foundation. K. Lorenz acknowledges funding by FCT, Portugal (PTDC/CTM-NAN/2156/2012, PTDC/FIS-NAN/0973/2012, Investigator FCT).

References

[1] Sekiguchi H, Takagi Y, Otani T, Matsumura R, Okada H and Wakahara A 2013 Jpn. J. Appl. Phys. **52** 08JH01
[2] Nishikawa A, Furukawa N, Kawasaki T, Terai Y and Fujiwara Y 2010 Appl. Phys. Lett. **97** 051113
[3] Park J H and Stekl A J 2006 Opt. Mater. **28** 859
[4] Wang K, Martin R W, Nogales E, Katchkanov V, O'Donnell K P, Hernandez S, Lorenz K, Alves E, Ruffenach S and Briot O 2006 Opt. Mater. **28** 797
[5] O'Donnell K P, Edwards P R, Kappers M J, Lorenz K, Alves E and Bockowski M 2013 Phys. Status Solidi **C 11** 662
[6] Roqan I S, O'Donnell K P, Martin R W, Edwards P R, Song S F, Vantomme A, Lorenz K, Alves A and Bockowski M 2010 Phys. Rev. **B 81** 085209
[7] Poplawsky J D, Nishikawa A, Fujiwara Y and Dierolf V 2013 Optics Express **21** 30633
[8] Wisniewski K, Jadwisienczak W, Thomas T and Spencer M 2009 J. Rare Earths **27** 667
[9] Li Z, Bang H, Piao G, Sawahata J and Akimoto K 2002 J. Crystal Growth **240** 382.
[10] Fleischman Z, Munasinghe C, Steckl A J, Wakahara A, Zavada J and Dierolf V 2009 Appl. Phys. **B 97** 607
[11] Mezdrogina M M, Danilovskii E Yu and Kuz'min R V 2011 Inorg. Mater. **47** 1450
[12] Henderson B and Imbush G F 1989 *Optical Spectroscopy of Inorganic Solids* (Oxford:Clarendon Press) chapter 8
[13] Abragam A and Bleaney B 1970 *Electron Paramagnetic Resonance of Transition Ions* (Oxford : Clarendon Press) p209 and p338

Journal of Physics: Conference Series **619** (2015) 012026 doi:10.1088/1742-6596/619/1/012026

Characterization of metal-functionalized flax orbitide as a new candidate for light-emitting semiconductor

Robert Bauer [1*], Paul Bazylewski [1], Pramodkumar Jadav [2], Jianheng Shen [2], Denis Paskal Okinyo-Owiti [2], Jian Yang [3], Gap Soo Chang [1*], Martin Reaney [2] and Ramaswami Sammynaiken [4*]

[1] Dept. of Physics and Engineering Physics, University of Saskatchewan, 116 Science Place, Saskatoon SK S7N 5E2, Canada
[2] Dept. of Plant Sciences, University of Saskatchewan, 51 Campus Drive, Saskatoon SK S7N 5A8, Canada
[3] Dept. of Pharmacy, University of Saskatchewan, 116 Thorvaldson Building, Saskatoon SK S7N 5C9, Canada
[4] Saskatchewan Structural Sciences Centre, 110 Science Place, Saskatoon SK S7N 5C9, Canada

E-mail: robert.bauer@usask.ca, gapsoo.chang@usask.ca, r.sammynaiken@usask.ca

Abstract. Organic materials display promise in numerous electronic applications, complimentary to traditional semi-conducting materials. Cyclolinopeptides show promise in light-emitting applications as an organic semiconductor. Photoluminescence measurements indicate charge transfer between the peptide and the metal, resulting in an increase in intensity of the emission from around the metal in the Cyclolinopeptide complex. Complementary X-ray absorption near-edge spectroscopy (XANES) shows a change in occupation of energy states in the peptide when complexed with the metal, indicating charge transfer, but peak positions show the peptide is not chemically changed by the metal. Combining X-ray emission and XANES provides element specific partial density of states, to estimate the element specific energy gap which is the proposed emission range for the peptide material. Organic light emitting diode devices have been fabricated, although no measurable emission has been seen as of yet. The devices have diode like current-voltage characteristics showing the peptide is semi-conducting with a threshold voltage of approximately 2.5 V.

1. Introduction

Organic electronics is a growing field driven by the low cost of fabrication and innate flexibility of organic materials. The cost reductions arise from the roll to roll processing, or printing fabrication potential of organic materials [1]. These methods reduce the amount of material utilized by employing bottom up fabrication, instead of the top down fabrication currently employed for mass produced electronics. The characteristics of the active organic material result from the chemical structure of the organic molecules rather than the dopant atoms in a silicon lattice, thereby eradicating the need for ion implantation of the active material. The vast number of suitable organic molecules and the potential for engineering their electronic properties, providing an almost endless supply of organic materials to be tested for electronic applications.

Cyclolinopeptides (CLs) are naturally occurring cyclic, hydrophobic and biologically active orbitides found in flaxseed, flax roots and flaxseed products [2, 3]. This group of orbitides comprise 8 or 9 proteinogenic amino acids fused at terminal amino and carboxyl groups to form stable cyclic structures [2]. The geometric arrangement of the atoms in the ring structures is greatly affected by environmental considerations, as well as functionalization and metal complexes. The effect of these considerations is a change in the twisting of the CL structure as seen by looking at the crystal structure, which is affected by the solvent from which the crystal was grown [4].

For photon emission from organic materials one must first look at the exciton (excited electron-hole pair) which undergoes recombination resulting in light emission. There are two main types of exciton pairs that are generally considered in emission of visible light from organic materials: singlet state, and triplet state excitons [5]. The difference between the two types of excited states is that a triplet state has a net electron spin of one and a singlet state has a net electronic spin of zero. This results from the orientation and phase of the spin of the excited electron with respect to its counterpart in the ground state. If the spin orientation or phase is the same for both excited and ground state electrons of the exciton then this is a triplet state, otherwise it is a singlet state exciton. Singlet state emission (fluorescence) is much more rapid because the transition does not require a change of spin orientation or phase of the electron. Triplet state emission (phosphorescence) is much slower and tends to result in a glow even after the device is turned off due to the lower probability of recombination of triplet state excitons [6]. Generally in organic electronics one can expect approximately 75% triplet state excitons and 25% singlet state excitons [7]. This causes the efficiency of emission from singlet state focused devices to suffer by limiting the proportion of singlet carriers, but benefits devices focused on triplet state emission [8]. This may be accomplished by adding a heavy atom (such as terbium (Tb) or europium (Eu)) to the system, which is a known method to enhance triplet state emission in organic systems. Organic materials have weak spin-orbital interaction because they are made up of materials that do not have a bound d-state. The heavy atom has larger spin orbit interaction of the system because they have a bound d-state and the spin orbit interaction is generally seen in the interaction between the p-states and partially occupied d-states of transition metals. Transition of triplet state excitons to singlet state excitons at metal sites creates localized sources of singlet excitons, increasing the exciton emission in the organic metal complexes; that could then cause a cascade of emission via stimulated emission [9].

2. Experiments and Discussion

Three techniques were employed to characterize CLs: photoluminescence (PL), XANES, X-ray emission spectroscopy (XES), and current-voltage (I-V) charcterization of organic light emitting diode (OLED) devices. PL measurements used a 240 nm ultraviolet light source and an Ocean Optics QE65000 spectrometer. XANES and XES spectra of carbon, oxygen, and nitrogen K-edge absorption and non-resonant K-α emission respectively, were collected at the Resonant Elastic and Inelastic X-ray Scattering (REIXS) beam line at the Canadian Light Source (CLS). OLEDs were fabricated using a standard four layer device structure. The four-layers are an emissive layer, a buffer layer, front and back electrodes. The front and back electrodes were indium tin oxide (ITO) and gold respectively. CLs complexed with various transition metals are used as the emissive layer. Between the front electrode and emissive layer poly(3,4-ethylenedioxythiophene) polystyrene sulfonate (PEDOT:PSS) is employed as a buffer. I-V characterization of OLEDs were preformed using a Keithley 2420 source measure unit.

PL measurements have been used to examine the emission properties of various CLs, containing several different transition metals. One such system is shown in Fig. 1 where one can see the change in the PL spectrum of Tb when it is complexed with cyclolinopeptide D (CLD). The orbitide CLD was chosen for this investigation because it contains tryptophan as one of the amino acids in the peptide structure; tryptophan is an excellent absorber of short wavelength electromagnetic radiation. Looking at Fig. 1 one can see the reflected UV light in the 240 nm region, with a large reflected peak in the Tb sample; when CL is added the reflected peak is greatly reduced. The reduction in reflected peak intensity is likely caused by absorption by the tryptophan in the CLD structure, but could also be caused by scattering in the CLD structure. Absorption is most likely due to the enhanced emission from Tb sites observed. Tryptophan also has an emission band in the 250-300 nm region, which does not appear in the collected spectrum, indicating charge transfer being the mechanism for energy transfer to the Tb rather than a secondary emission-absorption process. These interpretations are supported by looking at the interaction of Tb and Eu in other CL structures using XANES.

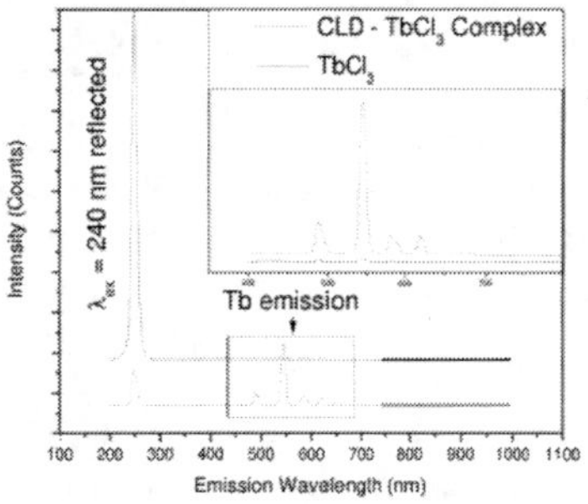

Figure 1: UV excited PL measurements of Tb and CLD Tb complex, showing enhancement of emission with CLD present indicating charge transfer.

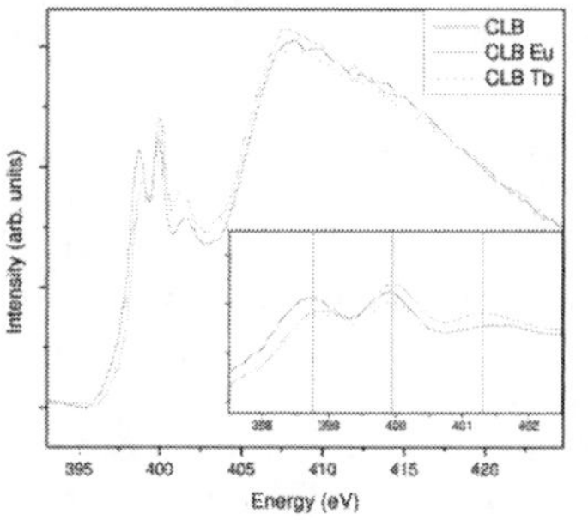

Figure 2: Nitrogen K-edge XAS of CLB and metal complexes, indicating chemical stability of CL and charge transfer in change in state occupation.

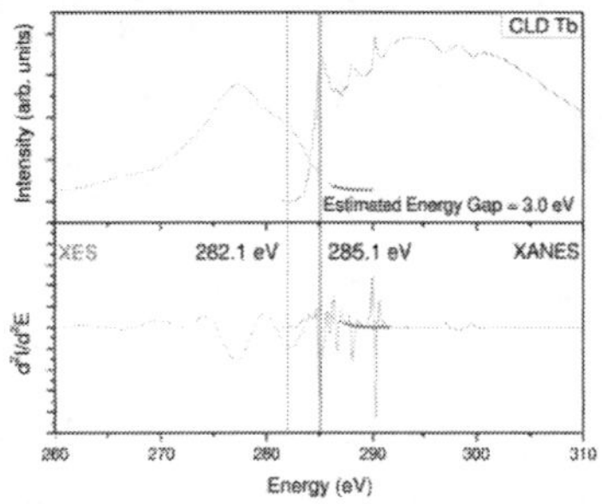

Figure 3: Carbon K-edge XES (left) and XAS (right) spectra to estimate element specific energy gap of in CLD Tb complex

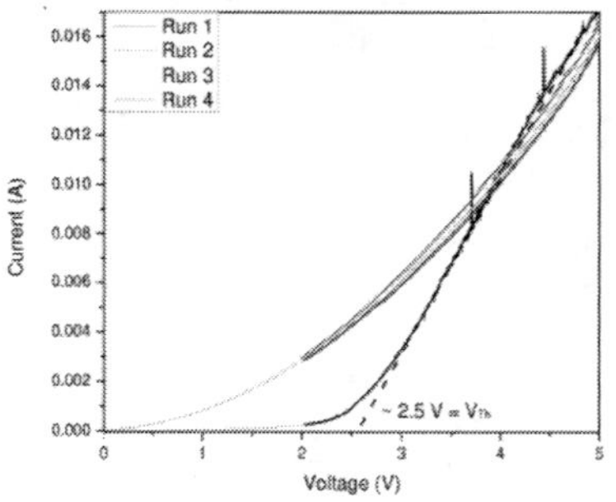

Figure 4: CLD terbium complex I-V characterization in runs of 4 successive tests, showing change in line shape.

Carbon and nitrogen K-edge XANES spectra (Fig. 2) indicate a change in the occupation of localized energy states attributed to bonding in the organic molecule. The energy location of peaks in Fig. 2 does not change between samples, meaning the chemical structure of the molecule is unchanged by metal addition to the system. The change in state occupation, seen as the variation in peak intensity, indicates charge transfer between the metal and CL. Combining the XANES and XES yields the element specific partial density of states as shown in Fig. 3; the element specific energy gap is estimated to get an idea of the maximum possible emission energy of 2.9 eV for the CLD.

I-V characterization of OLED devices was done to look at the turn on potential and relate emission to current. The I-V curves from one device provide the threshold voltage of approximately 2.5 V, even though no emission has been observed (Fig. 4). I-V characteristics show the typical diode line shape for the first run but subsequent runs immediately following show a significant change in the line shape. The line shape returns to its original shape after some time but changes again after the first run of the test, indicating a possible saturation of states resulting in a more conductive electronic structure.

3. Summary

In summary CLs are cyclic hydrophobic orbitides that occur in flax and show promise as light emitting materials. Charge transfer between CLD and a complexed metal was revealed by enhanced emission from the metal site under UV irradiation. This is supported by XANES measurements where intensity changes indicate a change in the occupation of states in the presence of the metal, while the peak positions remain unchanged indicating chemical stability. The enhanced emission indicates a possibility to utilize the plentiful triplet state excitons in organic materials for light emission.

Element specific energy gap estimation finds a maximum possible energy gap value (energy of emitted light) of 2.9 eV for CLD. OLED fabrication shows the semiconducting nature of CLD and a threshold voltage for devices made from CLD of approximately 2.5 V. There is a change in the line shape of I-V characteristic after the first test, but a return to a diode shape occurs when the system relaxes. The ability to tailor CLs for OLEDs by complexing with metals can be used to improve the performance of the devices for light emission, making CL-metal systems diverse and promising for OLED applications.

Acknowledgments

We gratefully acknowledge support from the Natural Sciences and Engineering Research Council of Canada (NSERC), Canada Foundation for Innovation (CFI), the Agriculture Development Fund (ADF), the Saskatchewan Structural Science Center (SSSC). Research described in this paper was performed at the Canadian Light Source, which is funded by the CFI, the NSERC, the National Research Council Canada, the Canadian Institutes of Health Research, the Government of Saskatchewan, Western Economic Diversification Canada, and the University of Saskatchewan.

References

[1] Forrest S R 2004 *Nature* **428** 911–918

[2] Okinyo-Owiti D P, Young L, Burnett P G G and Reaney M J T 2014 *Peptide Science* **102** 168–175

[3] Okinyo-Owiti D P, Burnett P G G and Reaney M J 2014 *Journal of Chromatography B* **965** 231 – 237

[4] Schatte G, Labiuk S, Li B, Burnett P G, Reaney M, Grochulski P, Fodje M, Yang J and Sammynaiken R 2012 *Acta Crystallographica Section E* **68** 50–51

[5] Griffiths D J 2005 *Introduction to quantum mechanics* (Pearson Prentice Hall)

[6] Kasha M 1947 *Chemical Reviews* **41** 401–419

[7] Baldo M A, O'Brien D F, Thompson M E and Forrest S R 1999 *Phys. Rev. B* **60**(20) 14422–14428

[8] Baldo M A, O'Brien D F, You Y, Shoustikov A, Sibley S, Thompson M E and Forrest S R 1998 *Nature* **395** 151–154

[9] Singh J 2007 *Phys. Rev. B* **76**(8) 085205

Journal of Physics: Conference Series **619** (2015) 012027 doi:10.1088/1742-6596/619/1/012027

The photoinduced birefringence and mass transport in azo compound K-D-2

K Klismeta, J Teteris

Institute of Solid State Physics, University of Latvia, 8 Kengaraga Str., LV 1063 Riga, Latvia

E-mail: k.klismeta@gmail.com

Abstract. Azobenzene containing compounds are among light polarization sensitive materials - the moieties may align relative to the electric field vector of light, leading to anisotropy and birefringence in the sample. Another phenomenon which can be observed in azo compounds under influence of light is macroscopic movement of the material. In this work photoinduced processes in low molecular weight organic glass - bis-azobenzene containing compound K-D-2 were experimentally studied. Birefringence was induced with linearly polarized laser light (473, 532 and 635 nm) and measured at 633 nm wavelength. Polarization holography with recording beam configuration $+45^0/-45^0$ was used to induce mass motion. Dependence of the surface relief depth on the recording laser wavelength in the visible spectrum (375 – 671 nm) was obtained. Formation of the SRG was observed with all used wavelengths and high birefringence values were obtained. Certain correlation between the absorption of the wavelength and photoinduced mass transport and birefringence is yet to be confirmed.

1. Introduction

Azobenzene containing compounds are among light polarization sensitive materials. Under influence of light azo compounds experience *trans-cis* isomerisation process [1]. As a result, the photochrome moieties may align relative to the electric field vector of light, leading to anisotropy in material and thus to birefringence [2]. Another phenomenon which can be observed in azo compounds is macroscopic movement of the material - if a lateral polarization modulation of light is present on the surface of the sample, a mass transport could occur [3]. One of the methods to induce mass motion is holographic recording where a periodical polarization pattern can be obtained at the submicron scale.

The photo-induced mass transport phenomenon allows formation of surface relief directly without chemical etching process. Materials which possess the ability of direct optically induced surface patterning may have potential application in various photoactive devices, optical information storage and the production of diffractive optical elements used in telecommunication (e.g. filters, Bragg reflectors) [4].

Low-molecular-weight organic glasses containing azobenzene moieties are the subject of intense current studies [5,6]. In contrast to functionalized azo-polymers, molecular glasses have the advantage of possessing a well-defined molecular structure and they exhibit more uniform physical properties. Recently, SRG formation in amorphous films containing diphenylamine based azochromophores using holographic recording by 532 nm laser has been reported [6]. In this work the efficiency dependence on the wavelength of the laser to photoinduced mass transport and birefringence in low molecular weight organic glass - bis-azobenzene containing compound K-D-2 was experimentally studied.

2. Experimental

The 2-(bis(4-(-(2-bromo-4-nitrophenyl)diazenyl)phenyl)amino)ethanol (referred to as K-D-2) was synthesized in Riga Technical University. The molecular structure of the compound is shown in figure 1. The absorption spectrum of the studied sample is shown in figure 5. The thickness of the film was determined using Veeco Dectak 150 surface profilometer and it was 150 nm.

Surface relief gratings were recorded using a holographic set-up shown in figure 2. Holographic recording was performed with eleven different lasers (L) with wavelengths 375 nm, 405 nm, 448 nm, 473 nm, 491 nm, 532 nm, 561 nm, 594 nm, 635 nm, 650 nm and 671 nm. The set-up was adjusted for each of the laser wavelengths so that the grating period was the same in all cases - 1 μm. The laser beam was split into two beams with equal intensities by a beam splitter (Bs). Using mirrors (M) the beams were directed to the sample (S) where they interfere. Half-wave plates (λ/2) were used to set the polarization states of the recording beams. Two orthogonally at +45/-45 degrees linearly polarized beams were used for recording thus the optimal conditions for surface relief grating formation was provided [7]. One period of the polarization modulation created on the surface of the sample is shown in figure 2. The intensity of the beams was $I_1=I_2=0.18$ W/cm^2 and the duration of the recording was 1 hour thus the dose of illumination received was E=1.30 kJ/cm^2 in all cases. The recording course was monitored by registering the 0th and 1st order diffracted beams of the laser with photodiodes (PD). The diffraction efficiency of the obtained grating was calculated as in equation (1)

$$\eta = \frac{I_1}{I_1 + I_0} \tag{1}$$

where I_0 is the intensity of the 0th order diffracted beam and I_1 is of the 1st order.

The profile of the produced SRG was examined by AFM.

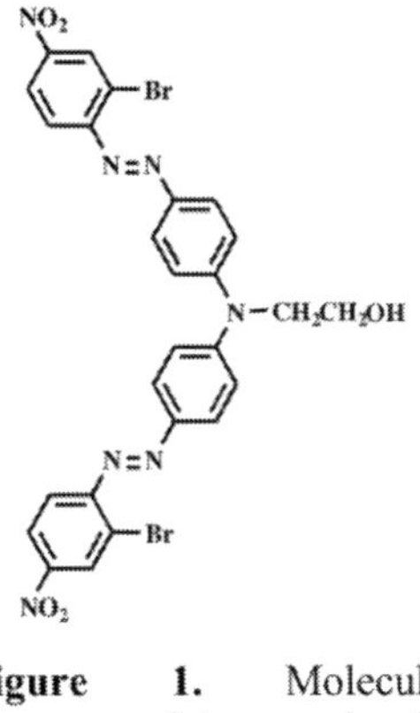

Figure 1. Molecular structure of low molecular weight organic glass K-D-2.

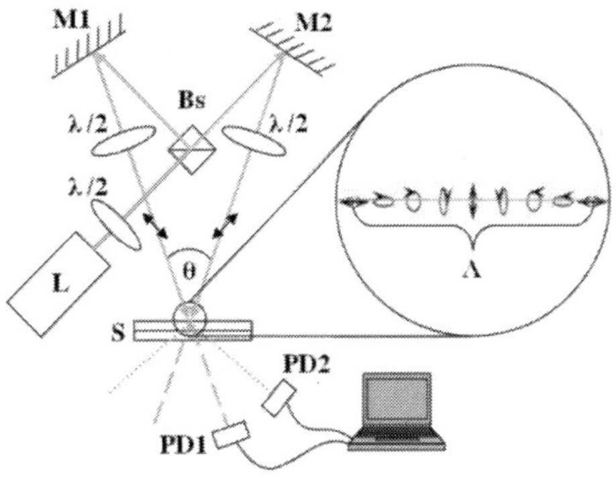

Figure 2. Experimental set-up for holographic recording. L – pump laser, λ/2 – half-wave plate, Bs – beam splitter, M – mirror, S – sample, PD - photodiode.

Photoinduced birefringence was measured using experimental set-up shown in figure 3. Experiments were carried out with three different lasers as pump lasers (L1) with wavelengths 473 nm, 532 nm and 635 nm to induce birefringence in the sample (S) and it was measured at 633 nm by a probe laser (L2). The polarization of the laser beams was set using half-wave plates (λ/2). The linearity of the polarization was improved by Glan-Taylor polarization cubes (P). Probe beam passes through the sample, a part of it is reflected to a photodiode to take into account the changes of transmittance regarding photobleaching but the rest of it passes through an analyzer (A). If a difference between the refractive indexes in orthogonal directions is present in the sample, the polarization plane of the probe beam shifts and a signal appears on the photodiode behind the analyzer. Birefringence then can be evaluated using formula (2)

$$\Delta n = \frac{\lambda}{\pi d} \arcsin\left(\sqrt{\frac{I}{I_0}}\right) \quad (2)$$

where I is the probe beam intensity passing through crossed polarizer and analyzer, Io – probe beam intensity passing through parallel polarizers (with transmittance changes taken into account), λ - probe wavelength and d is the thickness of the layer of material.
The intensity of the pump beam was I =0.12 W/cm^2 and the duration of the measurement was 1 hour thus the dose of illumination received in these experiments was E=0.44 kJ/cm^2. The intensity of the probe beam was much smaller compared to that of the pump beam to exclude its interaction with the material during the experiment.

A quarter-wave plate was used to determine whether the birefringence is negative or positive by using a phase shift compensating method.

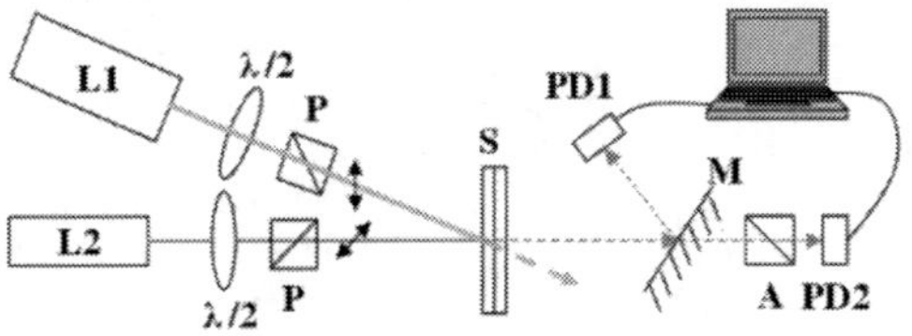

Figure 3. Experimental set-up for photoinduced birefringence measurements. L1 – pump laser, L2 - probe laser, λ/2 – half-wave plate, P - polarizer, A - analyzer, M – semitransparent mirror, S – sample, PD - photodiode.

3. Results and discussion

Figure 4 shows diffraction efficiency dependence on time for recording done with 653 nm, 405 nm and 375 nm lasers. The diffraction efficiency has not yet fully reached saturation after receiving the dose of 1.30 kJ/cm^2. Figure 5 shows the obtained values of modulation depth along with the absorption spectrum of the sample. It has to be mentioned that the modulation depth obtained with 491 nm, 532 nm, 561 nm and 594 nm lasers are likely not the maximal possible because the 300 nm exceed the thickness of the sample by 2 times, which simply means that there is not enough material to transport. AFM profiles of the obtained relief show that the substrate is already reached in these cases.

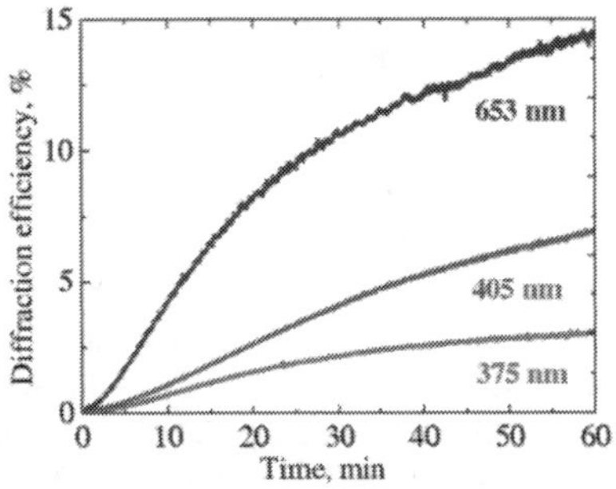

Figure 4. Diffraction efficiency in time for recording done with 653 nm, 405 nm and 375 nm lasers. The intensity of the beams was I_1=I_2=0.18 W/cm^2.

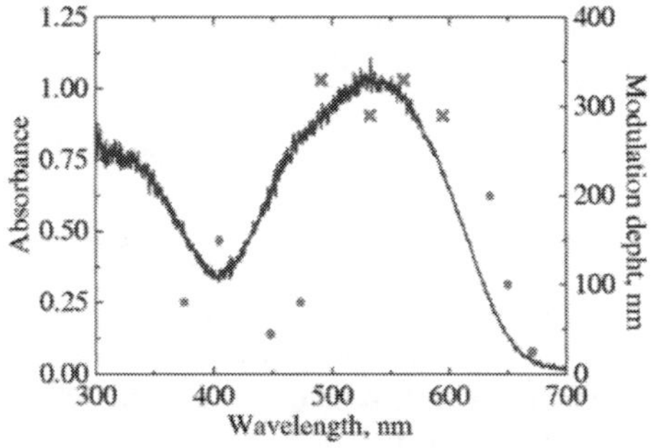

Figure 5. The absorption spectra of the sample (line) and modulation depth obtained with different recording laser wavelengths (points) at received dose of illumination E=1.30 kJ/cm^2.

Diffraction efficiency values correlate well with modulation depth, i.e., the gratings with deeper relief diffract better.

It can be seen in figure 5 that there is a correlation between the absorption of the wavelength and the efficiency of mass transport induced by this wavelength. Most likely it is related to the *trans-cis* isomerisation process of the azo compound and the corresponding absorption of each isomer [8]. It cannot be certainly concluded from these experiments whether particularly the stable *trans* configuration or the metastable *cis* configuration or the cycling process between the both of them is responsible for the mass transport.

Figure 8 shows the photoinduced birefringence measurements in time. It can be seen that there is a rapid change in the refractive index of the sample in the first seconds of the irradiation. That means the most of the moieties quickly orientate in presence of the light electric field thus making the material anisotropic. Birefringence values obtained in this sample are high - almost 0.2 and the birefringence was determined to be negative. It cannot be certainly concluded from these experiments whether there is a correlation between birefringence and surface modulation depth induced by one wavelength.

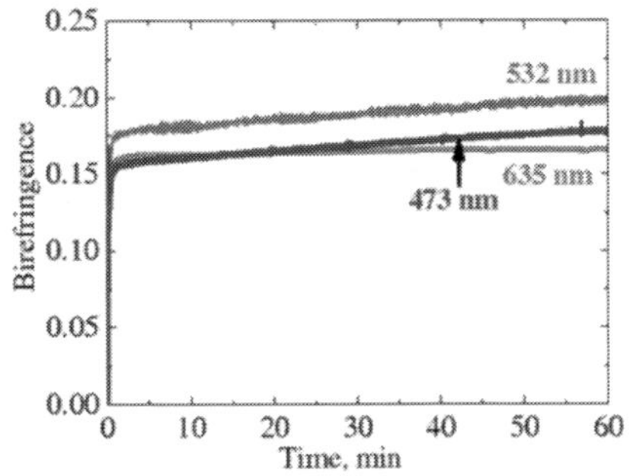

Figure 8. Birefringence values at 633 nm, induced by 473 nm, 532 nm and 635 nm lasers with beam intensity I =0.12 W/cm^2.

4. Conclusions

Formation of the SRG was observed with all used recording laser wavelengths with depth determined by AFM to be even greater than the thickness of the sample. Dependence of the surface relief depth on the recording laser wavelength in the visible spectrum was obtained but certain correlation between the absorption of the wavelength and the efficiency of the photoinduced mass transport is yet to be confirmed. High birefringence values were obtained in the sample with all laser wavelengths used to induce the anisotropy. Birefringence was concluded to be negative in all cases. Also certain correlation between the photoinduced mass transport and birefringence is yet to be confirmed.

Acknowledgements

The authors thank Prof. Andris Ozols for providing the K-D-2 films.

References

[1] Sudesh Kumar G and Neckers D C 1989 *Chem. Rev.* **89** 1915
[2] Todorov T, Nikolova L and Tomova N 1984 *Applied Optics* **23** 4309
[3] Rochon P, Batalla E and Natansohn A 1995 *Appl. Phys. Lett.* **66** 136
[4] Natansohn A and Rochon P 2002 *Chemical Reviews* **102** 4139
[5] Audorff H, Walker R, Kador L and Schmidt H-W 2009 *Proc.SPIE* **7233** 723300
[6] Ozols A, Kokars V, Augustovs P, Traskovskis K and Saharov D 2012 *IOP Conf.Series: Materials Science and Engineering* **38** 012011
[7] Jiang X L, Li L, Kumar J, Kim D Y, Shivshankar V and Tripathy S K 1996 *Appl. Phys. Lett.* **68** 2618
[8] Ojanen J and Rantala T T 2009 *The Open Chemical Physics Journal* **2** 37

Journal of Physics: Conference Series **619** (2015) 012028 doi:10.1088/1742-6596/619/1/012028

Fabrication of an interdigitated sample holder for dielectric spectroscopy of thin films

Mina Shenouda and Derek R. Oliver
Department of Electrical and Computer Engineering, University of Manitoba
Winnipeg MB Canada R3T 5V6

E-mail: Derek.Oliver@umanitoba.ca

Abstract: A planar interdigitated sample substrate has been designed to support a thin-film sample of a polymer while the frequency-dependent dielectric properties of the thin film are measured. Trenches for the electrodes were etched into a SiO_2/Si wafer surface. Chromium was used as an adhesion layer prior to thermal evaporation of copper for the body of the electrode. The device was placed in a standard probe station and the dielectric character was recorded as a function of frequency with an impedance analyser. Devices with 20 to 70 fingers were measured and the results compared to analytical and finite element simulations. At 1 kHz, the total capacitance of a typical 20-finger device was 8 pF. The capacitive contribution of the thin film due to the fringing field in the polymer was about 2% of the total capacitance of the fabricated structures.

1. Introduction

Dielectric spectroscopy may be used to characterize dielectric material character. For bulk samples, the material is generally clamped between planar electrodes and the dielectric character is extracted from measurements of the capacitance [1]. Despite considerable interest in thin film samples for a range of technology applications [2], these samples are not well-suited to the mechanical loading associated with this approach. The dielectric properties of thin films have been shown to differ from those measured for bulk materials [3]. However, the challenge associated with performing accurate dielectric measurements for thin films is centered on designing an appropriate sample holder [4].

Open ended coaxial probes [5] and short-terminated [6] probes demand additional fixtures as well as specific geometric dimensions of the sample to obtain accurate results. The free space method, in which a free-standing sample is placed between transmitting and receiving antennas and a loss measurement determined, has limited applicability as the sample width is generally less than the incident wavelength, thus edge diffraction has a considerable impact on the measurements [7]. The inability to produce free-standing samples makes approaches based on resonant cavities infeasible, as precisely-shaped samples are required [8]. Similarly, Fabry-Perot [9], coaxial [10] and split-post [11] techniques all rely on specific sample sizes and/or free-standing samples. A scanning near-field probing technique, sometimes termed scanning microwave microscopy [12], is limited to a sequence of pre-determined frequency measurements rather than the accessible frequency sweep of a network or impedance analyzer. The goal is to develop non-contact electrostatic force microscopy (EFM) to map dielectric variations with submicron resolution. The independence of the swept-frequency response to the (narrow) bandwidth of the EFM probe has already been demonstrated [13,14]

In this design, interdigitated electrodes were embedded in a substrate to provide a deposition plane for the thin film sample (figure 1). Well-separated contact pads provided contact locations for measurement probes.

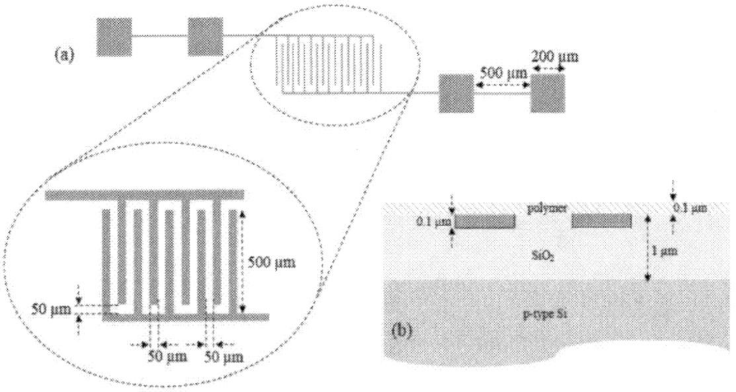

Figure 1. a) Plan view of the interdigitated sample holder. b) Cross-section of the embedded electrodes. The thickness of the p-type silicon wafer was 400 µm. The chromium adhesion layer is indicated at the electrode interface with SiO_2.

Analytical descriptions for the capacitances due to interdigitated electrodes fabricated in the SiO_2 layer of a SiO_2/Si wafer have been reported in the literature [15]. The complex capacitance, $C^*_{substrate}$, associated with fringing capacitance contribution due to the SiO_2 substrate is given by

$$C^*_{substrate} = \left(\frac{N-1}{2}\right)\left[\frac{1}{\varepsilon_{SiO_2}C(h_{SiO_2})} + \frac{1}{\varepsilon^*_{Si}}\left(\frac{1}{C(\infty)} - \frac{1}{C(h_{SiO_2})}\right)\right]^{-1} \quad (1)$$

where N is the number of interdigitated electrodes, $C(h_{SiO2})$ is the capacitance contribution due to the silicon dioxide, $C(\infty)$ arises from the infinite layer of air above the substrate when no polymer is present, ε^*_{Si} is the complex permittivity of silicon and $\varepsilon_{SiO2} = 4.5$ is the real part of the relative permittivity of SiO_2 [15]. The capacitance contribution from the polymer layer may be described [15]

$$C_{polymer} = (N-1)\varepsilon_0 \frac{K(k_{polymer})}{K(k'_{polymer})} L \quad (2)$$

where L is the finger length, K is an elliptical integral of the 1st kind and the parameters $k_{polymer}$, $k'_{polymer}$ are derived from the electrode geometry [15]. The direct capacitances between electrodes as well as between the nearest contact pad and the interdigitated electrodes are, respectively, expressed:

$$C_{elec} = 2(N-1)\varepsilon_0\varepsilon_{SiO_2}\frac{Lt}{s} \quad (3)$$

$$C_{elec-pad} \approx 2\varepsilon_0\varepsilon_{SiO_2}\frac{Lt}{H} \quad (4)$$

where t is the depth of the electrodes inside the SiO_2 substrate, s represents the spacing between the fingers and H is the spacing between the interdigitated electrodes and the neighbouring pad. The summation of these contributions yields the total capacitance between the contact pads:

$$C_{total} = Re(C^*_{substrate}) + C_{elec} + C_{elec\text{-}pad} + C_{polymer} \quad (5)$$

and predicts a linear relationship between the total capacitance, C_{total}, and the number of fingers, N.

2. Experimental

The capacitance of the system was estimated analytically using (1) – (5) and verified using a standard finite-element solver (COMSOL). The devices were fabricated on a 4" SiO_2/Si wafer, exploiting the doi:10.1088/1742-6596/619/1/012028

insulating properties of silicon dioxide ($10^{16}\Omega$.cm). The electrodes were patterned in a positive photoresist (HPR 504) using a shadow mask. Plasma etching (~ 20 nm/min) was used to create trenches in the SiO_2 layer. A 20 nm chromium adhesion layer was deposited and the remainder of the trench was filled with thermally-evapourated copper. The photoresist was dissolved in an ultrasonic bath containing acetone. For the capacitance measurements, the polymer film was deposited from a 2% wt/wt solution of polystyrene (PS) dissolved in toluene and was spin-coated at 1000 rpm for 30 seconds. The pads were exposed with a cotton swab dipped in toluene followed by surface cleaning using isopropyl alcohol (IPA).

A *KLA Tencor AS-500 Alpha-Step* profilometer was used for surface relief measurements to confirm the electrode trench etch depth and spin-coated polymer thickness. The difference between the deposited copper and SiO_2 surface was confirmed using contact atomic force microscopy (*Veeco DI3100* and a *Nanoscope IV controller*). An *Agilent 4294A* impedance anlayzer was used to obtain capacitance measurements in conjunction with an *Alessi REL 6100* probe station.

3. Results

The target polymer film thickness was 100 nm, 30 independent replications (stripping and redeposition) confirmed the film thickness was always in the range 100 ± 30 nm. The polymer contribution was estimated experimentally from the difference between a measurement of the interdigitated device without a before the polymer was spin-coated and a second measurement after the spin-coating (figure 2).

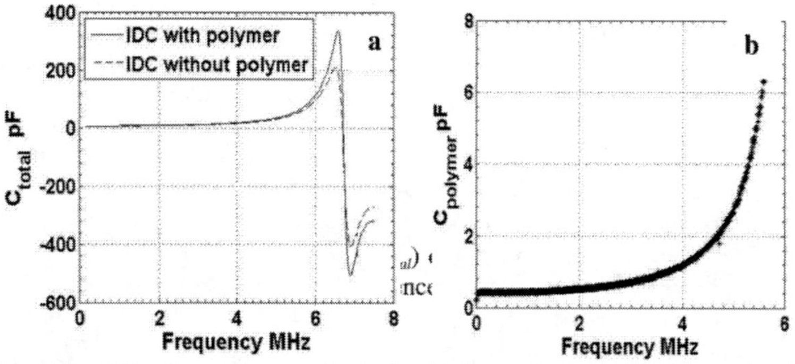

Figure 2: a) Total capacitance (C_{total}) of a 20-finger device with/without the thin polymer film; b) The difference between the 2 cases, i.e. $C_{polymer}$

The experimental data were also compared to the results obtained analytically and from finite element simulation The linearity of C_{total} in N (number of fingers) anticipated from (1) – (5) is clearly shown over the range N = 20 to N = 70 in figure 3.

4. Discussion

The data points in figure 3 represent an average of 5 independent measurements. Following each measurement, the polymer film was removed and a fresh film deposited. The maximum and minumum C_{total} values from each set of replicate data points were within 3% of the average. Thus, the uncertainty may be reasonably considered to lie within the scale of the data points shown in figure 3.

The resonant peak in figure 2a was associated with associated with the combination of cables and probes used in conjunction with the probe station and the impedance analyzer. These parasitc contribtions are also evident in the higher-frequency C_{total} measurements (figure 3b). The shallower gradient of the analytical calculation at 1 MHz was attributed to the quasi-static coplanar line model for interdigitated systems that formed the basis of these calculations [15].

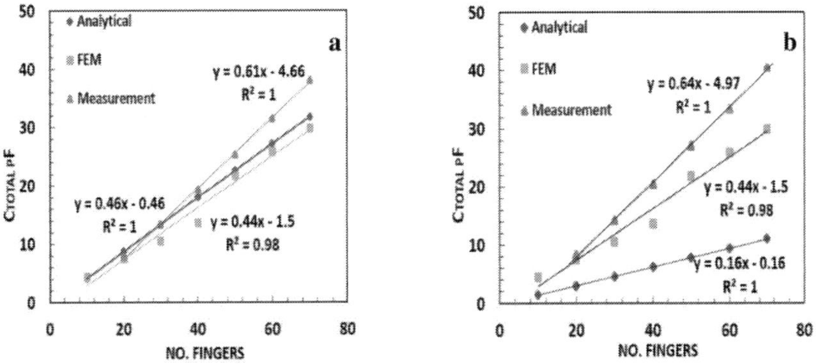

Figure 3: Comparison between the total capacitance obtained from analytical calculations, finite element (FEM) simulations and experimental measurements. The capacitance is expressed in pF. a) Data at 9.414 KHz; b) data at 1 MHz.

5. Conclusions

An interdigitated capacitor to measure the capacitance contribution and, in turn, the dielectric character of a thin polymer film has been designed and demonstrated. The capacitance contribution of a 100 nm polystyrene film was shown to be measureable using this technique.

Acknowledgments

This work was supported in part by the Natural Sciences and Engineering Research Council (NSERC) of Canada and the University of Manitoba. The work reported made use of surface characterization infrastructure in the Manitoba Institute for Materials.

References

[1] Kremer F and Schönhals A 2003 *Broadband Dielectric Spectroscopy* (Berlin:Springer)
[2] Poortmans J and Arkhibov V 2006 Thin film solar cells: fabrication, characterizaiton and applications (UK: John Wiley & Sons)
[3] Wasa K, Kitabatake M and Adachi H 2004 *Thin films materials technology: sputtering of compound materials* (Berlin:Springer)
[4] Venkatesh M S and Raghavan G S V 2004 *Biosyst. Eng.* **88** 1
[5] Krupka J 2006 *Meas. Sci. Technol.* **17** R55
[6] Wu M, Yao X and Zhang L 2000 *Meas. Sci. Technol.* **11** 1617
[7] Ghodgaonkar D, K Varadan V V and Varadan V K 1990 *IEEE Trans. Instrum. Meas.* **39** 387
[8] Chen L, Ong C K and Tan B T G 1999 *IEEE Trans. Instrum. Meas.* **48** 1031
[9] Asfar M N 1984 *IEEE Trans. Microw. Theory Tech.* **32** 1598
[10] Murata K, Hanawa A and Nozaki R 2005 *J. Appl. Phys.* **98** 084107
[11] Krupka J, Huang W T and Tung M J 2006 *Ferroelectrics* **1** 335
[12] Easton C D, Jacob M V and Krupka J 2007 *Meas. Sci. Technol.* **18** 2869
[13] Cheng K M, Weng Z, Oliver D R, Thomson D J and Bridges G E 2007 *J. Microelectromech. Syst.* **16** 1054
[14] Yahyaie I, Buchanan D A, Bridges G E, Thomson D J and Oliver D R 2012 *IEEE Trans. Ultrason. Ferroelectr. Freq. Control* **59** 1212
[15] Igreja R and Dias C J 2011 *Sensor Actuat. A: Phys.* **172** 392

Journal of Physics: Conference Series **619** (2015) 012029 doi:10.1088/1742-6596/619/1/012029

Influence of Gate Dielectrics, Electrodes and Channel Width on OFET Characteristics

V P Liyana*, **Stephania A M, K Shiju, and P Predeep**

Laboratory for Molecular Photonics and Electronics (LAMP), Department of Physics, National Institute of Technology, Calicut, Kerala, 673601, India.

*E-mail: ppredeep@gmail.com

Abstract. Organic Field Effect Transistors (OFET) possess wide applications in large area electronics owing to their attractive features like easy fabrication process, light weight, flexibility, cost effectiveness etc. But instability, high operational voltages and low carrier mobility act as inhibitors to commercialization of OFETs and various approaches were tried on a regular basis so as to make it viable. In this work, Poly 3-hexylthiophene-2,5diyl (P3HT) based OFETs with bottom-contact top-gate configuration using Poly vinyl alcohol (PVA) and Poly (methyl methacrylate) (PMMA) as gate dielectrics, aluminium and copper as source-drain electrodes are investigated. An effort is made to compare the effect of these dielectric materials and electrodes on the performance of OFET. Also, an attempt has been made to optimize the channel width of the device. These devices are characterised with mobility (μ), threshold voltage (V_T), on-off ratio (I_{on}/I_{off}) and their comparative analysis is reported.

1. Introduction

Organic Electronics has attracted a lot of attention lately as organic polymers mainly aromatic derivatives like thiophene [1] revealed moderate conductivity and mobility ($\sim10^{-3}cm^2V^{-1}s^{-1}$) due to their inherent optical and electrical properties which make them competent candidates in semiconductor industry. However, they still require improvement in manifolds to outperform silicon's ($\mu\sim10^2$-$10^3cm^2V^{-1}s^{-1}$) legacy as semiconductor material.

OFET has been undergoing incessant and intensive research to enhance performance which is expected to be attained by the rightful combination of all its components and interfaces. Gate dielectrics and electrodes are the elements that are under study and a total of three transistors with a permuted combination are built by employing solution based thin film technology. OFETs are fabricated with bottom contact top-gate. Regioregular Poly (3-hexylthiophene-2,5diyl) (P3HT), a popularly used organic semiconductor (p-type) is used as active layer as it is shown to have comparably good mobility ($\mu = 10^{-4}$-$10^{-1}cm^2V^{-1}s^{-1}$) [2] of its charge carriers.

This study involves assessment of prevalent materials and their comparison. For the same reason, Poly vinyl alcohol (PVA) (K=2) and Poly (methyl methacrylate) (PMMA) (K=3.5), a water soluble material and a non-polar solvent based material, respectively formed the two dielectric choices that are investigated in this study. Similarly, two commonly used and readily available metal contacts, namely aluminium and copper are tested for their compatibility with the semiconductor in use.

2. Materials and Methods

P3HT is dissolved in toluene without further purification to prepare a solution. Glass slides are used as substrates. Source-drain metallic layers are then coated on the substrate in a bell jar evacuated to a pressure under 1.5×10^{-5} mbar. Two devices are fabricated with aluminium (b.p. ~1,080 K) and one device with copper (b.p. ~1,600 K) as source/drain electrodes. RR-P3HT (degree of regioregularity ~ 99%) with a weight percentage of 10 mg/ml is drop cast onto the electrode with the help of a micro syringe in the dark to prevent decomposition. They are left undisturbed and screened from light to facilitate the alignment of the semiconductor between the electrodes. PVA (Mol. Wt 125,000 $gmol^{-1}$) with dielectric constant 2 is dissolved in water and PMMA (Mol. Wt 12,000 $gmol^{-1}$) with dielectric constant 3.5 is dissolved in acetate to obtain a clear solution. Both of these solutions were dispersed over the already prepared component films, employing the same procedure formerly adopted. The device with 4 mm channel width and aluminium electrode was coated with PMMA as dielectric and the rest are built with PVA. The final layer of gate was deposited with the same metals applying the same methods used earlier.

The OFET characterization is conducted in air at room conditions using Keithley 2400 Source Meter and Keithley 6517B Electrometer/High Resistance Meter. Electrical characterization is made by biasing the gate with a constant voltage (V_g), and a voltage is applied across drain and source (V_{sd}) with the latter grounded. V_{sd} is swept over a range of 60 V and the corresponding drain-source current (I_{sd}) is noted for the voltage range -20 V and 20 V of V_g in steps of 10 V.

3. Results and Discussion

The recorded values of I_{sd} are plotted against V_{sd} and V_g to obtain the transfer characteristics and transconductance (g_m) curve respectively. Further, $I_{sd}^{1/2}$ is also annotated on the g_m curve against V_g. The value g_m of is the slope of I_{sd} vs V_g curve. Charge carrier mobility μ ($cm^2\ V^{-1}s^{-1}$) is calculated from the slope of transconductance curve and the following relations [3].

$$\mu = (g_m L)/(WC_pV_{sd}) \tag{1}$$

From SEM images, the channel length (L) and thickness of the dielectric (t) are found to be 19 μm and 25 μm respectively. Threshold voltage (V_T) can be obtained by using the simple current equation in the linear regime.

$$I_{sd} = (\mu C_p W/L)\,((V_g-V_T)\,V_{sd}-V_{sd}^2/2) \tag{2}$$

3.1. Effect of Dielectric Materials

For gate dielectrics comparison, the devices with aluminium electrodes and dimensions as mentioned in table 1 are prepared with dielectric layer as PVA and PMMA for two separate devices. The transfer curve and the transconductance curve of the two devices are shown in figures 1(a), 2(a) and 1(b), 2(b) respectively.

In the case of FET when positive gate voltage increases, source drain current decreases which is due to the negative charges induced in the active channel with the help of positive gate bias voltages indicating p-type of FET operation in depletion mode.

Table 1. Various parameters of the two devices under study for dielectric comparison.

Parameters	Device 1 (PVA)	Device 2 (PMMA)
Channel Width, W (mm)	5	4
Mobility, μ(m2V-1s-1)	4.802x10-7	2.297x10-5
Threshold Voltage, VT (V)	9.968	-4.43
On-Off Ratio, Ion/Ioff	1.1785	6.348

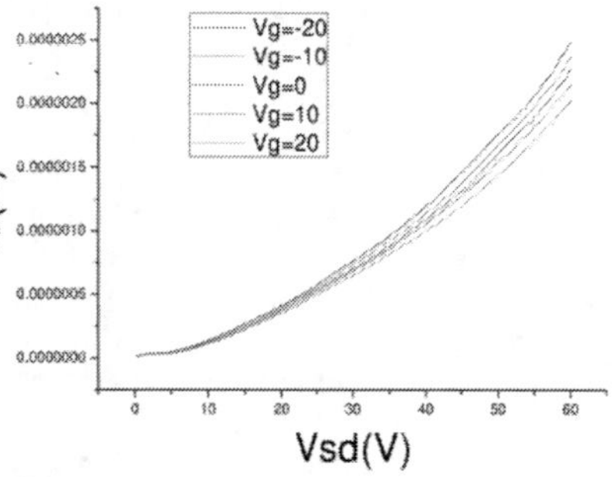

Figure 1(a). Relationship between Isd and Vsd

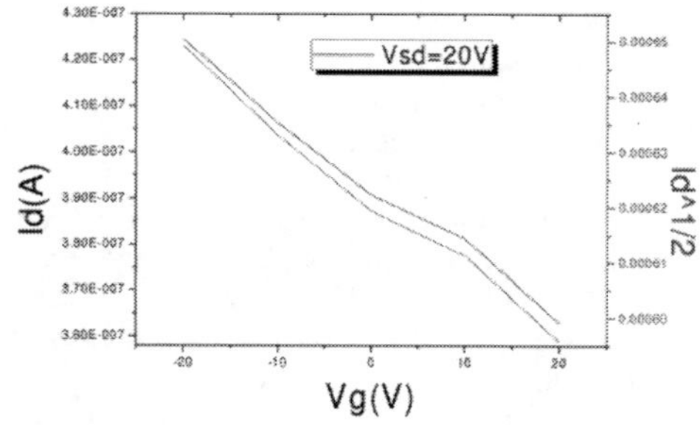

Figure 1(b). Transconductance curve

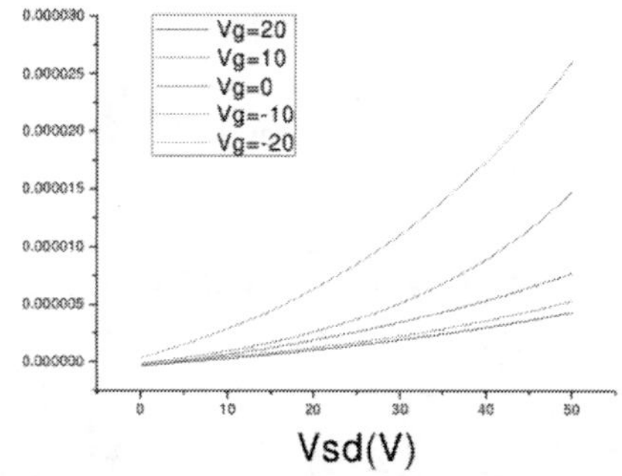

Figure 2(a). Relationship between Isd and Vsd

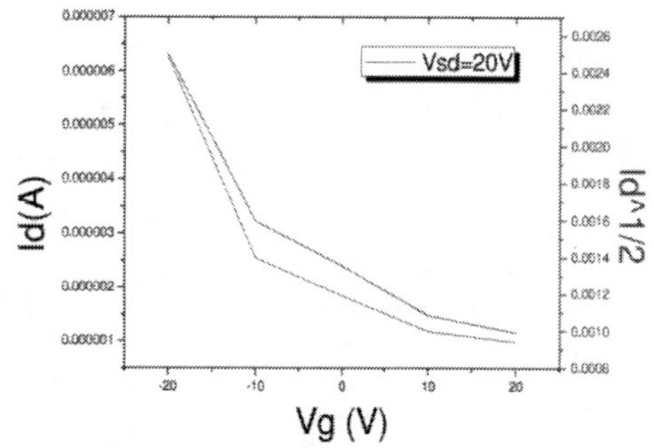

Figure 2(b). Transconductance curve

It can be seen from table 1 that performance of PMMA over PVA as a gate dielectric is discernible, as the corresponding device exhibited better OFET characteristics. The mobility of PMMA (μ=2.297x10^{5} $m^2V^{-1}s^{-1}$) exceeds that of PVA (μ=4.802x10^{-7} $m^2V^{-1}s^{-1}$) by a power of two. PMMA has intrinsic properties due to its molecular structure which makes it a good dielectric with insulating properties compared to its counterpart PVA. PMMA possesses better molecular packing than PVA. This results in the proximity of the functional groups which is essential for higher dipole moment. This in turn contributes to its higher dielectric constant, making it as a high capacitance dielectric (C_p= 1.239 $\mu F/m^2$) which is vital for a lower operating voltage. A reduction in operational voltage increases the mobile charge carrier density which improves the effective mobility ultimately leading to an increased current drive even at low voltage. Enhanced performance of device 2 also indicates that smaller channel width or a low aspect ratio enhances the performance of OFET and yields better output.

3.2. Effect of Electrode Material

Aluminium and Copper electrode devices fabricated having PVA as dielectric material is used for comparing the electrodes.

Table 2. Various parameters of the two devices for electrode study.

Parameters	Device 1 (Al)	Device 3 (Cu)
Channel Width, W (mm)	5	5
Mobility, μ ($m^2V^{-1}s^{-1}$)	4.802x10^{-7}	4.5167x10^{-4}
Threshold Voltage, V_T (V)	9.968	-86.435
On-Off Ratio, I_{on}/I_{off}	1.1785	1.3647

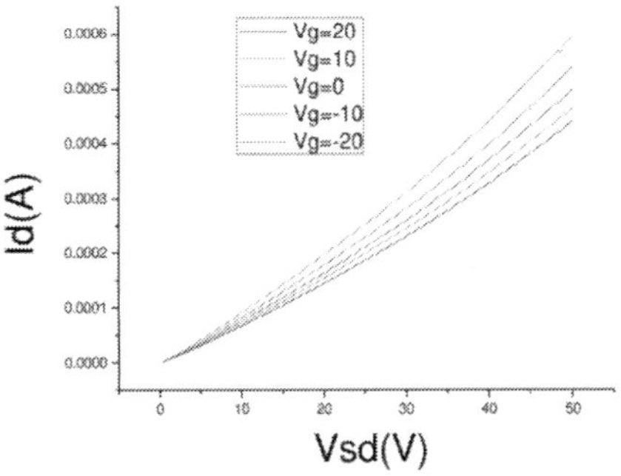

Figure 3(a). Relationship between I_{sd} and V_{sd}

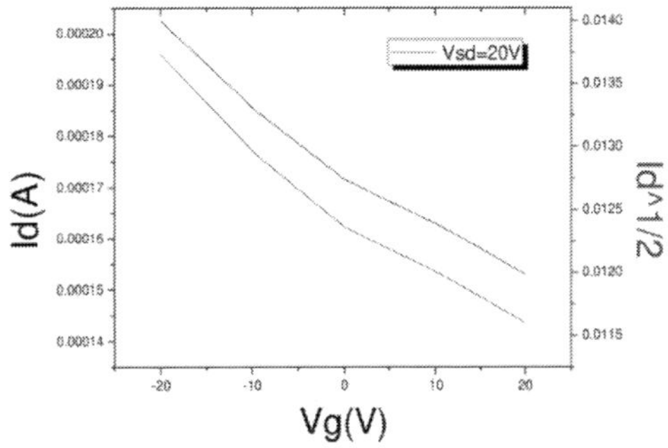

Figure 3(b). Transconductance curve of device

Referring to table 2, the results indicate that copper is better for electrode application in P3HT based OFET as this gives carrier mobility nearly 1000 times higher than aluminium. This will mean better compatibility of copper with the P3HT which again means better band alignment between the HOMO & LUMO (Highest Occupied Molecular Orbital and Lowest Unoccupied Molecular Orbital) of the organic semiconductor with the work function of the metal as illustrated in figure 4. Shift in energy levels of both the materials results in the formation of dipole barrier at the metal-semiconductor interface. The work function of copper (~4.7 eV) lies closer to the HOMO level of P3HT (~5.2 eV) [4] than aluminium (~4.08 eV) and is in concurrence with the observed results and provides a convincing explanation for the same. When gate source terminal is applied a negative bias voltage, I_{sd} would increase as shown in the transconductance curves in figure 3(b) and 1(b). Transfer curves are given in figure 3(a) and 1(a) for the two devices.

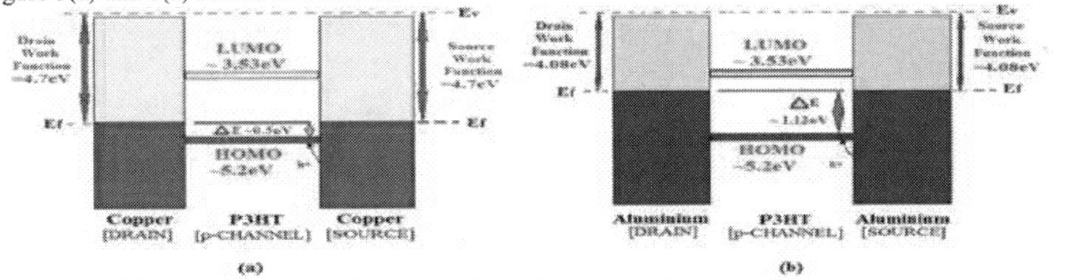

Figure 4. Energy level diagram of metal contact and semiconductor interface.

4. Conclusion

By the fabrication of hole conducting OFETs it is demonstrated that gate dielectric and electrodes have an effect on the performance of P3HT based OFET. It can be concluded that when P3HT is the semiconducting material, gate dielectric using PMMA shows better performance than PVA since former is showing 100 times better mobility and V_T, I_{on}/I_{off} are also in the preferred range.

References

[1] Tsumura A Koezuka H and Ando T 1988 Polythiophene field-effect transistor: Its characteristics and operation mechanism *Synthetic Metals* **25** 11-23

[2] Wang G, Swensen J, Moses D and Heeger A J 2003 Increased mobility from regioregular poly(3-hexylthiophene) field-effect transistors *Journal of Applied Physics* **93** 6137-6141

[3] S M Sze, Semiconductor devices physics and technology, 1936

[4] Young Min Nama, June Huh and Won HoJo 2010 Optimization of thickness and morphology of active layer for high performance of bulk-hetero junction organic solar cells *Solar Energy Materials & Solar Cells* **94** 1118–1124

Diffusion of excitons in materials for optoelectronic device applications

Jai Singh[1,a], **Monishka Rita Narayan**[b] , **David Ompong**[a]

[a]School of Engineering and IT, Faculty of EHSE, Charles Darwin University, Darwin, NT 0909, Australia

[b]Centre for Renewable Energy , Charles Darwin University, Darwin, NT 0909, Australia

jai.singh@cdu.edu.au

Abstract The diffusion of singlet excitons is known to occur through the Förster resonance energy transfer (FRET) mechanism and that of singlet and triplet excitons can occur through the Dexter carrier transfer mechanism. It is shown here that if a material possesses the strong exciton-spin-orbit-photon interaction then triplet excitons can also be transported /diffused through a mechanism like FRET. The theory is applicable to the diffusion of excitons in optoelectronic devices like organic solar cells, organic light emitting devices and inorganic scintillators.

1. Introduction

The performance of many optoelectronic devices depends on the diffusion, radiative recombination and dissociation of excitons. These devices include organic solar cells (OSCs) [1], organic light emitting devices (OLEDs) [2], scintillators [3], photo detectors, etc. In these devices excitons are either generated through photoexcitation or charge carrier injection by applying an electric field and after their generation they need to be transported within the device for a certain time or distance. Finally the photo generated excitons dissociate into free electron and hole pairs which are collected at the opposite electrodes of the device, e.g., in OSCs or the electro-generated excitons radiatively recombine by emitting photons, e.g., in OLEDs. In both cases the efficient diffusion of excitons is required for achieving the higher internal quantum efficiency. The exciton transfer can take place via the interaction between an excited molecule or chemical group, say A*, and a ground-state molecule or chemical group, say B, without emitting a photon. Excitons are usually generated in two spin configurations, singlet and triplet. Singlet excitons can recombine radiatively by emitting a photon whereas the radiative recombination of triplet excitons is spin forbidden. This distinction between the singlet and triplet spin configurations affects the way an exciton can move in organic solids from an excited molecule to another molecule in its ground state.

There are two well known models for transfer of excitations in materials. The first model is by Förster [4] and is called the Förster resonance energy transfer (FRET) mechanism where an excited molecule A^* gets de-excited and the energy released simultaneously excites another molecule B from its ground-state due to the Coulomb interaction between the electrons on these two molecules or chemical groups. The second model is due to Dexter [5], where an excited molecule or chemical group exchanges electrons between A^* and B to accomplish the non-radiative transfer process. Since the inception of their theories both models have been revisited by many in the literature and the simplest way to achieve the essence of both is by calculating the energy eigenvalue, $W_f(\mathbf{K},S)$ of a Frenkel exciton of spin S (= 0 for singlet and 1 for triplet) in the real crystal space, where the electron is excited in f state and hole in the ground state g of a molecule in organic solids, which is obtained as [6]:

$$W_f(\mathbf{K},S) = E_g + E_b(S) + L_f(\mathbf{K},S), \tag{1}$$

where E_g is the energy of the band gap, E_b (S) is the binding energy and $L_f(\mathbf{K},S)$ represents the transfer energy of excitons from one molecule to another and is given by:

$$L_f(\mathbf{K},S) = -N^{-1} \sum_{\mathbf{l},\mathbf{m}\neq\mathbf{l}} \exp[i\mathbf{K}\cdot(\mathbf{l}-\mathbf{m})] M^f_{\mathbf{l},\mathbf{m}}(S), \tag{2}$$

and

$$\begin{aligned} M^f_{\mathbf{l},\mathbf{m}}(S) =& < g,\mathbf{l}; f,\mathbf{m} \,|\, U \,|\, f,\mathbf{l}; g,\mathbf{m} > - \\ & (1-S) < g,\mathbf{l}; f,\mathbf{m} \,|\, U \,|\, g,\mathbf{m}; f,\mathbf{l} > \end{aligned}. \tag{3}$$

$M^f_{\mathbf{l},\mathbf{m}}(S)$ represents the transfer integrals between sites **l** and **m** due to the Coulomb potential $U = \frac{\kappa e^2}{r_{12}}$ of interaction between a pair of electrons (e); one in an excited state f and the other in the ground state g and it consists of two terms. The first term represents an e excited in state f on site **l** and another e in the ground state g on site **m** initially (right side of U) exchange their sites due to the Coulomb interaction, i.e., first e moves to the excited state of **m** and second moves to the ground state of **l** (left side of U). The second term, which vanishes for the case of triplet spin configuration (S = 1), represents where initially the first excited electron in state f on site **l** (right side of U) gets de-excited to the ground state g on site **l** in the final state (left side of U) and the second electron initially in the ground state g at site **m** (right side of U) gets finally excited to state f at site **m**(left side of U). The first term arises due to the exchange interaction between two electrons and it describes the Dexter's model of transfer of excitons applicable to both singlet and triplet excitons as shown on Fig. 1,

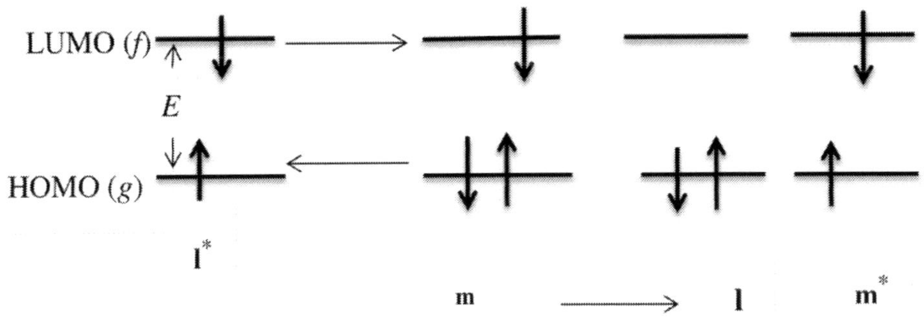

(a)

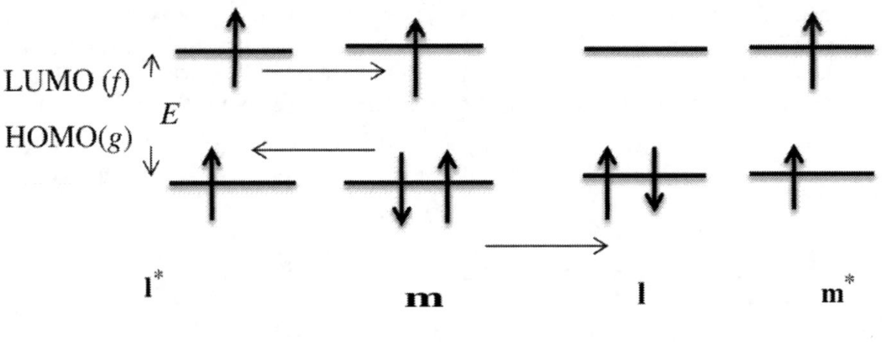

(b)

Figure 1. Dexter transfer mechanism due to the exchange interaction (first term of Eq. (3)): (a) for singlet excitons and (b) for triplet excitons

The second term of Eq. (3), which vanishes for triplet excitons, represents the resonance energy transfer as obtained in Förster's theory. In this case the excited atom at site $\mathbf{l}^*$ gets de-excited and transfers its energy to excite the molecule at site $\mathbf{m}^*$ as shown in Fig. 2. Förster's resonance

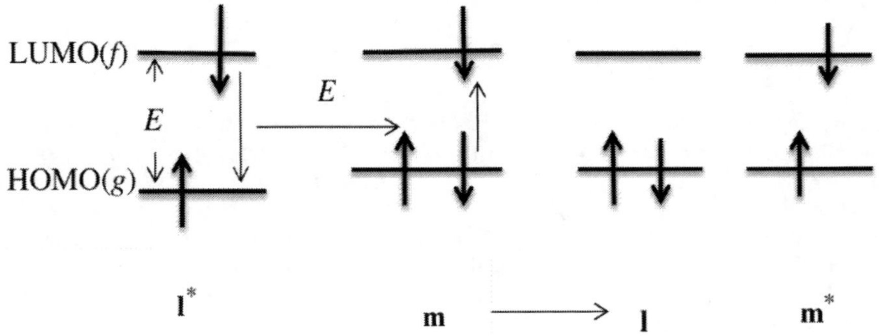

Figure 2. Schematic presentation of the Förster resonance energy transfer mechanism for singlet excitons (term 2 of Eq. (3)): excited molecule $\mathbf{l}^*$ gets de-excited and transfers its energy to excite the ground state molecule **m** (**m***gets excited).

energy transfer is applicable only to singlet excitons because although it is a non-radiative process it requires an excited molecule to get de-excited and the energy is transferred to excite another molecule. This excitation and de-excitation cannot take place if the excited state is a triplet because then such transitions would be spin forbidden, as it can also be seen from Eq. (3) that for $S = 1$ the second term vanishes. It may be noted that it is the term $L_f(\mathbf{K}, S)$ in Eq. (2) that contributes to the kinetic energy of an exciton with an effective mass in a lattice [6]. For non-crystalline solids the wave vector **K** is not relevant but the matrix element $M^f_{\mathbf{l},\mathbf{m}}(S)$ still has the same significance as described above [6-7].

However, recently the Förster resonance energy transfer (FRET) has been applied to interpret the photoluminescence kinetics in OLEDs [8] and inorganic scintillators [9], e.g., $CdWO_4$ and in both cases triplet Frenkel excitons are excited. In OLEDs the host or donor material is doped with compounds containing heavy metal atoms. The purpose of this paper is to illustrate that under the influence of strong exciton-spin-orbit-photon interaction, excitons either excited by photons (electro-magnetic field) or by injecting the charge carriers (electric field) as in OLEDs, FRET can be used to study the transfer of triplet excitons as well.

2. Effect of Spin-orbit interaction

The derivation of the time-dependent exciton-spin-orbit interaction is briefly described here and the details are published in [10]. For a pair of excited electron and hole in a solid with N atoms/molecules, the stationary spin-orbit interaction denoted by H_{so}^{sol} can be written as [10-11]:

$$\hat{H}_{so}^{sol} = -\frac{eg}{2\mu_x^2 c^2}\mathbf{s}_e \cdot (\mathbf{p}_e \times \sum_{n=1}^{N} \mathbf{E}_{ne}) + \frac{eg}{2\mu_x^2 c^2}\mathbf{s}_h \cdot (\mathbf{p}_h \times \sum_{n=1}^{N} \mathbf{E}_{nh}), \tag{4}$$

where μ_x is the reduced mass of an exciton (electron and hole), g is the gyromagnetic ratio (g = 2), $\mathbf{s}_e$ and $\mathbf{p}_e$ are the spin angular and orbital momenta of the electron, respectively, and $\mathbf{s}_h$ and $\mathbf{p}_h$ are those of the hole and $\mathbf{E}_{ne}$ and $\mathbf{E}_{nh}$ are the electric fields experienced by the electron and hole, respectively, due to the nucleus. When an electromagnetic radiation incidents on the sample, Eq. (4) modifies to:

$$\begin{aligned}\hat{H}_{so}^{sol} = &-\frac{eg}{2\mu_x^2 c^2}\mathbf{s}_e \cdot (\mathbf{p}_e + \frac{e}{c}\mathbf{A}_e) \times (-\frac{1}{c}\frac{\partial \mathbf{A}_e}{\partial t} - \sum_{n=1}^{N} \nabla V_{ne}) \\ &+ \frac{eg}{2\mu_x^2 c^2}\mathbf{s}_h \cdot (\mathbf{p}_h - \frac{e}{c}\mathbf{A}_h) \times (-\frac{1}{c}\frac{\partial \mathbf{A}_h}{\partial t} - \sum_{n=1}^{N} \nabla V_{nh})\end{aligned} \quad . \tag{5}$$

In obtaining Eq. (5), it is found that the influence of the magnetic field in the radiation vanishes [10-11]. Therefore, it is anticipated that the results are also applicable for the case of generation of electrons and holes by an electric field alone. The time-dependent exciton-photon-spin-orbit interaction extracted from Eq. (5) is thus obtained [10]:

$$\begin{aligned}\hat{H}_{so}^{sol(t)} = &-\frac{e^3 g\kappa}{2\mu_x^2 \varepsilon c^2}\sum_{\lambda,n}[\frac{Z_n}{r_{en}^2}(\frac{2\pi\hbar}{\varepsilon_0\varepsilon\omega_\lambda V})^{1/2}\sin\phi_{\lambda en}\, s_{ez} \\ &+ \sum_{\lambda,n}\frac{Z_n}{r_{hn}^2}(\frac{2\pi\hbar}{\varepsilon_0\varepsilon\omega_\lambda V})^{1/2}\sin\phi_{\lambda hn}\, s_{hz}]e^{-i\omega_\lambda t}c_\lambda^+\end{aligned} \quad , \tag{6}$$

Using the triplet spin configuration and the property of s_{ez} and s_{hz} spin operators as $s_{ez}a_e(\pm\frac{1}{2}) = \pm\frac{1}{2}\hbar a_e(\pm\frac{1}{2})$ and $s_{hz}d_g(\pm\frac{1}{2}) = \mp\frac{1}{2}\hbar d_g(\pm\frac{1}{2})$, we find that the triplet spin configuration of a triplet exciton gets flipped to the singlet spin configuration due to the spin-orbit interaction and the time-dependent operator in Eq. (6) becomes [10-11]:

$$\hat{H}_{so}^{sol(t)} \approx -\frac{2\hbar e^3 gZ\kappa}{\mu_x^2 c^2 \varepsilon r^2}\left(\frac{2\pi\hbar}{\varepsilon_0 \varepsilon V}\right)^{1/2} \sum_{\lambda} \frac{\sin\lambda}{\sqrt{\omega_\lambda}} e^{-i\omega\lambda t} \quad . \tag{7}$$
$$\times[\frac{1}{\sqrt{2}}(a_e(+1/2)d_g(-1/2)+a_e(-1/2)d_g(+1/2))]c_\lambda^\dagger$$

Thus the mechanism of the radiative recombination of triplet excitons through the time-dependent transition operator can be described in the following two steps:

1. The new operator is attractive for excitons so it attracts a triplet exciton to the heaviest atom as it is proportional to the atomic number. As the magnitude of attraction in inversely proportional to the square of the average distance between an electron and nucleus, only the nearest heavy nucleus will play the dominant role.

2. As soon as a triplet exciton interacts with such a spin-orbit-exciton-photon interaction, the spin gets flipped to a singlet configuration and exciton recombines radiatively by emitting a photon.

As the spin of the triplet exciton is flipped the electron and hole can recombine either radiatively to emit photons or get transferred non-radiatively by following Förster's resonance energy transfer mechanism. In this way the strong exciton-spin-orbit-photon interaction enables FRET mechanism to be applicable to triplet excitons as well.

3. Results and discussions

It is shown here that the effect of the time-dependent spin-orbit interaction on triplet excitons is that it flips the spin to singlet configuration and then they behave like singlet excitons as shown in Fig. 3.

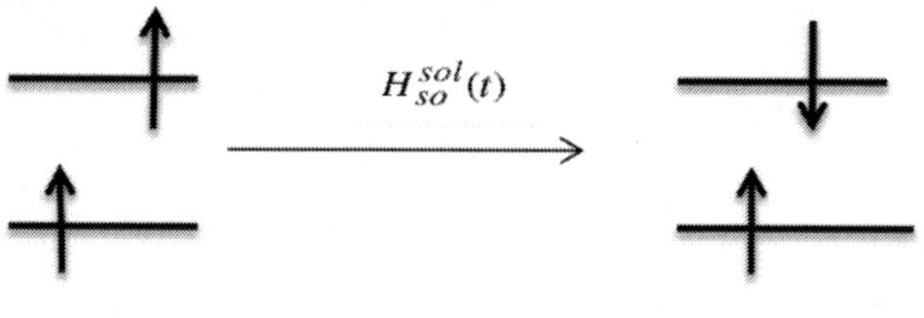

Figure 3. Effect of the time-dependent exciton-spin-orbit-photon interaction on triplet excitons: It flips the spin to singlet configuration.

In organic solids used for the fabrication of organic solar cells (OSCs) it has been observed that the conversion efficiency of OSCs gets enhanced if the donor is incorporated with heavy metal atoms [11-15]. As the time-dependent spin-orbit interaction in Eq. (4) is proportional to the atomic number *Z*, doping with heavy metal atoms enhances the effect of spin-orbit interaction and hence triplets are converted into singlets which diffuse faster through FRET than triplet excitons [11]. This results in a faster diffusion of excitons to the donor-acceptor interface resulting in faster dissociation and hence higher conversion efficiency. The triplet excitons transfer occurs through the Dexter mechanism which depends on the exchange interaction which is known to be a short range interaction [6, 8] and hence it is effective only at short distances, i.e., between nearest neighbours with overlapping electronic wave functions [8].

In OLEDs it is very well established that the emission efficiency gets enhanced when doped with heavy metal atoms like platinum (Pt), iridium (Ir) and palladium (Pd) [16,19], which has been attributed to the conversion of triplet excitons to singlets whose radiative recombinations are spin allowed [2, 20] as presented above. In inorganic scintillators, the transfer of triplet excitons in $CdWO_4$ can be better explained in terms of Förster's mechanism [9], which is based on the dipole-dipole interaction as is well known [8]. Inorganic solids have usually higher atomic numbers and hence stronger spin-orbit interaction which again converts triplets into singlets and then get transferred efficiently through the Förster mechanism.

4. Conclusions

It is shown that the time-dependent exciton-spin-orbit interaction, which depends on the atomic number, assists flipping the spin of triplet excitons to singlet spin configuration and hence the Förster mechanism can be applied for the transfer of triplet excitons as well. Present results can successfully explain many experimental results in OSCs, OLEDs and inorganic scintillators.

References

[1] M. R. Narayan and J. Singh, J. Appl. Phys. **114**, 073510 (2013)
[2] J. Singh, Phys. Status Solidi A **208**, No. 8, 1809 (2011)
[3] G. Bizarri, W. W. Moses, J. Singh, A. N. Vasil'ev, and R. T. Williams, J. Appl. Phys. **105**, 044507 (2009)
[4] T. Förster, Annalen. Phys. **2,** 55 (1948)
[5] D. L. Dexter, J.Chem.Phys. 21,836 (1953)
[6] J. Singh, *Excitation Energy Transfer Processes in Condensed Matter* (Plenum, N.Y., 1994)
[7] J. Singh and K. Shimakawa, *Advances in Amorphous Semiconductors* (Taylor & Francis, London, 2003)
[8] F. S. Steinbacher, R. Krause, A. Hunze, and A. Winnacker, Phys. Status Solidi A **209**, 340(2012).
[9] R. Lassner, N. Fedorov, R. Grigonis, S. Guizard, M. Krim, V. Makhov, S. Markov, V. Nagirnyi, V. Sirutkaitis, A. Vasil'ev, S. Vielhauer and I.A. Tupitsyna, J. Phys: Cond. Matt. **25**, 245901(2013)
[10] J. Singh, Phys. Rev. B**76**, 085205 (2007)
[11] M. R. Narayan and J. Singh, J. Appl. Phys. **114**, 154515 (2013)
[12] Z. Xu, B. Hu, and J. Howe, J. Appl. Phys. **103**, 043909 (2008)
[13] C.-M. Yang, C.-H. Wu, H.-H. Liao, K.-Y. Lai, H.-P. Cheng, S.-F. Horng,H.-F. Meng, and J.-T. Shy, Appl. Phys. Lett. **90**, 133509 (2007)
[14] G. L. Schulz and S. Holdcroft, Chem. Mater. **20**, 5351 (2008)
[15] J. Huang, J. Yu, Z. Guan, and Y. Jiang, Appl. Phys. Lett. **97**, 143301 (2010)
[16] C. Adachi, M. A. Baldo, M. E. Thompson, and S. R. Forrest, J. Appl. Phys. **90**, 5048 (2001)
[17] H. F. Wittman, R. H. Friend, M. S. Khan, and J. Lewis, J. Chem. Phys. **101**, 2693 (1994)
[18] V. Cleave, G. Yahioglu, P. L. Barny, R. H. Friend, and N. Tessler, Adv.Mater. **11**, 285(1999)
[19] A. Köhler and D. Beljonne, Adv. Funct. Mater. **14**, 11 (2004)
[20] J. Singh, H. Baessler and S. Kugler, J. Chem. Phys. **129**, 041103 (2008).

Spectroscopy of high index contrast Yb:Ta_2O_5 waveguides for lasing applications

A. Aghajani[1], G. S. Murugan[1], N. P. Sessions[1], V. Apostolopoulos[2] and J. S. Wilkinson[1]
[1]Optoelectronics Research Centre, Faculty of Physical Sciences and Engineering, University of Southampton, Southampton, UK SO17 1BJ
[2]School of Physics and Astronomy, Faculty of Physical Sciences and Engineering, University of Southampton, Southampton, UK SO17 1BJ

E-mail: aa15v07@soton.ac.uk

Abstract. Ytterbium-doped waveguides are required for compact integrated lasers and Yb-doped Ta_2O_5 is a promising candidate material. The design, fabrication and spectroscopic characterisation of Yb:Ta_2O_5 rib waveguides are described. The peak absorption cross-section was measured to be 2.75×10^{-20} cm^2 at 975 nm. The emission spectrum was found to have a fluorescence emission peak at a wavelength of 976 nm with a peak cross-section of 2.9×10^{-20} cm^2 and a second broad fluorescence band spanning from 990 nm to 1090 nm. The excited-state life time was measured to be 260 μs.

1. Introduction

Yb-doped optical materials have found widespread application for lasers emitting at wavelengths near 1 μm in bulk, disk and fiber configurations. Their main attractive properties include broad absorption near 0.98 μm, long excited-state lifetime, and broadband gain bandwidth operation. Lasing in Er-doped [1] and Yb-doped [2] alumina and Nd-doped [3] and Er-doped [4] tantalum pentoxide (Ta_2O_5) have been demonstrated, and here we explore the suitability of Ta_2O_5 as a waveguide host for Yb ions. Ta_2O_5 is a promising material for mass-producible, multifunctional, integrated photonic circuits on silicon, exhibiting excellent mechanical, electrical and thermal properties and good compatibility with complementary-metal–oxide–semiconductor (CMOS) technologies [5]. Its high refractive index ($n \approx 2.124$ at $\lambda \approx 980$ nm) [6] allows high index contrast between the silica waveguide cladding and core, providing for low-loss, tight bend radii thereby enabling the development of compact photonic circuits due to the strong confinement of the optical modes. As well as Ta_2O_5 excelling as host for rare-earth ions, it also provides a large third-order nonlinearity ($n_2 \approx 7.25\text{x}10^{-19}$ m^2/W at $\lambda \approx 980$ nm) [7] for all-optical processing.

In this work, the design and fabrication of 2.5 wt% Yb_2O_3 doped Ta_2O_5 rib waveguides on silicon using magnetron sputtering, photolithography and argon ion beam milling is presented. The spectroscopic properties of Yb^{3+} ions in Ta_2O_5 such as the absorption cross section (σ_{abs}), emission cross-section (σ_{em}), fluorescence spectrum and radiative lifetime were measured and currently the waveguides are being optimised for the realisation of compact waveguide lasers.

2. Channel waveguide design

Rib waveguides in Ta_2O_5 were designed to give single mode operation for wavelengths between 970 nm and 1100 nm, to cover the signal and pump wavelengths associated with typical Yb-doped materials. Waveguides were designed to be fabricated on oxidised silicon wafers and to have a silica cladding for symmetry and high index contrast, allowing tight modal confinement. A rib waveguide approach, as shown in figure 1, was adopted to allow single mode operation of channels with various widths to be readily fabricated using conventional photolithography, to maximise pump-signal overlap and to minimise waveguide loss resulting from any sidewall roughness. The rib height was chosen to be 1μm to ensure good confinement of the modes within the core and the initial design was selected using the approach of Soref et al. [8] and the waveguide properties fully determined using finite element analysis. While an etch depth of up to 500 nm could be used and still maintain monomode operation, an etch depth of 150 nm was chosen as a compromise between modal spotsize and loss due to sidewall roughness, leaving the outer slab thickness at 850 nm. With these constraints, waveguides with widths below 1.9 μm, readily achieved by conventional photolithography, were predicted to be monomode at a wavelength of 970 nm.

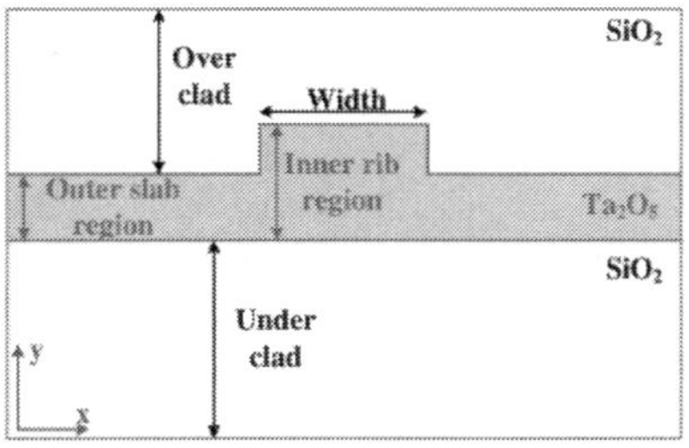

Figure 1. Cross-section of etched rib waveguide design

3. Fabrication of $Yb:Ta_2O_5$ waveguides

A 1 μm thick layer of $Yb:Ta_2O_5$, using RF magnetron sputter deposition system from a 150 mm wide circular powder-pressed Ta_2O_5 target doped with 2.5 wt.% of ytterbium oxide (~$6.2x10^{20}$ Yb atoms/cm^3) onto a 4 inch silicon substrate having a 2.5 μm thick thermally-grown silica overlayer. The deposition was performed in a vacuum chamber which had been pumped to a base pressure of 10^{-8} Torr. During the deposition process the chamber was filled with a mixture of oxygen and argon gases at flow rate of 5 sccm and 20 sccm, while the substrate was kept at a constant temperature of 200 °C and magnetron power of 300 W. The chamber was maintained at a pressure of 10 mTorr. These conditions are based on previous optimisation for Ta_2O_5 hosting erbium ions for 1.5μm emission to provide the lowest optical loss with an acceptable deposition rate [9]. Rib waveguides were realised using conventional photolithography and argon ion beam milling (IBM) to produce the waveguide design described in Section 2. Strips of photoresist of widths varying from 1 to 10 μm were defined in positive photoresist spun onto the deposited $Yb:Ta_2O_5$ film. The $Yb:Ta_2O_5$ film was then etched by 150 nm using an OIPT Ion Fab 300 Plus ion beam etch and deposition system. After etching and the removal of excess photoresist the wafer was immersed in a solution of 30% aqueous potassium hydroxide (KOH) at a temperature of 55 °C for 20 minutes to reduce surface roughness [9] introduced during the etching process. To reduce any oxygen deficiency and stresses introduced during the sputtering and etching processes the wafer was then annealed for two hours at 600 °C, with the furnace temperature ramped up at a rate of 3 °C per minute and ramped down at a rate of 2 °C per minute to room temperature [9].

To protect the waveguide layer from external influences and to create a symmetrical waveguide system, a 2 μm silica overcladding was deposited onto the patterned $Yb:Ta_2O_5$ film. The silica was sputtered using the same RF magnetron sputtering system used for the $Yb:Ta_2O_5$ layer. After the silica deposition, the wafer was again annealed at 600 °C to reduce oxygen deficiency and release the stresses introduced during the deposition process. To characterise the waveguides for their spectroscopic properties the wafer was diced and the end-facets were polished to optical quality.

4. Characterisation of spectral properties of Yb:Ta_2O_5

This section presents the procedures for determining spectral characteristics of the Yb:Ta_2O_5 waveguides such as the excited state lifetime, and the absorption and emission spectra, and the results obtained, leading to the deduced absorption and emission cross-sections. The excited-state lifetime and the absorption and emission cross-sections will enable the design of an efficient waveguide laser. These measurements were performed taking into account the effects of the waveguide configuration such as waveguide loss, the material/mode overlap factor and the potential for significant amplified spontaneous emission.

4.1. Absorption spectrum

The measurement of the absorption spectrum of the ytterbium ions in Ta_2O_5 was carried out to determine the absorption cross-section at the pump wavelength near 980nm. In a 3 mm long waveguide the spectrum was obtained by end-coupling broadband (700 – 1700 nm) white light from a tungsten halogen lamp into the polished end-facet of a waveguide using a monomode fibre. Light emerging from the waveguide was captured using a multi-mode fiber and fed into an optical spectrum analyser (OSA). Measurements were made with and without the device in place in order to quantify the spectral insertion loss, including coupling loss, propagation loss and Yb ion absorption.

The resulting spectrum is shown in figure 2a, where two prominent absorption bands of ytterbium at 935 and 975 nm are clearly visible. The peak absorption cross-section for ytterbium ions in Ta_2O_5 in the pump band occurred at 975 nm was calculated to be 2.75x10-20 cm^2, using an estimated concentration of the Ytterbium ions in the Ta_2O_5 of 6.2x1020 cm^{-3}, assuming that the composition of the film is the same as that of the sputtering target.

4.2. Fluorescence spectrum

The fluorescence spectrum for the Yb-doped Ta_2O_5 was captured using an experimental setup consisting of a 971 nm fiber Bragg grating-stabilised laser source butt-coupled via a single-mode fiber into an end-facet of a waveguide. The collection of the fluorescence was achieved using a multimode fiber positioned above the waveguide close to the input, and fed directly into an OSA. An estimate of the emission cross-section was deduced from the absorption cross-section using McCumber analysis [10, 11]. The accuracy of the deduced cross-section is poor at longer wavelengths due to the noise in the absorption spectrum and the nature of the McCumber transform, but is acceptable close to 977 nm, where the emission cross-section was estimated to be $\sigma_{em} = 2.90 \pm 0.73 \times 10^{-20}$ cm^2. This was used as a scaling factor to plot the emission cross-section spectrum from the fluorescence spectrum, as shown in figure 2b. The sharp peak at 971nm is due to pump radiation scattered by the waveguide.

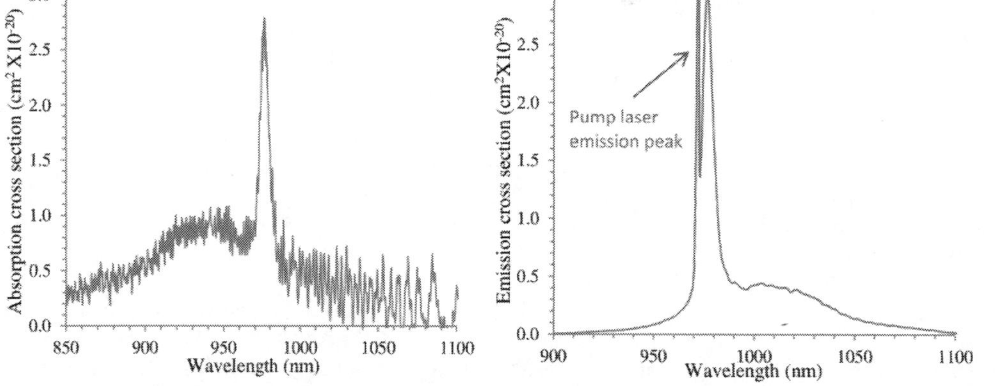

Figure 2a) Yb:Ta_2O_5 absorption cross-section spectrum **b)** Yb:Ta_2O_5 emission cross-section spectrum

4.3. Excited-state lifetime

The fluorescence lifetime was measured from a 10.8 mm long Yb:Ta_2O_5 waveguide sample from the same wafer as that used for the absorption and fluorescence characterisation. Light from the 971 nm pump laser diode was chopped at 200Hz using a mechanical chopper and focused into an end-facet of an Yb:Ta_2O_5 waveguide, with the pump power kept minimal to avoid significant amplified spontaneous emission (ASE). The light from the output of waveguide was passed through a set of long pass filters with a cut-off wavelength of 1 μm to eliminate the pump wavelength. The resulting fluorescence power was then detected using a silicon photodetector and amplifier connected to an oscilloscope. The time resolution of the system was approximately 20 μs and the lifetime was found to be 260 ± 30 μs.

5. Conclusion

In conclusion, we have studied the spectroscopic properties of Yb:Ta_2O_5 as a new material for waveguide laser applications, and determined the absorption and emission cross-sections and the excited-state lifetime. CMOS-compatible fabrication technology was used to realise rib waveguides on silicon substrates which operated in a single mode for wavelengths ranging between 970 nm and 1100 nm for waveguide widths smaller than 1.9 μm. The peak absorption cross-section was measured to be 2.75×10^{-20} cm^2 at 975 nm. The emission spectrum was found to comprise a sharp peak at a wavelength of 976 nm, with a peak emission cross-section of 2.9×10^{-20} cm^2, and a second broad peak which spanned from 990nm to 1090 nm. The upper state life time was measured to be 260 ± 30 μs. Yb:Ta_2O_5 shows promise as a material system for the development of advanced compact lasers with monolithically integrated components to add functionality. Optical structures such as ring resonators can be formed in this material system, exploiting the high index contrast, and nonlinear components for switching or modelocking can be realised, exploiting the high third order nonlinearity of tantalum pentoxide. Efforts are underway to demonstrate a mode-locked Yb:Ta_2O_5 waveguide laser.

References

[1] J.D.B. Bradley, R. Stoffer, L. Agazzi, F. Ay, K. Wörhoff and M. Pollnau 2010 *Opt. Lett.* **35** 73-75
[2] E.H. Bernhardi, H.A.G.M. van Wolferen, K. Wörhoff, R.M. de Ridder and M. Pollnau 2011 *Opt. Lett.* **36** 603-605
[3] B. Unal, M. C. Netti, M. A. Hassan, P. J. Ayliffe, M. D. B. Charlton, F. Lahoz, N. M. B. Perney, D. P. Shepherd, C.-Y. Tai, J. S. Wilkinson and G. J. Parker 2005 *J. Quantum Electron.* **41** 1565 - 1573
[4] A. Z. Subramanian, C. J. Oton, D. P. Shepherd and J. S. Wilkinson 2010 *Photon. Technol. Lett.* **22** 1571-1573
[5] C. Chaneliere, J.L. Autran, R.A.B. Devine and B. Balland 1998 *Materials Science & Engineering R-Reports* **22** 269-322
[6] A. Subramanian 2011 Tantalum pentoxide waveguide amplifier and laser for planar lightwave circuits *Ph.D. Thesis* University of Southampton
[7] C. Y. Tai, J. S. Wilkinson, N. M. B. Perney, M. C. Netti, F. Cattaneo, C. E. Finlayson and J. J. Baumberg 2004 *Opt. Express* **12** 5110 -5116
[8] R. Soref, J. Schmidtchen and K. Petermann, 1991 *J. Quantum Electron.* **27** 1971-1974
[9] A.Z. Subramanian, G.S. Murugan, M.N. Zervas, and J.S. Wilkinson 2012 *J. Lightwave Technol.* **30** 1455-1462
[10] D. E. McCumber 1964 *Phys. Rev.* **136** A954
[11] Y. Yu, Y. Huang, L. Zhang, Z. Lin and G. Wang 2013 *Plos One* **8** e54450

Mid infra-red hyper-spectral imaging with bright super continuum source and fast acousto-optic tuneable filter for cytological applications.

Mark Farries[1], Jon Ward[1], Stefano Valle[1], Gary Stephens[1], Peter Moselund[2], Koen van der Zanden[3], Bruce Napier[4]

[1]Gooch & Housego, Broomhill Way, Torquay. TQ2 7QL, UK,

[2]NKT Photonics A/S__Blokken 84__DK-3460 Birkerød__Denmark

[3]Xenics Infrared solutions, Ambachtenzone Haasrode 3227 + 3126__Ambachtenlaan 44__BE-3001 Leuven, Belgium

[4]Vivid Components Dr.-Rörig-Damm 22, 33102 Paderborn, Germany

mfarries@goochandhousego.com

Abstract. Mid-IR imaging spectroscopy has the potential to offer an effective tool for early cancer diagnosis. Current development of bright super-continuum sources, narrow band acousto-optic tunable filters and fast cameras have made feasible a system that can be used for fast diagnosis of cancer in vivo at point of care. The performance of a proto system that has been developed under the Minerva project is described.

1. Introduction

The possibility of using mid-ir light for medical diagnostics has been recognized for some years [1]. It has the potential to greatly improve the chances of early cancer diagnosis. However current systems are based on large expensive laboratory equipment that use light sources such as glow-bars, arrays of quantum cascade lasers or beams from synchrotron sources.

Journal of Physics: Conference Series **619** (2015) 012032 doi:10.1088/1742-6596/619/1/012032

There is a need for a mid-ir system that can be used to measure in vivo spectra at the point of care. The light source needs to be bright, broad-band, compact and modest cost. Additionally a spectrometer is required that has a high through put and fast scan rate so that spectra can be captured with cell level resolution on living samples. Also a sensitive infra-red camera would enable rapid capture of the spectra of many cells simultaneously.

The Minerva project was started in 2012 with the aim to develop technology solutions so that mid- ir spectroscopy can be used for medical point of care cancer diagnostics.

2. Mid-ir imaging spectrometer

The Mid-ir imaging spectrometer, which is schematically shown in fig 1. is made possible by the development of 3 key components that have been developed in the MINERVA project:- a fibre laser pumped super-continuum[2], an acousto-optic tunable filter (AOTF) and a sensitive fast IR camera.

The super-continuum output is in a single transverse mode that has a very high brightness. As a result the beam can be collimated through the high resolution AOTF that has a narrow (0.7°) acceptance angle. An advantage of the high-resolution mid-ir AOTF is that any wavelength can be tuned to in less than 1 ms, so that a customized spectra can be sent to the sample with minimal power loss. The high brightness beam is then projected onto a high resolution IR camera. As a result a 2.5mm square sample area can be analysed with 5μm spatial resolution, which is sufficient to resolve individual cells.

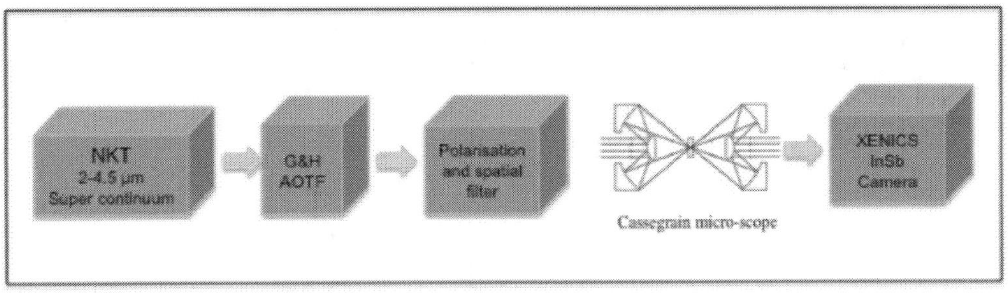

Figure 1 Schematic of mid-ir cytology system

Journal of Physics: Conference Series **619** (2015) 012032 doi:10.1088/1742-6596/619/1/012032

Figure 2 Bread-board mid ir hyper spectral cytology imaging system

3. System performance

The initial system performance of the breadboard system shown in Fig 2 is summarized below in table 1.

Paramater	**Value**
Super continuum power 2-4.5 μm	1.2 W
Bandwidth	2 nm
Mean Power per nm	2.0E-03
Transmission of AOTF	35%
Transmission of optical elements	34%
Power at sample	0.9 mW
Number of pixels	300k
Power per pixel	4.7E-10 W
Spatial resolution	5 μm
Integration time	1 ms
Camera Noise	2000 photons
Signal/noise	960
Minimum absorption detectable	0.0045dB

Table 1 Mid IR system performance

The performance of the short wavelength mid ir system was evaluated with a polymer film on a calcium fluoride substrate. Each image shown in Fig 3 has 300k pixels and was taken with a camera integration time of 200μseconds. The time to scan between any two wavelengths is less than 200μseconds so the limiting capture rate is governed by the frame grabber rate of 85 frames/second. Therefore a mid-ir hyper-spectral cube with 100 wavelength slices of 300k pixels each could be captured in 1.2 seconds.

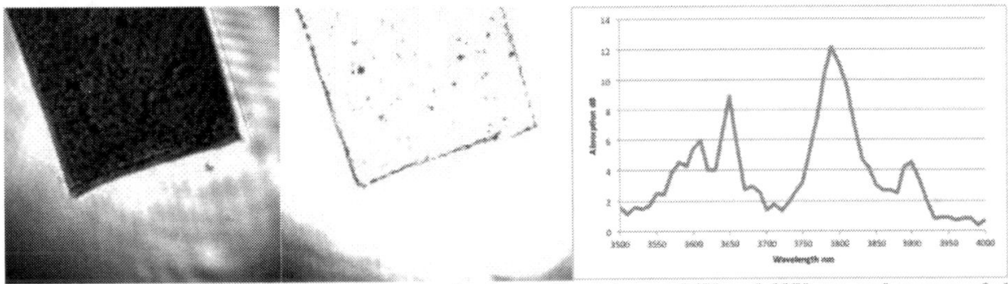

Figure 3 Examples of 300 k pixel spectral images of polymer film taken at 3650 and 3970 nm and spectra of single pixel

A sample spectra of one pixel is shown in Fig 3. Spectral classification software can then be used to identify regions of the sample that match the spatial and spectral characteristics of a particular material.

One significant advantage of the mid-ir spectral imaging system is the random wavelength tuning so that 100 spectral images may be taken of a sample at specific wavelengths that correspond to fingerprint markers of cancer cells without the need to scan through wavelengths that do not give useful information hence vastly speeding up diagnosis.

4. Conclusions

We have shown the efficacy of using mid ir light from a super continuum to rapidly identify features with spectral characteristics in the 2-4.5 μm wavelength region. While this wavelength region is of interest for diagnosing cancer and other medical conditions, far greater specificity can be obtained by extending the wavelength range into the finger print region. Work within the Minerva project is in progress to extend the wavelength range by developing super-continuum sources[3], AOTF's, IR cameras and fused fibre devices that will operate out to 12 μm.

5. Acknowledgements.

This work has been partly funded by the European Framework 7 programme.

References

[1] M.Jackson & H.Mantsch, "Biomedical infrared spectroscopy," pp. 311-340 in *Infrared Spectroscopy of Biomolecules* Toronto, Wiley-Liss (1996).

[2] I. Kubat, C.S. Agger, P.M. Moselund and O. Bang"Mid-infrared supercontinuum generation to 4.5 um in uniform and tapered ZBLAN step-index fibers by direct pumping at 1064 or 1550 nm"JOSA B 30, 2743 (2013)

[3] Agger, Christian ; Kubat, Irnis ; Møller, Uffe Visbech ; Bang, Ole ; Moselund, Peter M. ; Petersen, Christian ; Napier, Bruce ; Seddon, Angela ; Sujecki, Slawomir ; Benson, Trevor ; Farries, Mark ; Ward, Jon ; Lamrini, Samir ; Scholle, Karsten ; Fuhrberg, Peter "Numerical demonstration of 3-12μm supercontinuum generation in large-core step-index chalcogenide fibers pumped at 4.5μm" part of: Nonlinear Optics Technical Digest, pages: NW4A.09, 2013, Optical Society of America, Presented at: Nonlinear Optics (NLO 2013), Kohala Coast, Hawaii

Development of a new detection device using a glass clip emitting infrared fluorescence for laparoscopic surgery of gastric cancer

Shunko Albano Inada[1], Shingo Fuchi[2], Kensaku Mori[1], Junichi Hasegawa[3], Kazunari Misawa[4], Hayao Nakanishi[4]

[1]Information and Communications Headquarters/ Graduate School of Information Science, Nagoya University, Furo-cho, Chikusa-ku, Nagoya 464-8601 Japan.

[2]Department of Electrical Engineering and Electronics, Aoyama Gakuin University, 5-10-1 Fuchinobe, Chuo-ku, Sagamihara-shi,Kanagawa 252-5258, Japan.

[3]Department of Information Engineering, Chukyo University, 101 Tokodachi, Kaizu-cho, Toyota-shi, Aichi 470-0393, Japan.

[4]Division of Oncological Pathology, Aichi Cancer Center Research Institute, Kanokoden, Chikusa-ku Nagoya 464-8681, Japan.

sinada@mori.m.is.nagoya-u.ac.jp

In conventional method, to identify location of the tumor intraperitoneally for extirpation of the gastric cancer, charcoal ink is injected around the primary tumor. However, in the time of laparoscopic operation, it is difficult to estimate specific site of primary tumor. In this study we developed a glass phosphors was realized with Yb^{3+}, Nd^{3+} doped to Bi_2O_3-B_2O_3 based glasses, which have central emission wavelength of 1020 nm and 100 nm of FWHM. Using this glass phosphor, we developed a fluorescent clip and the laparoscopic fluorescent detection system for clip-derived near-infrared light. To evaluated clinical performance of a fluorescent clip and the laparoscopic detection system, we used resected stomach from the patients. Fluorescent clip was fixed on the gastric mucosa, and an excitation light (wavelength: 808nm) was irradiated from outside of stomach for detection of fluorescent through stomach wall. As a result, fluorescent emission from the clip was successfully detected. These results indicate that the glass fluorescent clip in combination with laparoscopic detection system is a very useful method to identify the exact location of the primary gastric cancer.

1. Introduction

Cancer is a leading cause of disease death worldwide, accounting for 8.2 million deaths in 2012. [1] Common causes of cancer death are derived from cancers such as lung, liver, gastric, colorectal, breast and oesophageal cancer. In the case of gastric cancer, 723,000 person were dead in 2012. [1] And 50% of the cases occurred in Eastern Asia. [2] Among several modalities for the treatment of gastric cancer such as surgery, radiotherapy and chemotherapy, surgical resection is the most reliable treatment method. Although open surgery has long been performed as a major treatment modality of gastric cancer, recently, laparoscopic surgery has been developed and is now recognized as an alternative surgical modality for gastric cancers because this method is low-invasive and allows good QOL for the patients, but the efficacy is equivalent to the open surgery. [3-6]

In a conventional laparoscopic surgery, charcoal ink is injected around the primary tumor to identify the location of the tumor in the stomach intraperitoneally. [7] However, in the time of laparoscopic operation, it is difficult to estimate exact site of primary tumor, because the injected charcoal diffusely spread to the area distant from the tumor in the stomach unlike colon cancer cases. Therefore, a broad area should be resected which results in a great inconvienience for the patients. To overcome this problem, we focused in the near-infrared wavelength of 1000 nm band which have high biological transmission. [8]

In this study, a glass phosphors was realized with Yb^{3+}, Nd^{3+} doped to Bi_2O_3-B_2O_3 based glasses, which have central emission wavelength of 1020 nm and 100 nm of FWHM (Fig. 1). [9] Using this glass phosphor with the dimension of 2x3x1 (mm), we developed a new fluorescent clip utilizing a commercially available hemostatic clip (HX-610-090L, Olympus Medical Systems CORP., Tokyo, Japan). For detection of the near-infrared light of the fluorescent clip, we developed the laparoscopic fluorescent detection system which is composed for rigid scope, two optical filters and InGaAs camera for optimizing 1000 nm near-infrared light.

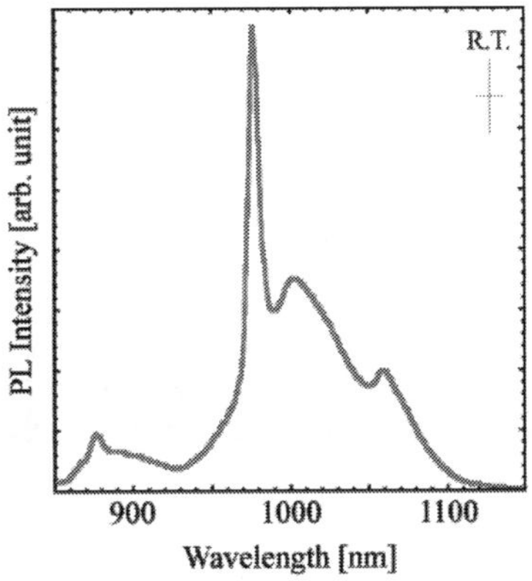

Fig. 1. PL spectrum of glass phosphor. Realized with Yb^{3+}, Nd^{3+} doped to Bi_2O_3-B_2O_3 based glasses.

2. Experimental Procedure

To evaluate clinical performance of a fluorescent clip and the laparoscopic detection system, we used resected stomach (thickness: 13 mm) from the patients. Fluorescent clip was fixed on the gastric mucosa, and an excitation light with wavelength of 808 nm (Laser Diode C9691-06, Hamamatsu Photonics K. K., Shizuoka, Japan) was irradiated from outside of the stomach to detect the fluorescence through stomach wall. The rigid scope of laparoscopic detection system was fixed 2 cm from the outside of stomach with 0 degrees of inclination. On the other hand, the excitation light was fixed 2 cm from the serosa-side of the stomach wall with 10 degrees of inclination. The excitation light was irradiating 2W of intensity. To examine the detection sensitivity of the laparoscopic detection system for the emission light from fluorescent clip, first a near-infrared light image was take when the fluorescent clip not covered with stomach wall. Next, near-infrared image was taken when the fluorescent clip is covered with stomach wall. For estimation of background fluorescence, near-infrared light image of the stomach without fluorescent clip was taken.

3. Results and Discussion

Figure 2 shows the result of fluorescent clip and the laparoscopic detection system using a resected stomach from the patient to evaluate the clinical performance. Specific fluorescent emission from the clip in the stomach was successfully detected under covering with stomach wall, although fluorescence emission was strongly reduced by the presence of the stomach wall. Furthermore, we confirmed that the detection sensitivity of this laparoscopic system for the fluorescent emission from the clip depends on the output power of the excitation light source.

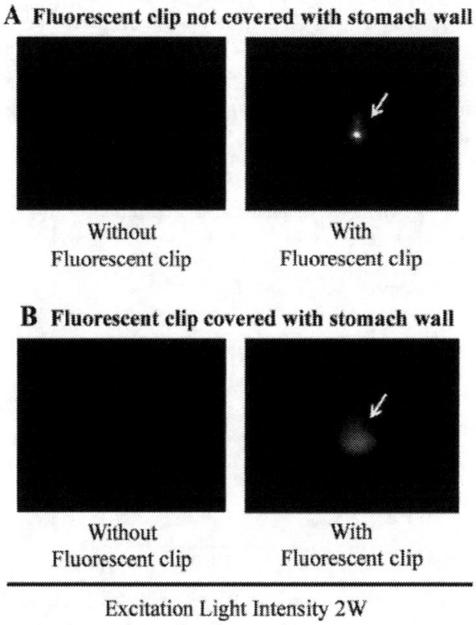

Fig. 2. Results of the fluorescent clip detection using humans resected stomach.

4. Conclusions

Stomach wall is very thick (13 mm) and contains a lot of blood and mucin as a barrier substance compared with colon. However, sufficient emission from the clip could be detected from the outside the stomach by our new laparoscopic fluorescent detection system. These results indicate that laparoscopic detection system in combination with the glass fluorescent clip has ability to identify the exact location of the primary gastric cancer intraperitoneally and suggest that the fluorescent clip and the laparoscopic fluorescent detection system is a very useful method to determine the resection margin of the primary gastric tumor in the laparoscopic surgery.

References

[1] Globocan 2012, IARC.
[2] Ferlay J, Shin HR, Bray F, et al.: GLOBOCAN 2008: Cancer Incidence and Mortality Worldwide in 2008. Lyon, France: IARC CancerBase No. 10.
[3] Hosono S, Arimoto Y, Ohtani H and Kanamiya Y.: Meta-analysis of short-term outcomes after laparoscopy-assisted distal gastrectomy. World J Gastroenterol. 2006; 12(47): 7676-83.

[4] Kitano S, Shiraishi N, Uyama I, Sugihara K and Tanigawa N.: A multicenter study on oncologic outcome of laparoscopic gastrectomy for early cancer in Japan. Ann Surg. 2007; 245(1): 68-72.

[5] Kim HH, Hyung WJ, Cho GS, Kim MC, Han SU, Kim W, Ryu SW, Lee HJ and Song KY.: Morbidity and mortality of laparoscopic gastrectomy versus open gastrectomy for gastric cancer: an interim report-a phase III multicenter, prospective, randomized trial (KLASS trial). Ann Surg. 2010; 251(3): 417-20.

[6] Katai H, Sasako M, Fukuda H, Nakamura K, Hiki N, Saka M, Yamaue H, Yoshikawa T and Kojima K.: Safety and feasibility of laparoscopy-assisted distal gastrectomy with suprapancreatic nodal dissection for clinical stage I gastric cancer: a multicenter phase II trial (JCOG 0703). Gastric Cancer. 2010; 13(4): 238-44.

[7] Kitamura K, Takahashi T, Yamaguchi T, Yamane T, Hagiwara A and Oyama T.: Identification, by activated carbon injection, of cancer lesion during laparoscopic surgery. Lancet. 1994 Mar 26; 343(8900): 789.

[8] R. R. Anderson and J. A. Parrish.: The optics of human skin. J. Inv. Der. 1981; 77(1): 13–19.

[9] Fuchi S, Sakano A and Takeda Y.: Wideband Infrared Emission from Yb^{3+} - and Nd^{3+} -Doped Bi_2O_3-B_2O_3 Glass Phosphor for an Optical Coherence Tomography Light Source. Jpn. J. Appl. Phys. 2008; 47(10):7932-35.

Journal of Physics: Conference Series **619** (2015) 012034 doi:10.1088/1742-6596/619/1/012034

Single Crystal Growth and Co-doping Effects of Lanthanum Substituted Gadolinium Pyrosilicate Scintillator

[1]Kei Kamada[1,2], Shunsuke Kurosawa[3], Rikito Murakami[3], Yuui Yokota[1], Jan Pejchal[1] Yuji Ohashi[3] and Akira Yoshikawa[1,2,3]

[1]New Industry Creation Hatchery Center, Tohoku University, 6-6-10 Aoba, Sendai 980-8579

[2]C&A corporation, 6-6-10 Aoba, Aramaki, Aoba-ku, Sendai 980-8579, Japan

[3]Institute for Material Reseach, Tohoku University, 2-1-1, Sendai,980-8577 Japan

E-mail: kamada@imr.tohoku.ac.jp

Abstract. Ca 100, 200, 500, 1000 and 3000 at. ppm co-doped $(Ce_{0.01}La_{0.3}Gd_{0.69})_2Si_2O_7$ single crystals were grown by the μ-PD method. Luminescence and scintillation properties such as radioluminescence spectra, light yield and decay time were evaluated. Expected Ce^{3+} 4f5d emission have been observed in 350-430nm peaking at 360nm. Emission intensity took maximum at the Ca 100at.ppm co-doped sample and decreased by increasing the Ca concentration. The Ca 100at.ppm co-doped sample showed the highest light yield of around 20,000 photons/MeV. Scintillation decay time was accelerated with increasing Ca concentration. Scintillation decay time of the Ca 0, 100, 200, 1000 and 3000 at. ppm co-doped samples were 66.5ns(21%) 468ns(79%) and 44.0ns(22%) 430ns(78%), respectively.

1. Introduction

Scintillator materials combined with photodetectors are used to detect high energy photons and accelerated particles in medical imaging techniques, high energy and nuclear physics detectors, high-tech industrial applications and most recently also in the advanced homeland security related techniques.[1] In the last two decades, great R&D effort brought several new material systems, namely the Ce-doped orthosilicates as Gd_2SiO_5 (GSO), Lu_2SiO_5 (LSO), $(Lu_{1-x}Y_x)_2SiO_5$ (LYSO), pyrosilicates based on $RE_2Si_2O_7$ (RE=Lu, Y,Gd) and most recently LaX_3 (X=Cl,Br) single crystal hosts[1–5].Recently, Ce:$Gd_2Si_2O_7$ (Ce:GPS) belonging to the pyrosilicate group has been investigated, and this crystal was found to have much higher light output and shorter decay time than Ce:GSO. [6–8] Toropov et al. found that the composition Gd_2O_3-2SiO_2 is not congruent in the Gd_2O_3–SiO_2 system,[9] and Kawamura et al. reported that the GPS phase can be grown by heavy Ce-doping (approximately10 mol %).[9] Furthermore, substituting Ce^{3+} site with La^{3+} whose ionic radius is similar to Ce^{3+} is also able to stabilize the congruent phase, and Ce:(La, Gd)$_2Si_2O_7$ (Ce:La-GPS) scintillator with a good energy resolution was developed [10].

More recently, the improvement of scintillation properties with co-doping divalent (e.g. Mg^{2+}, Ca^{2+}) for Ce:(Lu, Y)$_2SiO_5$ (Ce:LYSO) have reported [11]. The reason of improving light yield in Ca^{2+} co-doped Ce:LYSO systems was explained by decreasing the rate of electron capturing involved by O^{2-} defects [11]. Co-doped divalent introduce negative charge by substituting Lu^{3+} or Y^{3+} site, and it interact with the positive charge of O^{2-} defect [11]. Accordingly, this mechanism would prevent electron trapping [11].

[1] Kei Kamada, New Industry Creation Hatchery Center, Tohoku University, 6-6-10 Aoba, Sendai 980-8579

The aim of this work is to investigate Ca co-doping effects on luminescence and scintillation properties of Ce doped La-GPS. Ca co-doped Ce:$(La_{0.3}Gd_{0.7})_2Si_2O_7$ single crystal were grown by the micro-pulling down (μ-PD) method. Luminescence and scintillation properties were evaluated.

2. Experimental

2.1. Crystal growth procedure

A stoichiometric mixture of 4N La_2O_3, CeO_2, Gd_2O_3 and glass-SiO_2 powders (High Purity Chemicals Co.) was used as starting material. Ca^{2+} was co-doped in the rare-earth site of $(Ce_{0.01}La_{0.3}Gd_{0.69})_2Si_2O_7$. Ca co-doped $(Ce_{0.01}La_{0.3}Gd_{0.69})_2Si_2O_7$ single crystals were grown by the micro-pulling-down (μ-PD) method with an RF heating system. A schematic of the μ-PD growth apparatus is given in Refs. [11,12]. Typical pulling rates were 0.05–0.08 mm/min and the diameter was around 3 mm. Crystals were grown from an Ir crucible under N_2 atmosphere. The seed crystals were $(La_{0.3}Gd_{0.7})_2Si_2O_7$ crystals. Plates of 3mmϕ x 1mm were cut and polished for the measurements.

2.2. Measurements of scintillation properties

Radio-luminescence spectrum at room temperature was measured with the above spectrometer (EI FLS920) excited by 5.5MeV alpha rays from an 241Am source. To determine the light output, we obtained the pulse height spectrum of this crystal under excitation with 662 keV gamma rays from a ^{137}Cs source. The scintillation photons were detected using by a photomultiplier (PMT; Hamamatsu R7600U), and then the signals from the PMT were amplified (ORTEC 113), shaped (ORTEC 572A), and read out with a multi-channel analyzer (AMPTEK 8000A). The pulse height of $Bi_4Ge_3O_{12}$ (BGO) crystal, which has a light output of 8,500 photons/MeV was also measured as a reference. Additionally, its scintillation decay time excited by the gamma rays (^{137}Cs) was measured with the PMT and an oscilloscope (Tektronix TDS 3052B). The scintillation decay time was calculated from exponential approximation.

3. Results

3.1. Crystal growth procedure

Ca 100, 200, 500, 1000 and 3000 at. ppm co-doped $(Ce_{0.01}La_{0.3}Gd_{0.69})_2Si_2O_7$ single crystals were grown by the μ-PD method. Although $Gd_2Si_2O_7$ can not be grown from melt because of its incongruent melt composition, $(La_{0.3}Gd_{0.7})_2Si_2O_7$ became congruent melt by substituting La in GPS . Example photos are shown in Figure 1. The grown crystals were colorless transparent with 3 mm in diameter and 15-20 mm in length.

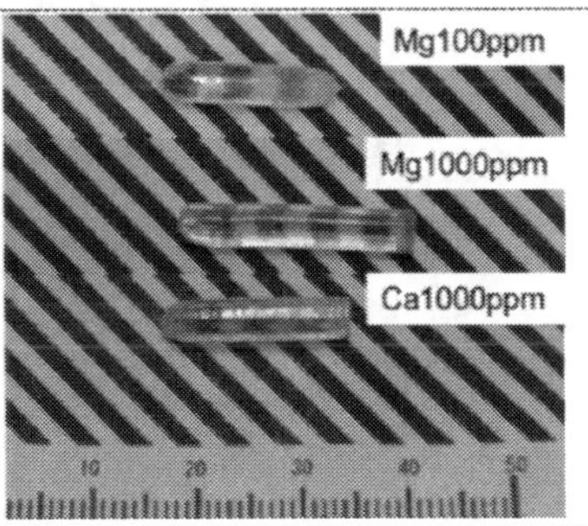

Figure 1. Example photographs of the grown Ca co-doped $(Ce_{0.01}La_{0.3}Gd_{0.69})_2Si_2O_7$ single crystals.

3.2. Luminescence and gamma-ray response measurement procedurembering

Figure 2 shows radioluminescence spectra of the Ca co-doped $(Ce_{0.01}La_{0.3}Gd_{0.69})_2Si_2O_7$ samples. Expected Ce^{3+} 4f5d emission have been observed in 350-430nm peaking at 360nm. Emission intensity took maximum at the Ca 100at.ppm co-doped sample and decreased by increasing the Ca concentration.

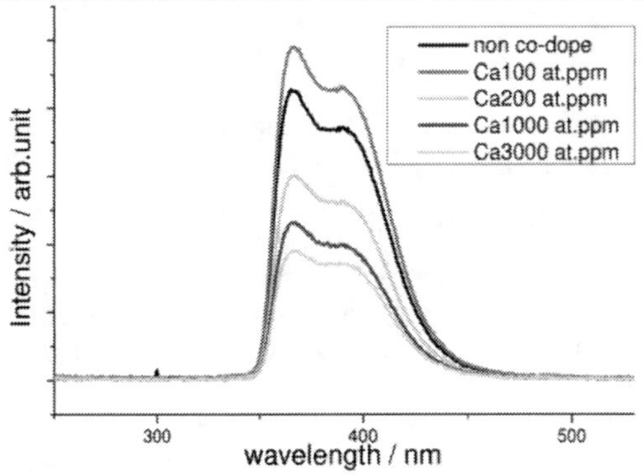

Figure 2. Radioluminescence spectra of the Ca co-doped $(Ce_{0.01}La_{0.3}Gd_{0.69})_2Si_2O_7$ samples.

Figure 3 exemplifies energy spectra of the Ca co-doped $(Ce_{0.01}La_{0.3}Gd_{0.69})_2Si_2O_7$ samples. The samples irradiated by ^{137}Cs, where 662 keV. The Ca 100at.ppm co-doped sample showed the highest light yield of around 20,000 photons/MeV considering with the emission wavelength of the samples and quantum efficiency of the PMT (45%@360nm, 32%@480nm). Light yield was decreased with increasing Ca concentration above 100 at.ppm. This result is good agreement with the result of radioluminescence spectra measurement. The Ca 100at.ppm co-doped sample showed the highest light yield. Light yields of the Ca 0, 200, 1000 and 3000 at. ppm co-doped samples were 18000, 14000, 14400 and 8000 photons/MeV, respectively.

Scintillation decay curves were obtained by using the PMT and digital oscilloscope and the sample was irradiated by ^{137}Cs (Figure 4). Scintillation decay time was accelerated with increasing Ca concentration. Scintillation decay time of the Ca 0, 100, 200, 1000 and 3000 at. ppm co-doped samples were 66.5ns(21%) 468ns(79%) and 44.0ns(22%) 430ns(78%), respectively.

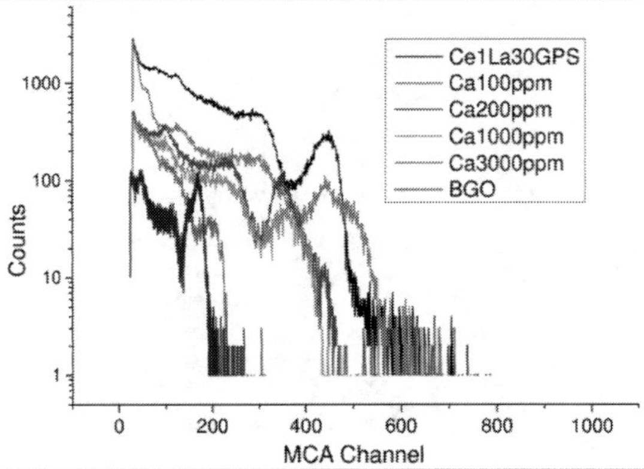

Figure 3. Energy spectra of Ca co-doped $(Ce_{0.01}La_{0.3}Gd_{0.69})_2Si_2O_7$ samples and BGO standard excited by a 662 keV gamma ray.

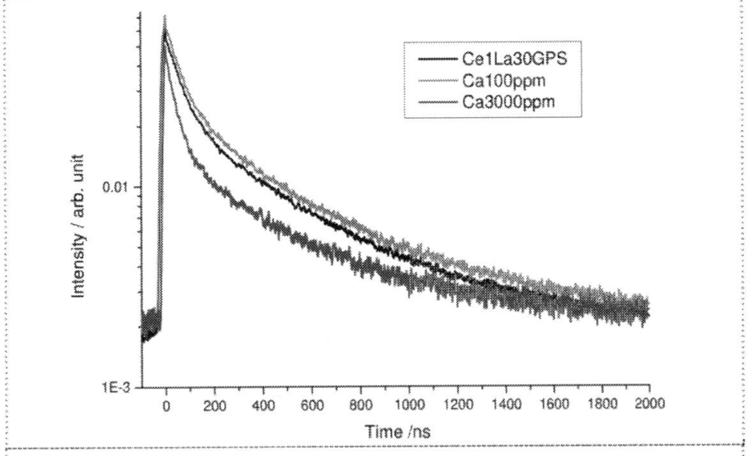

Figure 4. Decay curves of Nd1%:$Y_3Al_5O_{12}$ and Nd1%:$Y_3Ga_3Al_2O_{12}$ excited by a 662 keV gamma-ray using the PMT

4. Conclusion

Ca 100, 200, 500, 1000 and 3000 at. ppm co-doped $(Ce_{0.01}La_{0.3}Gd_{0.69})_2Si_2O_7$ single crystals were grown by the μ-PD method. Although $Gd_2Si_2O_7$ can not be grown from melt because of its incongruent melt composition, $(La_{0.3}Gd_{0.7})_2Si_2O_7$ became congruent melt by substituting La in GPS. Ce^{3+} 4f5d emission have been observed in 350-430nm Emission intensity took maximum at the Ca 100at.ppm co-doped sample and decreased by increasing the Ca concentration. The Ca 100at.ppm co-doped sample showed the highest light yield of around 20,000 photons/MeV. Light yield was decreased with increasing Ca concentration above 100 at.ppm. Scintillation decay time was accelerated with increasing Ca concentration. Scintillation decay time of the Ca 0, 100, 200, 1000 and 3000 at. ppm co-doped samples were 66.5ns(21%) 468ns(79%) and 44.0ns(22%) 430ns(78%), respectively.

5. References

[1] M. Nikl, "Scintillation detectors for X-rays",Meas. Sci. Technol. Vol.17, pp.R37-R54, Feb. 2006
[2] C. L. Melcher, et al.,IEEE Trans. Nucl. Sci., vol. 39, no. 4, pp.502–505, Aug. 1992.
[3] P. Lecoq, et al., IEEE Trans. Nucl. Sci., vol. 49, no. 4, pp. 1651–1654, 2002.
[4] M. Moszynski, et al.,Nucl. Instr. And Meth. A 385, pp. 123–131, Jan. 1997.
[5] K. S. Shah, et al.,IEEE Trans. Nucl. Sci., vol. 50, no. 6, pp. 2410–2413, Dec. 2003.
[6] Yagi, K. et al., IEEE Nuclear Science Symp. Conf. Rec., 2006, p. 1160.
[7] S. Kawamura, et al., Nucl. Instrum. Methods Phys. Res., Sect. A 583 (2007) 356.
[8] J. H. Kaneko, et al., Nucl. Instrum. Methods Phys. Res., Sect. A 529 (2004) 307.
[9] N. A. Toropov, et al.,Russ. Chem. Bull. 10 (1961) 497.
[10] A. Suzuk, et al., *Appl. Phys. Express*, **5**, 2012, 102601
[11] S. Blahuta *et al., IEEE Trans. Nucl. Sci.* Vol.60, No4, 2013, pp. 3134-3140

Nonstochiometry of $Lu_3Al_5O_{12}$ single crystal and its effects of on luminescence and scintillation properties

[1]**Kei Kamada**[1,2], **Shunsuke Kurosawa**[3] , **Yuui Yokota**[1], **Jan Pejchal**[1,] **Martin Nikl**[4], **Yuji Ohashi**[3] **and Akira Yoshikawa**[1,2,3]
[1]New Industry Creation Hatchery Center, Tohoku University, 6-6-10 Aoba, Sendai 980-8579
[2]C&A corporation, 6-6-10 Aoba, Aramaki, Aoba-ku, Sendai 980-8579, Japan
[3]Institute for Material Reseach, Tohoku University, 2-1-1, Sendai,980-8577 Japan
[4]Institute of Physics AS CR,16253 Prague, Czech Republic
E-mail: kamada@imr.tohoku.ac.jp

Abstract. Nonstoichiometric $Lu_3Al_{5+x}O_{12}$ (x= 0.05, 0.15, 0.35, 0, -0.05, -0.15, -0.35) crystals were grown by the μ-PD method. Luminescence and scintillation properties such as absorption, excitation and emission spectra, light yield and decay time were evaluated. Expected anti-site defect related host emission have been observed in 250-420nm. Emission intensity was increased by increasing the nonstochiometry. The x=-0.35 sample showed the highest light yield of around 12000 photons/MeV and slowest scintillation decay time of 1.96μs.

1. Introduction

Scintillator materials combined with photodetectors are used to detect high energy photons and accelerated particles in medical imaging techniques, high energy and nuclear physics detectors, high-tech industrial applications and most recently also in the advanced homeland security related techniques. Oxide materials based on garnet structure single crystals are promising candidates for scintillator applications because of well mastered technology developed for laser hosts and other applications, optical transparency and easy doping by rare-earth elements. For example, $Pr:Lu_3Al_5O_{12}$ (Pr:LuAG) has attracted attention because of its favorable combination of fast Pr^{3+} 5d-4f emission peaking at 311 nm, non-hygroscopic nature, high density (6.7 g/cm^3), high light output (around 20000 photon/MeV), very short decay time (20 ns), and good energy resolution (4.8% at 662 keV for a 10 × 10 × 10 mm LuAG:Pr sample and a Hamamatsu R1791 PMT) [1-3] Isostructural cubic garnet structures of YAG and LuAG form a solid solution and any intermediate mixed composition (i.e. $Y_xLu_{3-x}Al_5O_{12}$, for 0>x>3) can be grown. The solidification points of mixed compositions range between 2010 ^{0}C (for LuAG) and 1930 ^{0}C (for YAG) [4]. The garnet structure exhibits very flexible cation substitution and such substitution was used to prepare materials with tailored lattice constants [5] and can, in principle, be used for the preparation of highly substituted crystals with a homogeneous composition. However, at the same time, this readiness of solid solution formation in the garnet structure hints at the relative ease of intrinsic anti-site defect formation. Indeed, theoretical studies have shown that the anti-site defect, Y_{Al}, was found as the lowest energy intrinsic defect in both YAG [6-8] and all other REAGs, (RE ranging from Lu to Gd) [9]. Anti-site defect related host emission is

[1] Kei Kamada, New Industry Creation Hatchery Center, Tohoku University, 6-6-10 Aoba, Sendai 980-8579

observed around 280-360nm with decay time of few hundreds nano second. Anti-site defect can be enhanced by Sc doping in Al-site and Sc doped LuAG shows high light yield of 22500 photon/MeV with decay time of 610ns[10]

In this report, nonstochiometry of LuAG single crystal and its effects of on luminescence and scintillation properties were investigated. Nonstoichiometric $Lu_3Al_{5-x}O_{12}$ single crystals were grown by the micro-pulling down (μ-PD) method. Luminescence and scintillation properties were evaluated.

2. Experimental

2.1. Crystal growth procedure

A stoichiometric mixture of 4N Lu_2O_3 and α-Al_2O_3 powders (High Purity Chemicals Co.) was used as starting material. Nonstoichiometric $Lu_3Al_{5+x}O_{12}$ were grown by the micro-pulling-down (μ-PD) method with an RF heating system. A schematic of the μ-PD growth apparatus is given in Refs. [11,12]. Typical pulling rates were 0.05–0.12 mm/min and the diameter was around 3 mm. Crystals were grown from an Ir crucible under N_2 atmosphere. The seed crystals were <100> oriented LuAG crystals. Plates of 3mmϕ x 1mm were cut and polished for the absorption and luminescence spectra measurement.

2.2. Measurements of scintillation properties

Radio-luminescence spectrum at room temperature was measured with the above spectrometer (EI FLS920) excited by 5.5MeV alpha rays from an 241Am source. To determine the light output, we obtained the pulse height spectrum of this crystal under excitation with 662 keV gamma rays from a ^{137}Cs source. The scintillation photons were detected using by a photomultiplier (PMT; Hamamatsu R7600U), and then the signals from the PMT were amplified (ORTEC 113), shaped (ORTEC 572A), and read out with a multi-channel analyzer (AMPTEK 8000A). The pulse height of $Bi_4Ge_3O_{12}$ (BGO) crystal, which has a light output of 8,500 photons/MeV [13] was also measured as a reference. Additionally, its scintillation decay time excited by the gamma rays (^{137}Cs) was measured with the PMT and an oscilloscope (Tektronix TDS 3052B). The scintillation decay time was calculated from exponential approximation.

3. Results

3.1. Crystal growth procedure

Nonstoichiometric $Lu_3Al_{5+x}O_{12}$ (x= 0.05, 0.15, 0.35, 0, -0.05, -0.15, -0.35) crystals were grown by the μ-PD method. Example photos are shown in Figure 1. The grown crystals were colorless transparent with 2-3 mm in diameter and 15-23 mm in length.

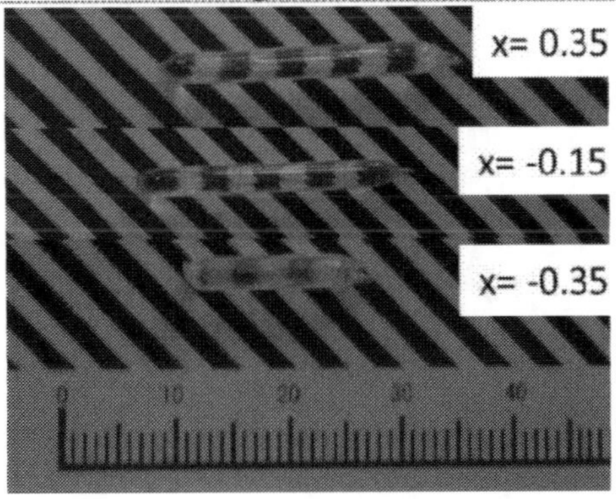

Figure 1. Example photographs of the grown $Lu_3Al_{5+x}O_{12}$ single crystals.

3.2. Luminescence and gamma-ray response measurement procedurembering

Figure 2 shows radioluminescence spectra of the $Lu_3Al_{5+x}O_{12}$ samples. Expected anti-site defect related host emission have been observed in 250-420nm. Emission intensity was increased by increasing the nonstochiometry. The x=+0.35 sample showed the highest emission intensity among the grown $Lu_3Al_{5+x}O_{12}$ samples.

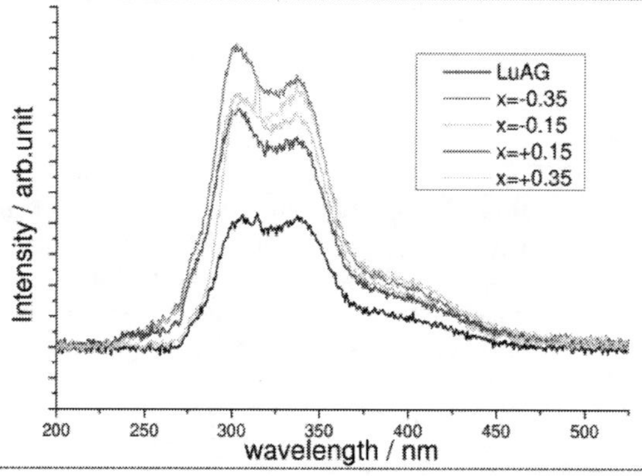

Figure 2. Radioluminescence spectra of the grown $Lu_3Al_{5+x}O_{12}$ crystals.

Figure 3 exemplifies energy spectra of the grown $Lu_3Al_{5+x}O_{12}$ samples irradiated by ^{137}Cs, where 662 keV. Light yield was increased with increasing the nonstochiometry. This result is good agreement with the result of radioluminescence spectra measurement. The x=-0.35 sample showed the highest light yield of around 12000 photons/MeV considering with the emission wavelength of the samples and quantum efficiency of the PMT (44%@330nm, 30%@480nm). Light yields of x= 0, 0.05, 0.15, 0.35, -0.05, -0.15, -0.35 were 8000, 8600, 9200, 12000, 8200, 9000 and 11000 photons/MeV, respectively.

Scintillation decay curves were obtained by using the PMT and digital oscilloscope and the sample was irradiated by ^{137}Cs (Figure 4). Scintillation decay time became slower with increasing the nonstochiometry. Scintillation decay time of x= 0, 0.05, 0.15, 0.35, -0.05, -0.15, -0.35 were 1.10, 1.32, 1.43, 1.96, 1.59, 1.62 and 1.88μs, respectively.

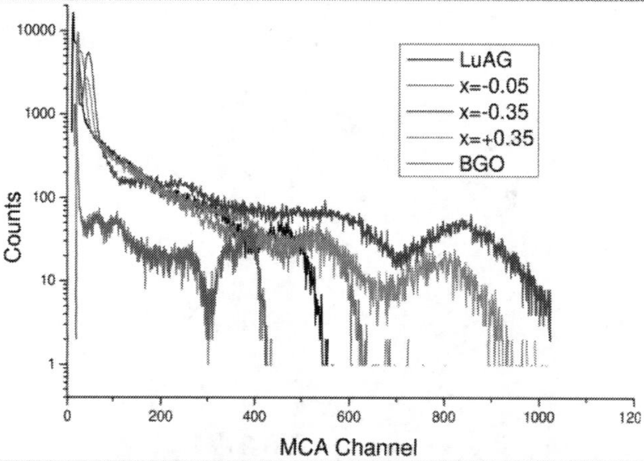

Figure 3. Energy spectra of the grown $Lu_3Al_{5+x}O_{12}$ crystals and BGO standard excited by a 662 keV gamma ray.

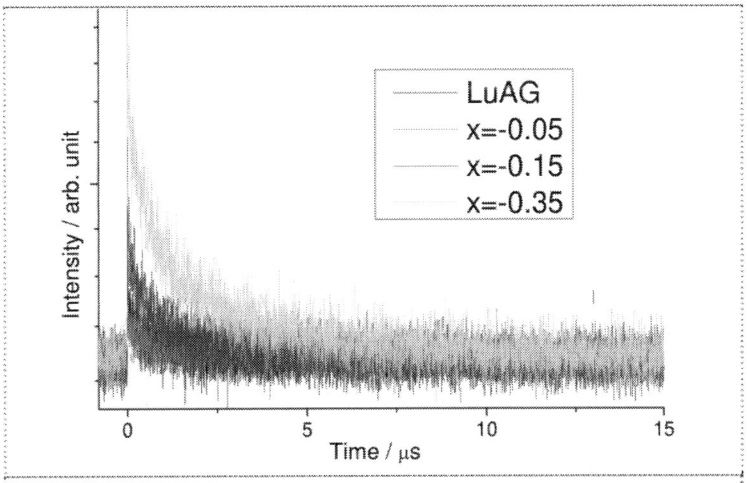

Figure 4. Decay curves of Nd1%:$Y_3Al_5O_{12}$ and Nd1%:$Y_3Ga_3Al_2O_{12}$ excited by a 662 keV gamma-ray using the PMT

4. Conclusion

Nonstoichiometric $Lu_3Al_{5+x}O_{12}$ (x= 0.05, 0.15, 0.35, 0, -0.05, -0.15, -0.35) crystals were grown by the μ-PD method. The grown crystals were colorless transparent with 2-3 mm in diameter and 15-23 mm in length. Expected anti-site defect related host emission have been observed in 250-420nm. Emission intensity was increased by increasing the nonstochiometry. Light yield was increased with increasing the nonstochiometry. The x=-0.35 sample showed the highest light yield of around 12000 photons/MeV. Scintillation decay curves were obtained by using the PMT and digital oscilloscope and the sample was irradiated by ^{137}Cs. Scintillation decay time became slower with increasing the nonstochiometry. Scintillation decay time of x= 0, 0.05, 0.15, 0.35, -0.05, -0.15, -0.35 samples were 1.10, 1.32, 1.43, 1.96, 1.59, 1.62 and 1.88μs, respectively.

5. References

[1] W. Drozdowski, et al., IEEE. Nucl. Trans. Sci. 55(4) (2008) 2420-2424
[2] K. Kamada , et al., IEEE Trans. Nucl. Sci., 56 (2009) 570-573
[3] L. Swiderski, et al., IEEE Trans. Nucl. Sci. 56, (2009) 2499-2505
[4] Y. Kuwano, et al., J. Cryst. Growth 260 (2004) 159.
[5] D. Mateika, et al., J. Cryst. Growth 102 (1990) 994.
[6] M. Kuklja, J. Phys.: Condens. Matt. 12 (2000) 2953.
[7] M. Nikl , et al., Phys. Stat. Sol. (b), vol. 181, pp.R10 , 2000
[8] M. Nikl, et al., Appl. Phys. Lett. Vol.88, pp. 141916-141918, Apr. 2006
[9] C.R. Stanek, et al., Nucl. Instr. Meth. Phys. Research A 579 (2007) 27.
[10] A. Yoshikawa, et al. J. Crystal Growth 205 (1999) 305-316.
[11] V.I. Chani, et al. J. Crystal Growth 204 (1999) 155-162.
[12] M. Moszynski, et al. IEEE Trans. Nucl. Sci. 44 (1997) 1052.

Growth and scintillation properties of Ce doped $Gd_2Si_2O_7/SiO_2$ eutectics

[1]Kei Kamada[1,2], Shunsuke Kurosawa[1,3], Rikito Murakami[3], Yuui Yokota[1], Jan Pejchal[1] Yuji Ohashi[3] and Akira Yoshikawa[1,2,3]
[1]New Industry Creation Hatchery Center, Tohoku University, 6-6-10 Aoba, Sendai 980-8579
[2]C&A corporation, 6-6-10 Aoba, Aramaki, Aoba-ku, Sendai 980-8579, Japan
[3]Institute for Material Reseach, Tohoku University, 2-1-1, Sendai,980-8577 Japan
E-mail: kamada@imr.tohoku.ac.jp

Abstract. Ce:$Gd_2Si_2O_7/SiO_2$ eutectic was grown by the μ-PD method. The square-shape sample with a side of 5 mm and a length of 15 mm was obtained. Two phases of orthorhombic $Gd_2Si_2O_7$ and SiO_2 was observed. Rod-phase was SiO_2 and matrix phase was $Gd_2Si_2O_7$. Ce^{3+} 4f5d emission have been observed at 400nm. The sample showed light yield of around 16,000 photons/MeV. Scintillation decay time was 46.3ns(21%) 249ns(79%).

1. Introduction

A number of halide and oxide scintillator materials have been developed and they are applied to various fields such as astronomy, medical imaging, and homeland security. Although halide scintillators, such as Tl:NaI, Tl:CsI and Ce:$LaBr_3$ [1], have high light outputs of more than 30,000 photons/MeV, they have hygroscopic nature. In the X-ray imaging, indirect flat panel detector (FPD) composed of a scintillator layer such as CsI:Tl and photo-sensor array requires improvements in their spatial resolution. But, CsI:Tl columnar films reduce the spatial resolution, because the thicker it becomes the more light scattering occurs [2]. Recently submicron-diameter phase-separated scintillator fibers (PSSFs) were reported and they possessed both the properties of an optical fiber and a radiation-to-light conversion. The PSSFs were fabricated using a directionally solidified eutectic (DSE) system. The DSE systems have been discovered in various materials for many applications [3-6]. In PSSFs, the light emitted from the scintillator fibers is confined and transported along the fiber direction by a total reflection mode as shown in Fig.1, so that high-resolution radiation imaging can be achieved. CsI/NaCl [6] and GAP/Al_2O_3[5] have been already reported as PSSFs.

Ce:$Gd_2Si_2O_7$ (Ce:GPS) had a good light output of 30,000 photons/MeV and FWHM energy resolution of 6.0% at 662 keV at room temperature. However, according to the previous report [7-10], Ce:GPS crystals need to be grown with heavy Ce-doping (approximately, 10 mol%) in order to stabilize the crystal growth process. Such high Ce-concentration would lead to lower light output because of self-absorption or concentration quenching. The main reason is the fact that $Gd_2Si_2O_7$ is not congruently melt. When the chemicals with stoichiometric composition is melted, initial melt has apatite phase.

[1] Kei Kamada, New Industry Creation Hatchery Center, Tohoku University, 6-6-10 Aoba, Sendai 980-8579

Scintillator materials combined with photodetectors are used to detect high energy photons and accelerated particles in medical imaging techniques, high energy and nuclear physics detectors, high-tech industrial applications and most recently also in the advanced homeland security related techniques. Recently, Ce:$Gd_2Si_2O_7$ (Ce:GPS) belonging to the pyrosilicate group has been investigated, and this crystal was found to have much higher light output and shorter decay time than Ce:GSO. [6–8] Toropov et al. found that the composition Gd_2O_3-2SiO_2 is not congruent in the Gd_2O_3–SiO_2 system,[9] and Kawamura et al. reported that the GPS phase can be grown by heavy Ce-doping (approximately10 mol %).[9] Furthermore, substituting Ce^{3+} site with La^{3+} whose ionic radius is similar to Ce^{3+} is also able to stabilize the congruent phase, and Ce:(La, Gd)$_2Si_2O_7$ (Ce:La-GPS) scintillator with a good energy resolution was developed [10-13].

In this study, Ce 1mol% doped $Gd_2Si_2O_7/SiO_2$ eutectics were grown by the micro pulling down (m-PD) method and their DSE system has been investigated. Investigation of obtained eutectic structure , luminescence and scintillation performances were performed.

2. Experimental

2.1. Crystal growth procedure

A stoichiometric mixture of 4N CeO_2, Gd_2O_3 and glass-SiO_2 powders (High Purity Chemicals Co.) was used as starting material. Ce:$Gd_2Si_2O_7/SiO_2$ eutectics were grown by the micro-pulling-down (μ-PD) method with an RF heating system at the eutectic composition (25.6mol%$Gd_{1.98}Ce_{0.02}O_3$_74.4mol%SiO_2). Here, the volume ratio is about $Gd_{1.98}Ce_{0.02}Si_2O_7$:$SiO_2$=0.809:0.191, theoretically. Typical pulling rates were 0.3-0.8mm/min. Crystals were grown from an Ir crucible under N_2 atmosphere. A Ir rod was used as a seed. Plates of 5mm x 5mm x 0.5mm were cut and polished for the measurements. The eutectic phase structure was investigated by back scattered electron image (BEI). The phases in the eutectic crystal determined by powder XRD patterns.

2.2. Measurements of scintillation properties

Radio-luminescence spectrum at room temperature was measured with the above spectrometer (EI FLS920) excited by 5.5MeV alpha rays from an 241Am source. To determine the light output, we obtained the pulse height spectrum of this crystal under excitation with 662 keV gamma rays from a ^{137}Cs source. The scintillation photons were detected using by a photomultiplier (PMT; Hamamatsu R7600U-200), and then the signals from the PMT were amplified (ORTEC 113), shaped (ORTEC 572A), and read out with a multi-channel analyzer (AMPTEK 8000A). The pulse height of $Bi_4Ge_3O_{12}$ (BGO) crystal, which has a light output of 8,500 photons/MeV was also measured as a reference. Additionally, its scintillation decay time excited by the gamma rays (^{137}Cs) was measured with the PMT and an oscilloscope (Tektronix TDS 3052B). The scintillation decay time was calculated from exponential approximation.

3. Results

3.1. Crystal growth procedure

Ce:$Gd_2Si_2O_7/SiO_2$eutectic was grown by the μ-PD method. The square-shape sample with a side of 5 mm and a length of 15 mm was obtained (Fig.1). The BEI image of transverse cross-section of the eutectic is shown in Fig. 2. The phases were determined as orthorhombic $Gd_2Si_2O_7$ and SiO_2glass by the EDX measurement and powder XRD pattern of the grown eutectic. Rod-phase was $SiO_2$and matrix phase was $Gd_2Si_2O_7$according to the EDX measurement result. The diameter of SiO_2scintillation fiber was approximately 700 nm (Fig.2) and dispersed well on the transverse cross-section.

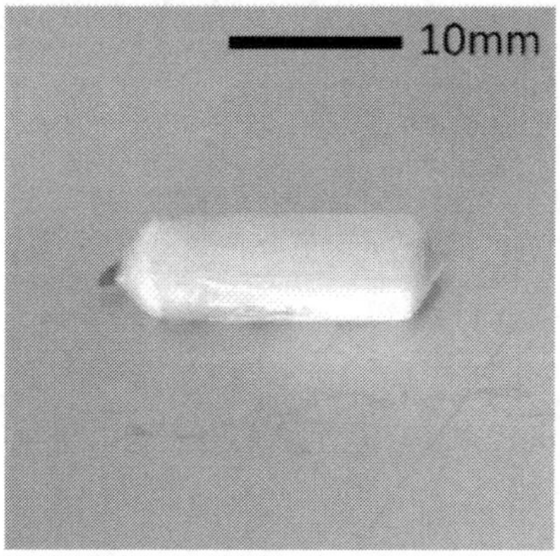

Figure 1. Photograph of the grown $Ce:Gd_2Si_2O_7/SiO_2$ eutectic

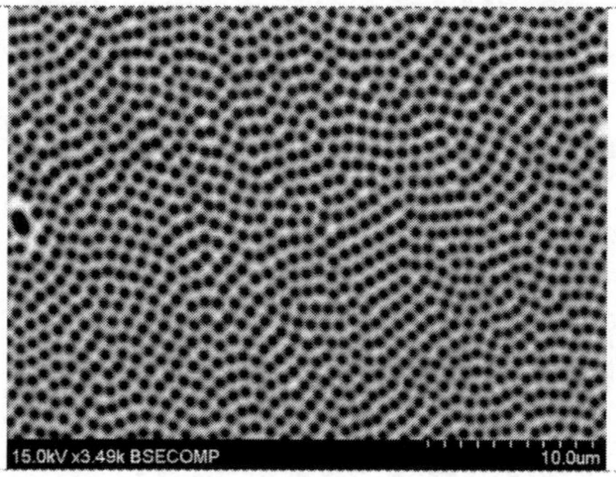

Figure 2. BEI of transverse cross-section of the grown $Ce:Gd_2Si_2O_7/SiO_2$ eutectic. White phase was $Ce:Gd_2Si_2O_7$ and brack phase was SiO_2.

3.2. Luminescence and gamma-ray response measurement procedurembering

Figure 3 shows radioluminescence spectra of the $Ce:Gd_2Si_2O_7/SiO_2$ eutectic samples. Expected Ce^{3+} 4f5d emission have been observed in 350-440nm peaking at 400nm. Figure 4 exemplifies energy spectra of the $Ce:Gd_2Si_2O_7/SiO_2$ eutectic samples. The samples irradiated by ^{137}Cs, where 662 keV. The sample showed light yield of around 14,000-16,000 photons/MeV considering with the emission wavelength of the samples and quantum efficiency of the PMT (45%@390nm, 32%@480nm). Scintillation decay curves were obtained by using the PMT and digital oscilloscope and the sample was irradiated by ^{137}Cs (Figure 5). Scintillation decay time was 46.3ns(21%) 249ns(79%).

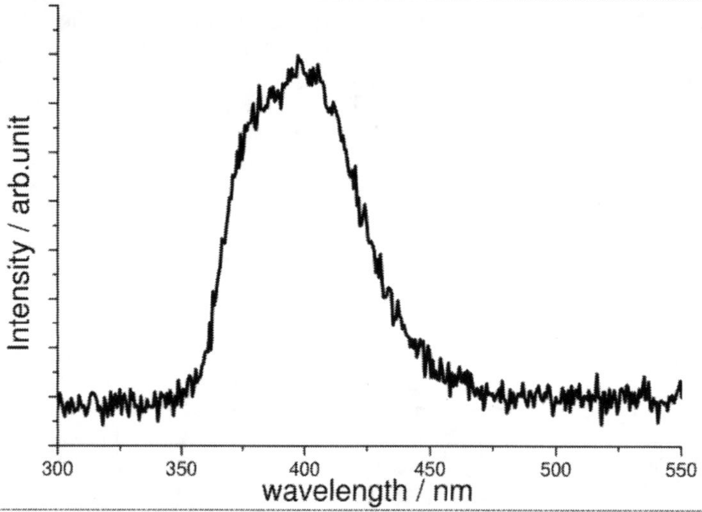

Figure 3. Radioluminescence spectra of the $Ce:Gd_2Si_2O_7/SiO_2$ eutectic sample.

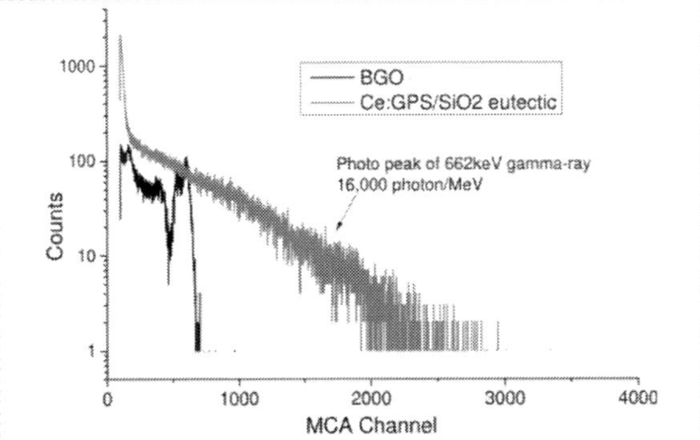

Figure 4. Energy spectra of the Ce:$Gd_2Si_2O_7/SiO_2$ eutectic sample and BGO standard excited by a 662 keV gamma ray.

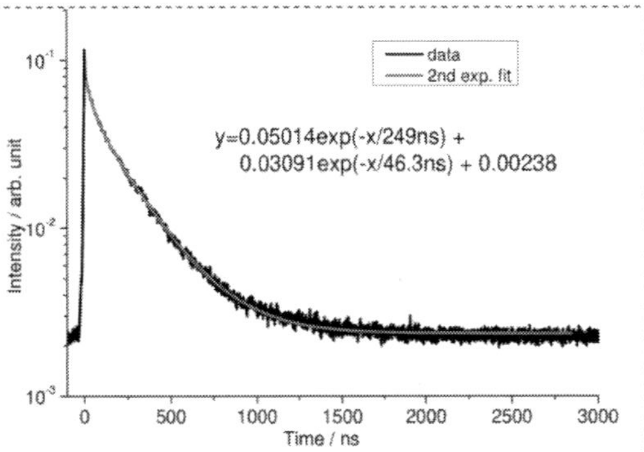

Figure 2. Scintillation decay curve of the Ce:$Gd_2Si_2O_7/SiO_2$ eutectic sample excited by a 662 keV gamma ray.

4. Conclusion

The sub-micron diameter SiO_2 glass fibers surrounded with Ce:GPS matrix from directionally solidified eutectics were fabricated. In this crystal, the length of longest fiber of SiO_2 along the growth direction was about 200 μ m or so. Expected Ce^{3+} 4f5d emission have been observed in 350-440nm peaking at 400nm. The sample showed light yield of around 16,000 photons/MeV. Scintillation decay time was 46.3ns(21%) 249ns(79%). By using DSE system of Ce:$Gd_2Si_2O_7/SiO_2$ eutectic, high resolution X-ray imaging can be expected.

5. References

[1] M. Nikl, "Scintillation detectors for X-rays",Meas. Sci. Technol. Vol.17, pp.R37-R54, Feb. 2006
[2] C. L. Melcher, et al.,IEEE Trans. Nucl. Sci., vol. 39, no. 4, pp.502–505, Aug. 1992.
[3] P. Lecoq, et al., IEEE Trans. Nucl. Sci., vol. 49, no. 4, pp. 1651–1654, 2002.
[4] M. Moszynski, et al.,Nucl. Instr. And Meth. A 385, pp. 123–131, Jan. 1997.
[5] K. S. Shah, et al.,IEEE Trans. Nucl. Sci., vol. 50, no. 6, pp. 2410–2413, Dec. 2003.
[6] Yagi, K. et al., IEEE Nuclear Science Symp. Conf. Rec., 2006, p. 1160.
[7] S. Kawamura, et al., Nucl. Instrum. Methods Phys. Res., Sect. A 583 (2007) 356.
[8] J. H. Kaneko, et al., Nucl. Instrum. Methods Phys. Res., Sect. A 529 (2004) 307.
[9] N. A. Toropov, et al.,Russ. Chem. Bull. 10 (1961) 497.
[10] A. Suzuk, et al., *Appl. Phys. Express*, **5**, 2012, 102601
[11] S. Kurosawa et al., Nucl. Instrum. Methods Phys. Res., Sect. A 744 (2013) 30
[12] S. Kurosawa et al., Radiation Detectors And Their Uses Proceedings of The 27th Workshop on Radiation Detectors and Their Uses in KEK, accepted.
[13] S. Blahuta *et al., IEEE Trans. Nucl. Sci.* Vol.60, No4, 2013, pp. 3134-3140

Twenty Years of Laser Cooling of Solids

Galina Nemova[*1] **and Raman Kashyap**[1,2]

[1]Department of Engineering Physics & [2]Department of Electrical Engineering, Polytechnique Montréal, P.O. Box 6079, Station Centre-ville, Montréal, Canada, H3C 3A7

E-mail: galina.nemova@polymtl.ca

Abstract. Laser induced cooling of solids or optical refrigeration is an area of optical science investigating interaction of light with condensed matter. This addresses a very important practical issue: design and construction all optical solid-state cryocoolers, which are compact devices, free from mechanical vibrations, moving parts, or fluids. They are based on reliable diode pump technology and in the most part free from electromagnetic interference in the cooled area. The optical cryocooler has a broad range of applications such as in the development of biomedical sensing, magnetometers for geophysical sensors and other sensors, satellite instrumentations where compactness and the lack of vibration are key parameters. The operation of these devices is based on *anti-Stokes fluorescence* also known as *luminescence upconversion*, in which light quanta in the red tail of the absorption spectrum are absorbed in a material from a pump laser and by adding thermal energy, blue-shifted photons are spontaneously emitted. Laser cooling of solids can be realized in rare-earth doped low phonon energy glasses and crystals as well as in direct band gap semiconductors. Both of these areas are very interesting and important and are discussed in this article.

1. Introduction

The term "laser cooling" must not be confused with the cooling of lasers by external means. This type of cooling is *with* a laser. Laser cooling of atoms and ions, for example, Doppler cooling [1] or Sisyphus cooling [2] are well known today. This area of science has progressed immensely in the last two decades facilitating the observation of Bose-Einstin condensates and many related phenomena and resulted in Nobel prizes in physics in 1997 [3] and 2001 [4]. It is worth remembering, that nearly half a century before Doppler cooling of atoms was observed, cooling of solids by anti-Stokes fluorescence was proposed and predicted by Peter Pringsheim in 1929 [5]. It has been shown that some materials emit light at shorter wavelengths than that with which the material is illuminated due to thermal interactions accompanied by phonon absorption with the excited atoms. Typically, this can be realized in rare-earth (RE) doped low phonon energy hosts (figure 1a) as well as in direct band gap semiconductors (figure 1b). Laser cooling of solids is similar to laser cooling of atoms and ions: photons in the red tail of the absorption spectrum are absorbed from a nominally monochromatic source followed by anti-Stokes fluorescence of blue-shifted photons with higher energy. This process is well known as anti-Stokes fluorescence rather than Stokes fluorescence in which the emitted wavelength is longer than the absorbed one. Since anti-Stokes fluorescence involves the emission of higher energy photons than those which are absorbed, the extra energy has to be extracted from the solid-state lattice in the form of the phonons, the quanta of vibrational energy that generates heat. The net anti-Stokes fluorescence can cause removal of energy from the material illuminated and results in its refrigeration. In the case of laser cooling in RE doped hosts the ground and the first excited manifolds are involved in the cooling cycle. In cooling in semiconductors, the conduction and valence bands of direct band gap semiconductors are involved (figure 1). The idea of cooling solids with anti-Stokes fluorescence stimulated an active discussion between Vavilov and Pringsheim concerning the

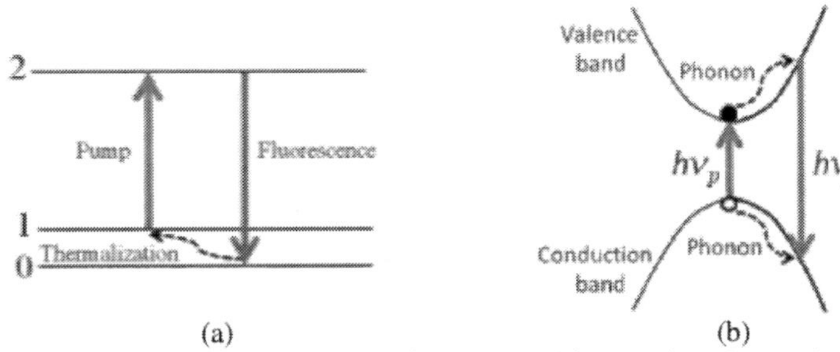

(a) (b)

Figure 1. Cooling cycles in a RE doped host (a) and in a semiconductor (b) with $h\nu_p$ absorbed energy followed by emission of an up-converted luminescence photon at $h\nu_f$.

thermodynamic aspect of this process. Vavilov believed [6, 7] that optical cooling of solids by anti-Stokes fluorescence contradicts the second law of thermodynamics. Vavilov argued that the cycle, which includes excitation and fluorescence, is a reversible one and the energy yield of greater than unity would be equivalent to the complete transformation of heat to work. Pringsheim contended [8] that in anti-Stokes cooling, a monochromatic and unidirectional input beam transforms into isotropic broadband fluorescence, and therefore this process must be irreversible, akin to Feynmann's experiment with a water sprinkler [9]. In 1946, Landau [10] concluded this controversy by providing a thermodynamic theory of the process by considering the entropy of the incident and scattered light. He has proved that the fluorescence carries much higher entropy than the beam of the pump laser; it is possible for the energy yield to exceed unity.

2. Laser cooling in rare earth doped low phonon glasses and crystals

2.1. Short history of laser cooling in RE-doped hosts

An effort to realise experimentally optical refrigeration of solids has a long history. In 1950, Kastler [11] and in 1961, Yatsiv [12] suggested the use of RE doped solids for optically cooling for the first time. The main advantage of RE ions is the optically active 4*f* electrons shielded by the filled 5*s* and 5*p* outer shells, which limit interaction with the lattice surrounding the RE-ion and suppress non-radiative decay. Hosts with low phonon energy such as fluoride glasses or crystals can provide high quantum efficiency by diminishing non-raditive decay. In 1968, Kushida and Geusic [13] in an attempt to cool a Nd^{3+}:YAG crystal observed a reduction of heating.

For the first time *net* laser cooling of solids with anti-Stokes fluorescence was realized experimentally only in 1995 by Epstein's research team at Los Alamos National Laboratory (LANL). In this proof-of-concept experiment, an ytterbium-doped fluorozirconate ZrF_4-BaF_2-LaF_3-AlF_3-NaF-PbF_2 (ZBLANP) glass sample of volume 43 mm^3 was cooled by only 0.3 K below room temperature [14]. Since the first experimental observation optical refrigeration has been realized also with Yb^{3+},

Tm^{3+} and Er^{3+} ions doped in the wide variety of low phonon energy glasses and crystals (see for e.g. reviews [15-17]). The record temperature drop for thulium ions is 24K in a Tm^{3+}:ZBLAN sample starting from room temperature [18]. In this experiment the transition between $^3H_6 \rightarrow {}^3F_4$ levels of Tm^{3+} ions was involved in the cooling cycle. The record for the temperature drop for the erbium ions is only 6K staring from room temperature in $Er^{3+}:Cs_2NaYCl_6$ [19]. In this experiment the transition between $^4I_{15/2} \rightarrow {}^4I_{9/2}$ levels of Er^{3+} ions was used in the cooling cycle. Ytterbium ions are the most suitable for laser cooling applications. They have only one excited manifold free from excited state absorption, which can be a source of undesirable heat generation in the system. It is not surprising that the today's record temperature of optical refrigeration of 93K±1K starting from room temperature is reached with ytterbium ions. This temperature has been achieved with 10% Yb^{3+}:YLF sample in 2014 [20]. All these experiments have been done within vacuum chambers with only a radiative heat load. In 2013, Yb^{3+}:YAG sample was cooled in air for the first time with a temperature drop 8.8K starting from room temperature [21]. The progress in laser cooling of RE-doped glasses and crystals is summarized in figure 2. As seen in figure 2, considerable progress has been made in recent years in laser cooling of RE-doped crystals. The long-range order in crystals in comparison with glasses results in a decrease in the inhomogeneous broadening and increases the quantum efficiency of cooling.

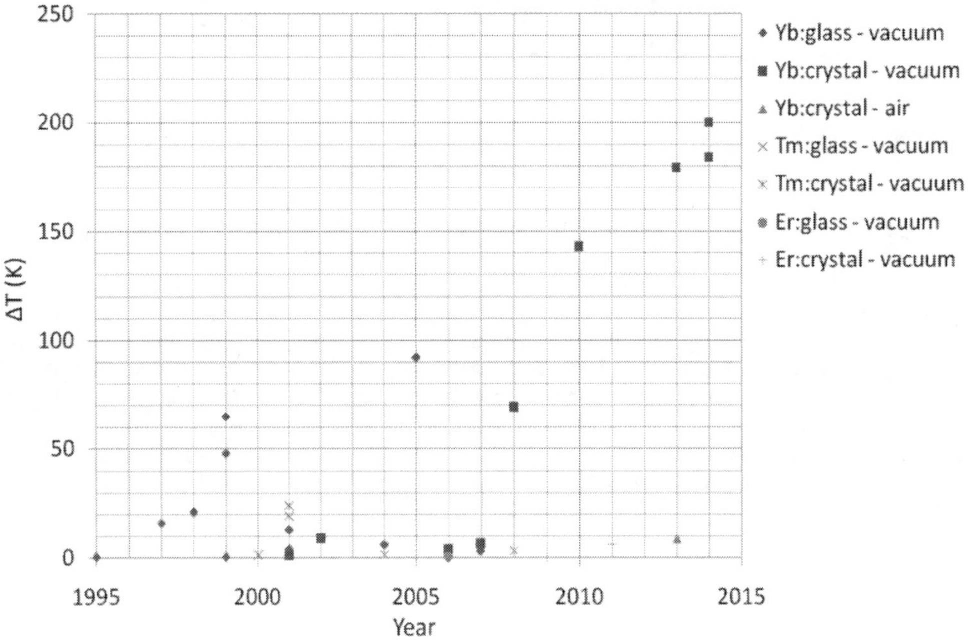

Figure 2. Progress in laser cooling of rare earth doped glasses and crystals. ΔT is the temperature drop starting from room temperature.

3. Laser cooling in direct band gap semiconductors

As shown in the previous part of the paper, the progress in laser cooling of RE-doped glasses and crystals has been slow but good; however, laser cooling of semiconductors is more attractive in a number of applications. Iindistinguishable charge carriers in the Fermi-Dirac distributions allow semiconductors to be cooled to lower temperatures than RE-doped materials. Due to Boltzmann distribution, the population of the highest energy levels of the ground manifold in the RE-doped systems decreases dramatically as soon as the temperature is lowered. The cooling cycle in RE-doped hosts ceases when the Boltzmann constant times the lattice temperature, k_BT, becomes comparable to the width of the ground state manifold. No such limitation exists in undoped semiconductors where temperatures as low as 10K may be achieved [22]. Semiconductor coolers provide more efficient absorption of the pump radiation. The radiative recombination rates in semiconductors are much faster than in RE ions. Optical cryocoolers based on a semiconductor can be easily integrated with electronic and optical devices. However, significant challenges remain, since the high refractive index of semiconductors leads to energy trapping within the solid causing reabsorption and unwanted heating.

3.1. Short history of laser cooling in semiconductors

There are several theoretical approaches for analysing laser cooling in semiconductors. The approach based on a rate-equation theory, in which the effects of nonradiative, radiative and Auger recombination are taken into account, were presented in 1996 by Oraevsky [23]; and in 1997 by Gauck and colleagues [24]. In 1997, Rivlin and Zadernovsly proposed an energy balance theory for semiconductors [25]. The possibility of excitonic luminescence in the limit of vanishing electron-hole pair density was taken into account, but not of excitonic absorption. In 2004, Sheik-Bahae and Epstein extended a rate-equation theory to analyze laser cooling of bulk GaAs based on a microscopic theory for the luminescence and absorption spectra [26]. They took into account the effect of luminescence reabsorption. The main feature of the rate-equation approach is its simplicity. This theory neglects the change of carrier distribution with temperature. It is valid only for a small change in the temperature. In 2004 and 2005, comprehensive theoretical studies of laser cooling of semiconductors was carried out by a research team at Kirkland Air Force Base [27, 28]. The change in the carrier distribution with temperature as well as temperature diffusion of phonons was calculated. It was shown that the lattice and the carriers can have different temperatures varying in space and time.

In 2005, a four-step model for spatially selective laser cooling of carriers in undoped semiconductor AlGaAs/GaAs/AlGaAs quantum wells was developed in a paper [29]. In 2007, Li developed a cooling theory for semiconductor quantum wells at the Hartree-Fock level [30]. In these papers, no indication that quantum wells are better suited to optical refrigeration than bulk semiconductors is found under the assumption of quasi-thermal equilibrium.

In 2008, Eliseev theoretically investigated heavily doped semiconductors [31]. It was shown that the minimum temperature, which can be reached in such a semiconductor, is about 60-120K depending on the dopant concentration.

Most of the papers devoted to laser cooling in semiconductors considered the cooling process with GaAs (group III-V semiconductor). However in 2013, in a surprising but exciting advance, the first *experimental* observation of laser cooling with semiconductor CdS (group II-VI semiconductor) *nanobelts* was announced by Zhang and colleagues [32]. A net cooling by about 40K starting at 290 K was reported. It was shown that strong exciton - longitudinal optical phonon (LOP) coupling in CdS nanobelts permits the resonant annihilation of *multiple* LOPs in the luminescence up-conversion processes, such that in a CdS structure, each cooling cycle can remove more heat from the sample than in a GaAs sample.

3.2. Principles of laser cooling in semiconductor nanobelts

Let us consider an undoped semiconductor pumped with a laser at frequency ν_p. The rate equation for the electron-hole pair density, N, is given by the relationship [26]:

$$\frac{dN}{dt}=\frac{\alpha(\nu_p,N)}{h\nu_p}I-AN-BN^2-CN^3+(1-\eta_e)BN^2, \tag{1}$$

where $\alpha(\nu_p,N)$ is the interband absorption coefficient. η_e is the probability that an emitted photon escapes from the semiconductor without being reabsorbed. The recombination process consists of linear nonradiative recombination with the rate AN, a "bimolecular" radiative recombination with the rate BN^2, and an Auger nonradiative recombination with the rate CN^3. A, B and C are decay coefficients, which characterize the fundamental properties of semiconductors. The cooling efficiency can be presented as [26]:

$$\eta_{cool} = \eta_{ext}\frac{\tilde{\nu}_f}{\nu_p} - 1, \quad (2)$$

where η_{ext} is the external quantum efficiency described by the relation:

$$\eta_{ext} = \frac{\eta_e BN^2}{AN + \eta_e BN^2 + CN^3}. \quad (3)$$

Let us take the first derivative of the external quantum efficiency (3) with respect to the electron-hole pair density N. The optimum electron-hole pair density, N_{opt}, for the maximum external quantum efficiency can be determined as

$$N_{opt} = \sqrt{\frac{A}{C}}. \quad (4)$$

From (3) and (5) the external quantum efficiency can be obtained at N_{opt}:

$$\eta_{ext}(N_{opt}) = 1 - 2\frac{\sqrt{AC}}{\eta_e B}. \quad (5)$$

In relation (5) the nonradiative recombination coefficient, A and the extraction efficiency η_e are functions of the thickness of the sample. *Surface* nonradiative recombination, A_s dominates coefficient, A [33]. The surface nonradiative recombination A_s is a function of the thickness of the sample, d, and the surface recombination rate, S:

$$A_s = \frac{S}{d}. \quad (6)$$

Following paper [33] and considering $\eta_e = \eta_0\exp(-\alpha_{eff}d)$ one can obtain

$$\eta_{ext}(N_{opt}) = 1 - 2\frac{\sqrt{2SC}}{\sqrt{dB\eta_0\exp(-\alpha_{eff}d)}}. \quad (7)$$

For CdS with the values $B = 10^{-11}$ cm^3/s, $C = 10^{-30}$ cm^6/s, $S = 250$ cm/s, $\alpha_{eff} = 4.6$ μm^{-1}, and $\eta_0 = 0.996$ the maximum external quantum efficiency can be achieved with a thickness of the sample of 110nm. The source of the decreased external quantum efficiency, η_{eff}, is an increase in the surface nonradiative recombination. A decrease in η_{eff} of the thick nanobalts is due to the reabsorption process. It is expected that in the near future progress in the development of laser cooling of semiconductors will be associated first of all with cooling of nanosize semiconductor samples.

4. Conclusions

We have reviewed the progress in the very interesting and promising area of laser physics known as *laser induced cooling of solids* or *optical refrigeration*. Recent developments in our understanding of the science stimulated by a number of applications such as an all optical solid state cryocooler, which is free from mechanical vibrations, moving parts and liquids, is progressing fast. A significant advantage of these devices is that they use compact long lived diode pump lasers. Thermoelectrical coolers (TEDs) are free from mechanical vibrations as well, but present day TECs can reach only ~180K; although recent developments may lower this limit if material can be better engineered. Mechanical coolers can reach as low as ~10K, but they are large and cause mechanical vibrations, which are undesirable characteristics for a number of applications. Optical cryocoolers may be beneficial for satellite instrumentation and small sensors, where compactness and the lack of vibrations are important factors. Optical solid state cryocooler have already reached an extremely low temperature of 93K±1K in a small sample of a crystal, but the problem of scaling still remains a challenge. An optical cooler based on a semiconductor should reach as low as ~10K. The realization of laser cooling with semiconductors is a significantly more difficult task than laser cooling in RE doped glasses or crystals. At the present time only nanobelt semiconductor samples have been cooled. Several methods to improve the cooling process have been proposed. Multipassing of the pump beam with an extra-cavity as well as with the intra-cavity have been proposed [34]. Development of new artificially engineered structures with suitable for laser cooling parameters may prove to be very promising for the future design and implementation of all optical cryocoolers.

References

[1] Hänsch T W, Schawlow A L 1975 *Opt. Commun.* **13** 68-69
[2] Dalibard J, Cohen-Tannoudji C 1985 *J. Opt. Soc. Am. B* **2** 1707-1720
[3] Chu S, Cohen-Tannoudji C & Philips W D 1997 *Nobel Prize in Physics* http://nobelprize.org
[4] Cornell E A, Ketterle W & Weiman C E 2001 *Nobel Prize in Physics* http://nobelprize.org
[5] Pringsheim P 1929 *Z. Phys.* **57** 739-746
[6] Vavilov S I 1945 *J. Phys. USSR* **9** 68-73
[7] Vavilov S I 1946 *J. Phys. USSR* **10** 499-501
[8] Pringsheim P 1946 *J. Phys. USSR* **10** 495-498
[9] Gleick J 1993 *Genius: Richard Feynman and modern physics* In Princeton, *Forward of backward* (Abacus Press) pp 106-108
[10] Landau L D 1946 *J. Phys. USSR* **10** 503-506
[11] Kastler A 1950 *J. Phys. Radium* **11** 255-265
[12] Yatsiv S 1961 Anti-Stokes fluorescence as a cooling process *Advances in Quantum Electronics* (ed. Singer, J.R.) pp 200-213, Columbia Univ. Press, New York
[13] Kushida T, Geusic J E 1968 *Phys. Rev. Lett.* **21** 1172-1175
[14] Epstein R I, Buchwald M I, Edwards B C, Gosnell T R, Mungan C E 1995 *Nature (London)* **377** 500-502
[15] Sheik-Bahae M, Epstein R I 2009 *Laser Photon. Rev.* **3** 67-84
[16] Nemova G, Kashyap R 2010 *Rep. Prog. Phys.* **73** 086501:1-20
[17] Seletskiy D V, Hehlen M P, Epstein R I, Sheik-Bahae M 2012 *Adv. Opt. Photon.* **4** 78-107
[18] Hoyt C W 2003 Laser cooling in thulium-doped solids. *PhD Thesis* University of New Mexico
[19] Hasan Z, Qiu Z, Lynch J 2011 *Proc. of SPIE* **7951** 79510E:1-7
[20] Melgaard S D, Albrecht A D, Hehlen M P, Seletskiy D V, Sheik-Bahae M, dx.doi.org/10.1364/CLEO_QELS.2014.FTh4D.4
[21] Soares de Lima Filho E, Nemova G, Loranger S, Kashyap R 2013 *Opt. Express* **21** 24711-24720
[22] Rupper G, Kwong N H, Binder R 2006 *Phys. Rev. Lett.* **97** 117401:1-4
[23] Oraevsky A N 1996 *J. Russ. Laser Res.* **17** 471-479
[24] Gauck H, Gfroerer T H, Renn M J, Cornell E A, Bertness K A 1997 *Appl. Phys. A., Mater. Sci. Process.* **64** 143-147
[25] Revlin L A, Zadernovsky A A 1997 *Opt. Commun.* **139** 219-222
[26] Sheik-Bahae M, Epstein R I 2004 *Phys. Rev. Lett.* **92** 247403:1-4
[27] Huang D, Apostolova T, Alsing P M, Cardimona D A 2004 *Phys. Rev. B* **70** 033203:1-4
[28] Apostolova T, Huang D, Alsing P M, Cardimona D A 2005 *Phys. Rev. A* **71** 013810: 1-6
[29] Huang D, Apostolova T, Alsing P M, Cardimona D A 2005 *Phys. Rev. B* **72** 195308:1-8
[30] Li J 2007 *Phys. Rev. B* **75** 155315:1-10
[31] Eliseev P G 2008 *Opto-Electron. Rev.* **16** 199-207
[32] Zhang J, Li D, Chen R, Xiong Q 2013 *Nature* **493** 504-508
[33] Li D, Zhang J, Xiong Q 2013 *Opt. Express* **21** 19302-19310
[34] Heeg B, Rumbles G 2002 *J. Appl. Phys.***91** 3356-3362

Journal of Physics: Conference Series **619** (2015) 012038 doi:10.1088/1742-6596/619/1/012038

X-ray Induced Luminescence Spectroscopy of Samarium Doped Barium Sulfate Prepared by Sintering Method

T Kumeda[*1]**, K Maeda**[1]**, Y Shirano**[1]**, K Fujiwara**[1]**, K Sakai**[2]**, T Ikari**[1]

[1] Faculty of Engineering, University of Miyazaki, 1-1 Gakuen-Kibanadai-Nishi, Miyazaki City 889-2192, Japan
[2] Cooperative Research Center, University of Miyazaki, 1-1 Gakuen-Kibanadai-Nishi, Miyazaki City 889-2192, Japan

E-mail: tc13006@student.miyazaki-u.ac.jp

Abstract. X-ray induced luminescence (XL) properties of phosphor materials made of samarium doped barium sulfate have been investigated. The samples were prepared by sintering method heated at 900-1250 °C for 3 hours in air from the mixture of $BaSO_4$ and Sm_2O_3. The concentration of Sm were prepared from 0.01-6 at.%. In as-prepared sample, the Sm^{3+} was detected by photoluminescence (PL). The PL intensity is maximum about 2 at.% with Sm, and then starts decreasing. The PL intensity showed concentration quenching. The XL observed Sm^{2+} and Sm^{3+} ions. The XL was shown from the sample sintered up to 1200 °C. The XL intensity increased with Sm concentration up to 1 at.%. The intensity was almost constant larger than 1 at.% Sm. These concentration dependences is different since the X-ray energy absorbed to the host material at once, and the energy transferred to both Sm^{3+} and Sm^{2+} ions. Sm doped $BaSO_4$ is found a host for XL phosphor materials.

1. Introduction

X-ray imaging techniques are used in such as medical fields and non-destructive testing. Various materials such as $BaSO_4$ and $SrSO_4$ have been developed for X-ray detection materials [1]. Indirect imaging method for obtaining X-ray images using fluorescent materials is required high sensitivity phosphor and, large area and high resolution [2]. In addition, an observation of X-ray induced luminescence (XL) has a merit of real-time measurement.

Rare earth ions exhibit high intensity and stable optical property in host materials. Samarium (Sm) ions emitted in the red region with high efficiency, and it can be coupled with Si photo-detector. Samarium ions were divalent and trivalent state in phosphor. The photoluminescence from Sm^{2+} ions are more efficient than that of Sm^{3+} ions [3]. After X-ray irradiation, some electrons can be trapped by Sm^{3+} ions, leading to the formation of Sm^{2+} ions while holes are trapped by other defects [4]. However, divalent ions are usually unstable as compared to trivalent. Following points are required to host material of X-ray phosphor of guest ions: a good solubility, a low phonon energy material and contain a heavy atom. In our previous paper [5, 6], we have studied X-ray phosphor of Sm -doped BaS. However Sm^{2+} ions were not presented in the material. In this paper, we choose the other host material, $BaSO_4$, because the emission from Sm^{2+} ions has been reported in the photoluminescence (PL) [7]. We study the optical and X-ray luminescence properties of Sm -doped $BaSO_4$.

2. Experimental

2.1. Sample preparation
Samples were made by the mixture of barium sulfate ($BaSO_4$, 99.9%) and samarium oxide (Sm_2O_3, 99.9%) compounds. The concentration of Sm were prepared from 0.01-6 at.%. These were weighted with a total weight of 3 g. These were ground in an agate mortar. The mixtures were synthesized by solid state reaction method at 900-1250 °C for 3 hours using an electric furnace in air. The obtained bulk samples were polished to a disk shape with, about 2 mm thickness and 5 mm diameter.

2.2. Measurements and characterization
The samples were characterized by X-ray diffraction method (XRD) obtained by PANalytical, X'Pert Pro using a CuKα radiation (45 kV, 40mA, λ = 0.154 nm) with a Ge monochromator at room temperature. PL spectra were measured in the range of 530-770 nm at room temperature using TRIAX320 Imaging Monochromator (Jobin yvon) with a cooled CCD detector. The PL excitation source used a 405 nm laser diode operating at a power of 24 mW. The PL intensity was calculated from the integrated intensity. The intensity of each sample was calibrated to a laboratory standard sample to allow a meaningful comparison.

The X-ray luminescence (XL) was measured by following method. The X-ray source was that for the same source as XRD measurement. The X-ray irradiate to 45 degrees of incidence to a sample. A digital camera (Canon EOS50D) was set up parallel to the sample plate surface. The red light-emitting from the sample during the X-ray radiation was taken as a photograph in the dark condition. XL measurements were carried out in situ. An expose time is in the range of 1 to 8 sec. Emission by Sm^{3+} and Sm^{2+} ions were measured by using a band pass filter with a transmission wavelength is 600nm and 700nm, respectively. After that, a luminescent image is electrical developed to be proportional to light intensity in image analysis software on PC. The XL intensity was calculated from the average of emission intensity of the sample surface. Finally, the XL intensity of the sample was determined by the slope of the light intensity versus the exposure time. XL spectra were measured with a monochromator (Ocean Optics, USB4000-UV-VIS) of which focal length is 6.8 cm through an optical fiber.

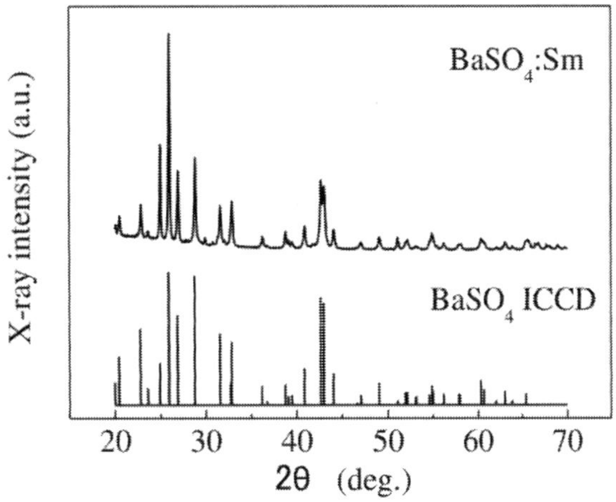

Figure 1. X-ray diffraction patterns of $BaSO_4$ ceramics samples doped (upper curve) 2 at.% of Sm and (lower curve) ICCD data base of card number 01-080-0512 .

Journal of Physics: Conference Series **619** (2015) 012038 doi:10.1088/1742-6596/619/1/012038

3. Results and discussions

3.1. Characterization of samples

The X-ray diffraction of Sm doped $BaSO_4$ sample was shown in Figure 1. From the analysis of the XRD pattern of Sm doped $BaSO_4$ compound, it was found that the sample belongs to the orthorhombic structure at room temperature in correspondence with the ICCD data base of card number 01-080-0512. It was found that the sample is a single phase of $BaSO_4$.

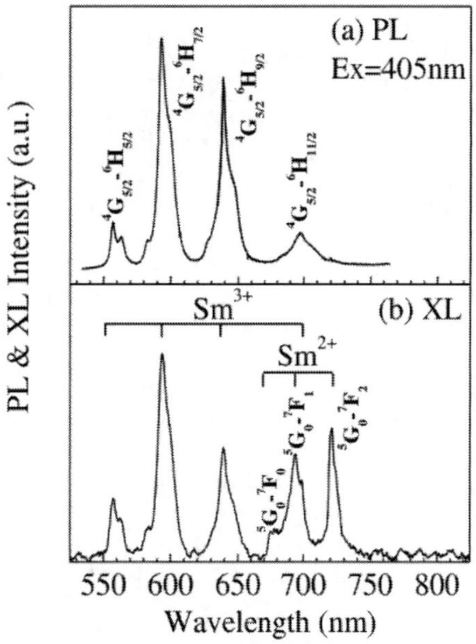

Figure 2. PL spectra excited at 405 nm light is shown in (a), and XL spectra is shown in (b).

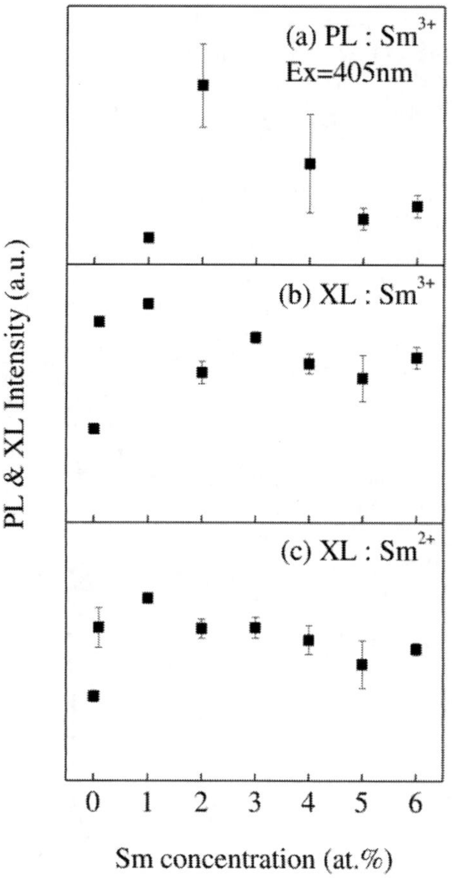

Figure 3. PL and XL intensity of $BaSO_4$:Sm ceramics as function of Sm concentration.

3.2. Photoluminescence

Figure 2 (a) showed the PL spectra of $BaSO_4$ doped Sm 2at.% excited at 405 nm. Four luminescence bands are obtained at around 557, 593, 639, 698 nm. These correspond to $^4G_{5/2}$-^6H_J (J: 5/2, 7/2, 9/2, 11/2) transitions of Sm^{3+} [8]. All bands from Sm^{3+} consist of many peaks. The spectral bands from Sm^{2+} where are at 680, 694 and 721 nm [9] have not been observed. The PL intensity strongly depended on synthesized temperature. The PL intensity gradually increased with synthesized temperature increasing. The PL intensity at synthesized temperature of 1250 °C was ten times larger than that of 900 °C .

The Sm concentration dependence of PL intensity excited at 405 nm is shown in Figure 3 (a). The PL intensity increased with Sm concentration up to 2 at.%, and then started decreasing with Sm. The phenomenon of these Sm concentration dependences seems as known effect of the concentration quenching [10].

3.3. X-ray luminescence

The red XL light could be seen in the dark during exposing X-ray. The X-ray luminescence was obtained from Sm doped $BaSO_4$ sample sintered up to 1200 °C. The XL spectrum of $BaSO_4$ including 2 at.% of Sm are shown in Fig. 2 (b). The spectra had a typical four luminescence bands of Sm^{3+} presented in the PL spectrum, previously. Also, three luminescence bands are obtained at around 680, 694, 721 nm. These correspond to $^5D_0$-^7F_J (J: 0, 1, 2) transition of Sm^{2+} [10]. Some Sm^{3+} ions are reduced into Sm^{2+} ions by X-ray irradiation [7].

The Sm concentration dependence of XL intensity is shown in Figure 3 (b) and (c). The XL intensity increased with Sm concentration up to 1 at.%. The intensity was almost constant larger than 1 at.% Sm. The dependence of XL did not show concentration quenching different to that of PL. The energy of excited light corresponds to the energy difference between $^6H_{5/2}$ and $^6P_{3/2}$ in Sm^{3+} ion [11]. Then, the light could directly excite Sm^{3+} ions. In the case of XL [3], the X-ray energy absorbed to the host material at once, and the energy transferred to both Sm^{3+} and Sm^{2+} ions. The mechanism is indirectly excitation. It seems that these concentration dependence is dissimilar since the difference in the emission mechanism of PL and XL. Sm doped $BaSO_4$ is expected as the XL phosphor.

4. Conclusions

We have investigated the optical properties and X-ray luminescence of $BaSO_4$ ceramics doped Sm. All samples sintered up to 1200 °C showed X-ray luminescence. The PL spectral bands were identified to the transitions in Sm^{3+} ions. The XL spectral bands were identified to the transitions in Sm^{3+} and Sm^{2+} ions. Some Sm^{3+} ions are reduced into Sm^{2+} ions by X-ray irradiation. The dependence of XL did not show concentration quenching. Sm doped $BaSO_4$ is found a host for XL phosphor materials.

References

[1] Yamashita N, Yamamoto I, Kiyotake N 1985 *Jpn. J. Appl. Phys.* **24** pp 1174-1180
[2] Lanca L and Silva A 2009 *Radiography* **15** pp 58-62
[3] Koughia C, Edgar A, Varoy C R, Okada G, Seggern H von, Belev G, Kim C Y, Sammynaiken R and Kasap S 2011 *J. Am. Ceram. Soc.* **94** pp 543-550
[4] Wang J, Li Y, Huang Y, Shi L and Seo J H 2010 *Mater. Chem. Phys.* **120** pp 598-692
[5] Maeda K, Tsudome R, and Ido M 2011 *Phys. Status Solidi* C **8** pp 2692-2695
[6] Maeda K, Kawaida N, and Tsudome R 2012 *Phys. Status Solidi* C **9** pp 2271-2274
[7] Stefani R, Maia A D, Teotonio E E S, Monteiro M A F, Felinto M C F C and Brito H F 2006 *J. Solid Status Chem.* **179** pp 1086-1092
[8] Sidike A, Kusachi I and Yamashita N 2003 *Phys. Chem. Minerals* **30** pp 478-485
[9] Edgar A, Varoy C R, Koughia C, Tonchev D, Belev G, Okada G, Kasap S O, Seggern H von and Ryan M 2009 *Opt. Mater.* **31** pp 1459-1466
[10] Lei B, Man S Q, Liu Y and Yue S 2010 *Mater. Chem. Phys.* **124** pp 912-915
[11] Venkatramu V, Babu P, Jayasankar C K, Troster Th, Sievers W and Wortmann G 2007 *Opt. Mater.* **29** pp 1429-1439

Luminescent properties of $Gd_3(Al,Ga)_5O_{12}$ crystal co-doped with Ce and M^{4+}

Shunsuke Kurosawa[1,2], Kei Kamada[2,3], Jan Pejchal[2,4], Yasuhiro Shoji[1,3], Yuji Ohashi[1], Yuui Yokota[2] and Akira Yoshikawa[1,2,3]

[1]Institute for Material Reseach, Tohoku University, 2-1-1, Sendai,980-8577 Japan
[2]New Industry Creation Hatchery Center, Tohoku University, 6-6-10 Aoba, Sendai 980-8579
[3]C&A corporation, 6-6-10 Aoba, Aramaki, Aoba-ku, Sendai 980-8579, Japan
[4]Institute of Physics AS CR, 10 Cukrovarnicka, Prague 6, 16200, Czech Republic

E-mail: kurosawa@imr.tohoku.ac.jp

Abstract. Since the light output of Zr^{4+} co-doped $Ce:Gd_2SiO_5$ scintillator improved compared with the $Ce:Gd_2SiO_5$ one, the light output and other scintillation properties of M^{4+}/Ce^{3+} co-doped $Gd_3Al_2\ Ga_3O_{12}$ (GAGG), where M = Zr, Hf, were investigated. $(Gd_{1-x}, Ce_{0.01}, M_x)_3Ga_3Al_2O_{12}$ (M=Zr, Hf, x = 0.001 and 0.0002) crystals were grown by the micro-pulling down (μ-PD) method. We found these samples had longer decay time constants and the light outputs of these co-doped samples were degraded compared with that of the conventional Ce:GAGG.

1. Introduction

Scintillation materials have been investigated and applied in many fields such as homeland security, medical imaging system, oil well logging and astronomy [1, 2]. Some Compton cameras consist of scintillators, and can be applied in medical imaging or astronomy [3-8]. Ce-doped $Gd_3Al_2\ Ga_3O_{12}$ (Ce:GAGG) was found to have high light-yield (~ 46,000 photon/MeV) and good energy resolution (~ 4.9 % at 662 keV, FWHM) [9 - 11]. Ce^{3+}-dopant is crystallographically substituted for Gd^{3+} in the dodecahedral site coordinated by eight oxygen atoms. The scintillation emission is related to the 5d-4f transition of Ce^{3+} with an emission wavelength of approximately 500 to 550 nm [9].

Shimura et al. reported that $Ce:Gd_2SiO_5$ (Ce:GSO) doped with around 200 ppm Zr^{4+} has shown maximum light output which is about 20% larger than that of conventional GSO:Ce [12]. On the other hand, light output of $Ce:(Lu, Y)_2SiO_5$ (Ce:LYSO) co-doped with divalent cation (e.g. Mg^{2+}, Ca^{2+}) was improved compared with the conventional Ce:LYSO scintillator [13].
However, the light output of the Ca^{2+}/Ce^{3+} co-doped GAGG was degraded [14]. Up to now, effect of codoping with tetravalent cation has not been sufficiently investigated for Ce: GAGG. In this paper, we investigate the light output and other scintillation properties of M^{4+}/Ce^{3+} co-doped GAGG, where M = Hf.

2. Materials and Methods

(Gd_{1-x}, $Ce_{0.01}$, M_x)$_3Ga_3Al_2O_{12}$ (M= Hf, x = 0.001 and 0.0002) crystals were grown by the micro-pulling down (μ-PD) method, and the powders of (α-Al_2O_3, Ce_2O, β-Ga_2O_3, Gd_2O_3, HfO_2) of 99.99% purity were used as the crystal materials. The growth atmosphere was N_2 (flow, 1 atm) and an Ir crucible was used. To grow the crystal, $Lu_3Al_5O_{12}$ crystal was used as the seed, and the pulling rate was approximately 0.1 mm/min. Moreover, a conventional co-dopant-free Ce:GAGG crystal (x=0) was also grown by the same way as the above.

The crystals were cut and polished to the 1 mm thickness for the evaluation of their optical and scintilaltion propeties. Transmittance spectra from 200 to 700 nm were measured with a spectrophotometer V-530 (JASCO). Photo-luminessence emission spectra were obtained with a spectrofluorometer (FLS920, Edingurgh Instrument) and Xe lamp as an excitation source.

Radio-luminescence spectra excited by 5.5-MeV alpha rays from an ^{241}Am soruce were measured with the same spectrofluorometer used for the photo-luminescence measurement. To determine the light output and energy resolution, we obtained the pulse height spectra of these crystals under excitation with 662-keV gamma rays from a ^{137}Cs source. Here, the scintillation photons were detected by a photomultiplier (R7600U-200, Hamamatsu). Additionally, we used an oscilloscope (TDS 3052B, Tektronix) to measure the PMT output in order to estimate the decay times of these samples irradiated with the gamma rays (^{137}Cs). The decay curves were fitted with the following equation:

$$I(t) = A_1 \exp[-t/\tau_1] + A_2 \exp[-t/\tau_2] + c, \quad (1)$$

where A_i and τ_i denote the amplitude and decay time (i = 1, 2, $\tau_1 < \tau_2$). The term c indicates the background.

3. Results and discussions

We succeeded in growing the (Gd_{1-x}, $Ce_{0.01}$, Hf_x)$_3Ga_3Al_2O_{12}$ (x = 0.001 and 0.0002) crystals as shown in Fig 1. The polished samples with thicknesses of 1mm had good transmittance of over 70%. Emission maximum peaks for the all samples including the conventional Ce: GAGG were located at the similar wavelength positions.

Figure 2 shows 5.5-MeV alpha-ray excited radio-luminescence spectra of (Gd_{1-x}, $Ce_{0.01}$, Hf_x)$_3Ga_3Al_2O_{12}$, where x = 0.0002, 0.001, and conventional Ce:GAGG. These samples had also the similar radio-luminescence spectra when compared to the conventional Ce: GAGG. These results were the same as the results of the photo-luminessence emission spectra.

From the pulse height spectra of these crystals, all the samples co-doped with Hf^{4+} had lower light output than the conventional Ce:GAGG by 10-20% as shown in Fig. 3, Moreover, the co-doped samples had similar decay time (primary) components of roughly 100 – 110 ns to the conventional one (~120 ns) as shown in Fig. 4. Meanwhile, co-doped samples had secondary decay time components of 400 – 420 ns, although intensity ratios, defined as $A_2/(A_1 + A_2)$, were only 5%. Thus, some defects or some traps could be generated by these tetravalent cations, and light outputs of these samples were degraded slightly.

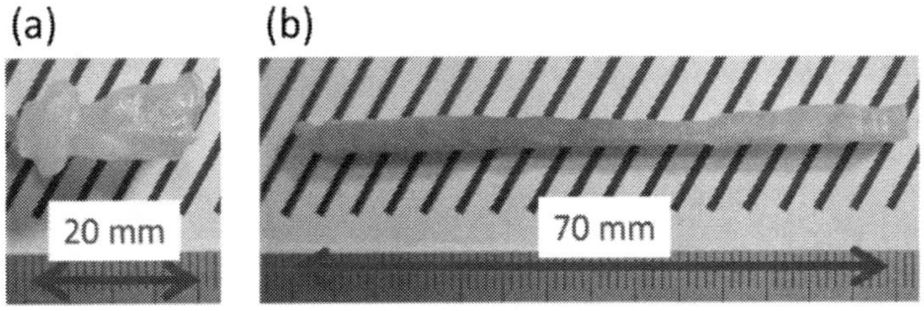

Fig. 1 Photograph of (Gd_{1-x}, $Ce_{0.01}$, Hf_x)$_3Ga_3Al_2O_{12}$, where x= 0.0002(a) and 0.001(b).

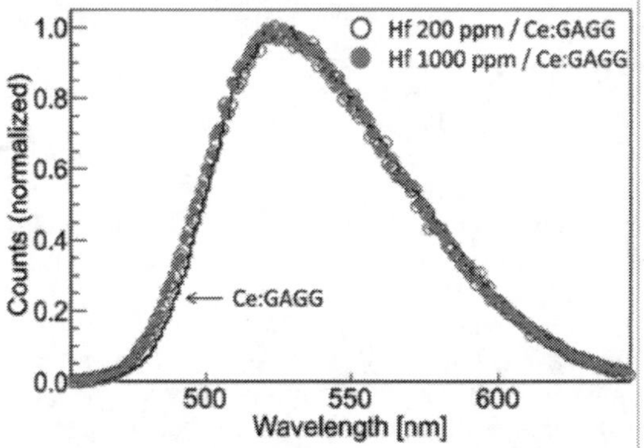

Fig. 2 Radio-luminescence spectra of $(Gd_{1-x}, Ce_{0.01}, M_x)_3Ga_3Al_2O_{12}$, where (M, x) = (Hf, 0.0002, open circles), (Hf, 0.001, closed circles) and (none, 0, solid line) excited by 5.5-MeV alpha rays from an ^{241}Am source..

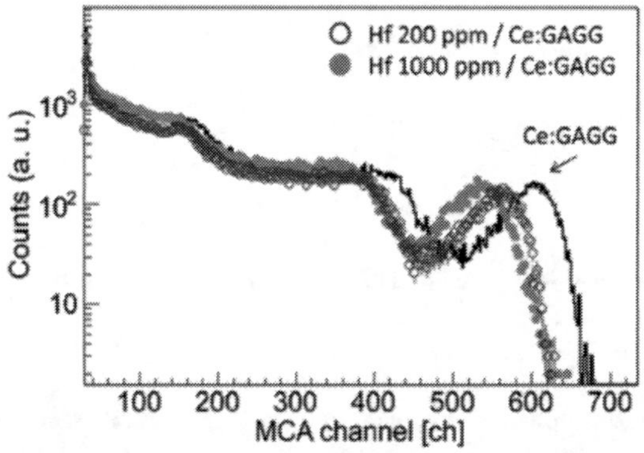

Fig. 3 Pulse height spectra of $(Gd_{1-x}, Ce_{0.01}, M_x)_3Ga_3Al_2O_{12}$, where (M, x) = (Hf, 0.0002, open circles), (Hf, 0.001, closed circles) and (none, 0, solid line) irradiated with gamma rays from a ^{137}Cs source..

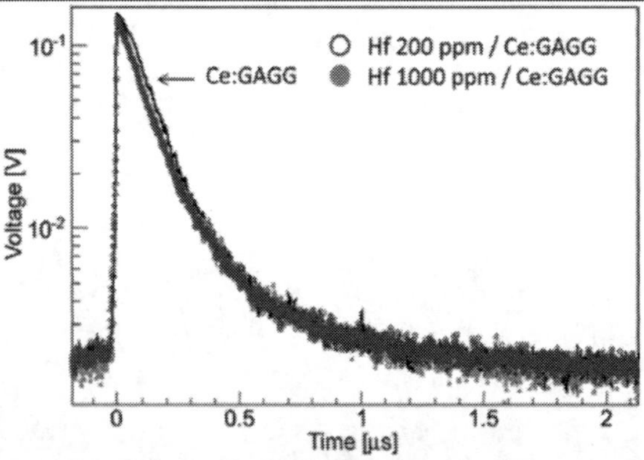

Fig. 4 Decay time corves of $(Gd_{1-x}, Ce_{0.01}, M_x)_3Ga_3Al_2O_{12}$, where (M, x) = (Hf, 0.0002, blue open circles), (Hf, 0.001, red closed circles) and (none, 0, black solid line) irradiated

3. Conclusions

$(Gd_{1-x}, Ce_{0.01}, Hf_x)_3Ga_3Al_2O_{12}$ (x = 0.001 and 0.0002) crystals were grown by the micro-pulling down method, and their emission spectra were almost the same as that of the conventional Ce-doped $Gd_3Al_2Ga_3O_{12}$ (Ce:GAGG). However, these samples had longer decay time constants and their light output of was degraded compared with the conventional Ce:GAGG.

Acknowledgements

This work is partially supported by (i) Japan Society for the Promotion of Science (JSPS) Research Fellowships for Young Scientists (S. Kurosawa), (ii) Development of Systems and Technology for Advanced Measurement and Analysis, Japan Science and Technology Agency (JST), (iii) Adaptable & Seamless Technology Transfer Program through Target-driven R&D (A-STEP), JST, (iv) the Association for the Progress of New Chemical Technology, (v) The Murata Science Foundation, (vi) Nippon Sheet Glass Foundation for Materials Science and Engineering. In addition, we would like to thank following persons for their support: Mr. Yoshihiro Nakamura in Institute of Multidisciplinary Research for Advanced Materials (IMRAM)/Tohoku University, and Mr. Hiroshi Uemura, Ms. Keiko Toguchi, Ms. Megumi Sasaki, Ms. Yuka Takeda and Ms. Kuniko Kawaguchi of IMR/Tohoku University.

References

[1] Trotta C., et al. 2007 *Nucl. Instrum. Methods Phys. Res., Sect. A.* **577** 604.
[2] Melcher C. L., Schweitzer J.S., Manente R.A., Peterson C.A., 1991 *Journal of Crystal Growth* **109** 37.
[3] Tanimori T., Kubo H., Miuchi K., Nagayoshi T., Orito R., Takada A., Takeda A., Ueno M. 2004 *New Astronomy Reviews, Volume* **48**, 263
[4] Kubo H., Hattori K., Kabuki S., Kurosawa S., Miuchi K., Nagayoshi T., Nishimura H., Okada Y., Orito R., Sekiya H., Takada A., Tanimori T., Ueno K. 2006 IEEE Nuclear Science Symposium Conference Record **1** 406.
[5] A. Takada, Kubo H., Nishimura H., Ueno K., Hattori K., Kabuki S., Kurosawa S., Miuchi K., Mizuta E., Nagayoshi T., Nonaka N., Okada Y., Orito R., Sekiya H., Takeda A., Tanimori T. 2011 *The Astrophysical Journal* **733** 13.
[6] Kabuki S., Hattori K., Kohara K., Kunieda E., Kubo A., Kubo H., Miuchi K., Nakahara T., Nagayoshi T., Nishimura H., Okada Y., Orito R., Sekiya H., Shirahata T., Takada A., Tanimori T., Ueno K. 2007 *Nucl. Instrum. Methods Phys. Res., Sect. A.* **580** 1031.
[7] Kurosawa S., Kubo H., Hattori K., Ida C., Iwaki S., Kabuki S., Kubo A., Kunieda E., Miuchi K., Nakahara T., Nishimura H., Parker J.D., Takada A., Takahashi M., Tanimori T., Ueno K. 2010 *Methods Phys. Res., Sect. A.* **623** 249
[8] Kurosawa S., Kubo H., Hattori K., Ida C., Iwaki S., Kabuki S., Miuchi K., Nishimura H., Okada Y., Parker J. D., Takada A., Takahashi M., Tanimori T., Ueno K., Yanagida Y. 2009 *IEEE Trans. Nucl. Sci.* **56** 3779
[9] Kamada K., Yanagida T., Endo T., Tsutumi K., Usuki Y., Nikl M., Fujimoto Y., Fukabori A., Yoshikawa A. 2012 *J. Cryst. Growth* **352** 88.
[10] Kamada K., Yanagida T., Pejchal J., Nikl M., Endo T., Tsutumi K., Yoshikawa A. 2011 *J. Phys. D: Appl. Phys.*, **44**, 505104.
[11] Kamada, K., Endo, T., Tsutumi, K., Yanagida, T., Fujimoto, Y., Fukabori, A., Yoshikawa, A., Pejchal, J., Nikl, M., 2011, Cryst. Growth Des. **11** 4484.
[12] Shimura N., Kamada, M. ; Gunji, A. ; Yamana, S. ; Usui, T. ; Kurashige, K. ; Ishibashi, H. ; Senguttuvan, N. ; Shimizu, S. ; Sumiya, K. ; Murayama, H. 2004 *2004 IEEE Nuclear Science Symposium Conference Record* **5** 2720.
[13] Blahuta, S. ; Bessiere, A. ; Viana, B. ; Dorenbos, P. ; Ouspenski, V. 2013 *IEEE Trans. Nucl. Sci.* **60** 3134.
[14] Donnald, S.B. ; Tyagi, M. ; Rothfuss, H. ; Fang Meng ; Hayward, J.P. ; Koschan, M. ; Melcher, C.L., 2013 *IEEE Trans. Nucl. Sci.***60**. 4002.

Photoluminescence of Valence-Changed Samarium Ions in Sintered $BaSO_4$ Induced by X-ray Radiation

K Maeda[*1], **T Kumeda**[1], **K Arimura**[1], **K Sakai**[2], **T Ikari**[1]

[1] Faculty of Engineering, University of Miyazaki, 1-1 Gakuen-Kibanadai-Nishi, Miyazaki 889-2192, Japan
[2] Cooperative Research Center, University of Miyazaki, 1-1 Gakuen-Kibanadai-Nishi, Miyazaki 889-2192, Japan

E-mail: t0b153u@cc.miyazaki-u.ac.jp

Abstract. The conversion of ionic valence induced by X-ray irradiation can possibly be applied in X-ray imaging devices. Here, we report the X-ray irradiation time dependence of the valence change of Sm ions doped in a $BaSO_4$ matrix. The samples were prepared by sintering a mixture of $BaSO_4$ and Sm_2O_3 powder. The as-prepared sample consisted mainly of Sm^{3+}. Sm^{2+} and Sm^{3+} were detected by using the photoluminescence (PL) spectral shapes. After X-ray irradiation (λ = 0.154 nm), the PL intensity of the Sm^{2+} increased proportionally to the X-ray irradiation time from 4 to 1000 s. The intensity began to saturate at more than 1000 s. The intensity of Sm^{3+} was constant regardless of the increasing Sm^{2+} intensity.

1. Introduction

X-ray-induced phenomena, such as photoionization, have been studied to develop high-sensitivity X-ray detectors [1, 2, 3]. The ions that have undergone valence changes can be observed after irradiation and are detected by luminescence techniques, which are highly sensitive, non-destructive, and simple. Optically stimulated luminescence (OSL) and thermoluminescence (TL) are widely used for measuring radiation doses [4]. TL intensity is initially proportional to the radiation dose [5]. Valence changes caused by X-ray irradiation can be detected by conventional photoluminescence (PL). Sm^{2+} shows strong luminescence in the near-infrared region from the $^5D_0$ to ^7F_j (j = 0, 1, 2) transitions. The wavelength of Sm^{2+} luminescence closely matches that of the Si detectors used for digital X-ray imaging. The host material for the luminescent ions is important because the efficiency of the luminescence depends on the rare earth ion sites in the crystalline structure. $BaSO_4$ compounds are suitable for X-ray irradiation because of their high equivalent absorption coefficient (Z_{eff} = 45) [5]. Many studies have reported the TL and OSL properties of $BaSO_4$ with Eu and Mn as dopants [5, 6]. In this paper, we detected the X-ray-induced ionic valence change from Sm^{3+} to Sm^{2+} in $BaSO_4$ by conventional PL because the method is simple and highly sensitive. The valence change of the Sm ions and the irradiation time dependence of the transition were determined from the PL intensity.

2. Experimental

Sm-doped $BaSO_4$ samples were prepared from $BaSO_4$ and Sm_2O_3 powders weighed with 0.1–2 at. % Sm. The mixed powders were sintered for 3 h in an electric furnace. The fabrication conditions for each sample were: sample A1, $BaSO_4$:Sm 1 at. %, sintered at 1200 °C in air; sample A2, $BaSO_4$:Sm 1 at. %, sintered at 1000 °C in air; sample B1, $BaSO_4$:Sm 1 at. % Sm sintered at 1200 °C in a reducing atmosphere (95% Ar + 5% H_2) The cylindrical sintered sample was cut and polished before X-ray irradiation. X-rays were obtained from a conventional XRD apparatus (X'Pert Pro MRD, PANalytical) operated at 40 kV and 45 mA with a Cu target. The exposure time was changed from 4 to 3600 s for each sample.

The PL and PL excitation (PLE) spectrum were measured by an RF-5300PC spectrometer (Shimadzu). The PL intensity was obtained by using a conventional monochromator (TRIAX320, HORIBA Jobin Yvon) at room temperature. The excitation light source for PL was a diode laser operating at a wavelength of 405 nm. The PL intensity was calculated from the area of the specific PL band for Sm^{3+} and Sm^{2+}.

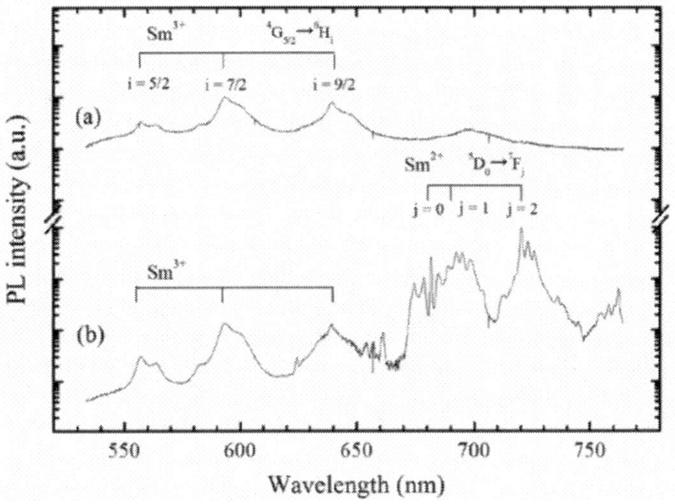

Figure 1. PL spectra of $BaSO_4$:Sm 1 at. % (sample A1) sintered at 1200 °C in air. (a) Before X-ray irradiation and (b) after 480 s X-ray irradiation.

3. Results and discussion

3.1. Photoluminescence

Figure 1(a) and (b) show PL spectra of sample A1 (1 at. % Sm-doped $BaSO_4$ sintered at 1200 °C in air) before and after X-ray irradiation for 480 s, respectively. The PL peaks observed in Fig. 1(a) at 557, 593, and 639 nm originated from the Sm^{3+} transitions $^4G_{5/2}$ to $^6H_{5/2}$, $^6H_{7/2}$, and $^6H_{9/2}$, respectively [7]. After irradiation, the peaks at 680, 692, and 720 nm corresponded to the Sm^{2+} transitions $^5D_0$ to $^7F_0$, $^7F_1$, and $^7F_2$, respectively. The intensities of the corresponding Sm^{2+} were weak or absent for other samples in the same batch before irradiation. The results confirmed a strong photoionization effect. However, irradiated samples showed relatively weak Sm^{3+} peaks that were similar before and after irradiation.

3.2. Photoluminescence excitation spectra

Figure 2(a) shows the PLE spectrum for sample A1 (1 at. % Sm-doped $BaSO_4$ sintered at 1100 °C in air), before X-ray irradiation. The excitation spectrum was obtained by monitoring the emission peak of Sm^{3+} ($^4G_{5/2}$ to $^6H_{7/2}$) at 591 nm. The strong narrow peaks at 343, 360, 373, and 400 nm were assigned to the Sm^{3+} transition from $^6H_{5/2}$ to $^4H_{9/2}$, $^4D_{3/2}$, $^6P_{7/2}$, and $^6P_{3/2}$, respectively [8]. Figure 2(b) and (c) shows the PLE spectrum for the sample A1 irradiated for 480 s, and sample B1 ($BaSO_4$ doped 1 at. % Sm sintered at 1200 °C in Ar/H_2), unirradiated. These spectra were obtained by monitoring the Sm^{2+} peak ($^5D_0$ to $^7F_2$) at 721 nm. PLE spectra of Sm^{2+} shows a broad band at 300–550 nm around 325 nm, hereafter called the 4f5d band, which corresponds to the ground state $4f^6$ level to $4f^55d^1$ band transition [9].

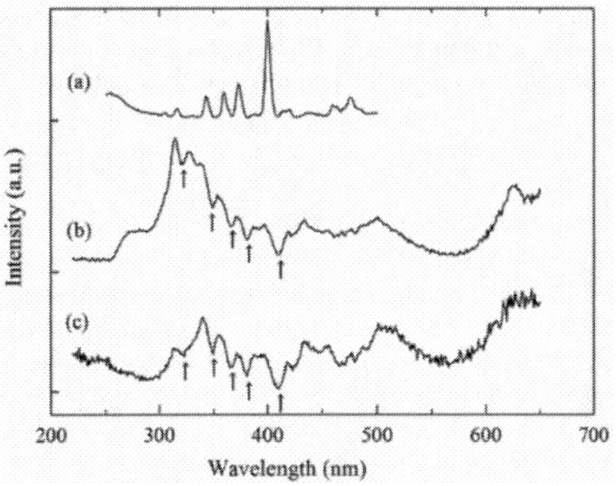

Figure 2 Excitation spectra of $BaSO_4$:Sm 1 at. %. (a) Sample A1, before irradiation, λ_{ex} = 591 nm for Sm^{3+}. (b) Sample A1, after 480 s irradiation, λ_{ex} = 721 nm for Sm^{2+}. (c) Sample B1, unirradiated, λ_{ex} = 721 nm for Sm^{2+}.

Figure 2(b) and (c) show the effect of X-ray irradiation and preparation under a reducing atmosphere on Sm^{2+} excitation spectra. The difference in spectral shape is caused by the condition of Sm^{3+} rather than by that of Sm^{2+}. The maximum peak position of the 4f5d band is 315 nm in sample A1 and 340 nm in sample B1. The 4f5d band of Sm^{2+} contains some dips (arrows in Figure 2(a) and (b)) at 320, 349, 366, 381, and 410 nm, and the 410 nm dip is the deepest for sample A1. These wavelengths correspond to the bands of the excitation spectra of Sm^{3+} in Figure 2(a) and the maximum is the 405 nm peak. This indicates that the Sm^{3+} absorption overlapped with PLE spectrum of the 4f5d band. Thus, the difference in the PLE spectral shape arises from the absorbed amount of Sm^{3+}. PLE spectrum of Sm^{2+} between generated by X-ray radiation and prepared in reducing atmosphere are not significantly difference.

3.3. Irradiation time dependence of photoionization

Figure 3 shows the log-log plot of the normalized PL intensities as a function of X-ray irradiation time. Figure 3(a) and (b) show the intensity of the Sm^{2+} peak for samples A1 and A2, respectively. The intensities for the X-ray irradiated samples increased linearly with time. The linear range of normalized PL intensity spanned two orders of magnitude of irradiation time. The good linearity is useful for contrast in X-ray imaging. The intensity saturated at irradiation times of more than 1000 s. The samples containing other Sm concentrations showed similar results.

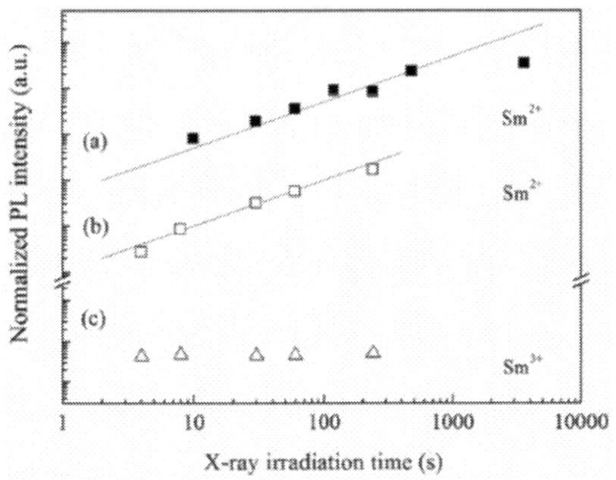

Figure 3. Sm ion emission intensity as a function of X-ray irradiation time. λ_{ex} = 405 nm. (a) Sample A1, Sm^{2+}, (b) sample A2, Sm^{2+}, and (c) sample A2, Sm^{3+}.

Figure 3(c) shows that the intensity of the Sm^{3+} peak for sample A2 was constant within experimental error. The X-ray penetration depth for $BaSO_4$ is about 7 μm, calculated from the mass absorption coefficient. The amount of Sm^{3+} in the region between the surface of sample and the penetration depth of the X-ray decreases because of the conversion to Sm^{2+} with increasing X-ray radiation time. The penetration depth of the excitation light for PL was much larger than that for X-rays because the energy of the excited light is lower than the optical absorption edge of $BaSO_4$. The depth of the light could not be evaluated because the scattering effect is very large around the grain boundaries of $BaSO_4$. Thus, the PL intensity from Sm^{3+} appears constant.

4. Conclusions

$BaSO_4$ samples doped with Sm were prepared by sintering. After X-ray irradiation for 4 to 1000 s, strong PL intensity from Sm^{2+} was detected. The X-ray irradiation time dependence of the PL intensity was linearly proportional to the amount of valence converted Sm^{2+}. The linearity of the conversion spanned more than two orders of magnitude of irradiation time. The linearity is suitable for X-ray imaging detector. We would like to develop wider linearity materials for optimizing the fabrication conditions.

References

[1] Marcazzo J, Santiago M, D'Angelo C, Furetta C and Caselli E 2010 *Nucl. Instr. Meth. Phys. Res. B* **268** pp 183-186
[2] Koughia C, Edgar A, Varoy C R, Okada G, Seggern H von, Belev G, Kim C Y, Sammynaiken R and Kasap S 2011 *J. Am. Ceram. Soc.* **94** pp 543-550
[3] Maeda K, Kawaida N, and Tsudome R 2012 *Phys. Status Solidi* C **9** pp 2271-2274
[4] Dixon R L and Ekstrand K E 1974 *J. Lumin.* **8** pp 383-390
[5] Gonzales P R, Furetta C, Calvo B E, Gaso M I and Cruz-Zaragoza E 2007 *Nucl. Instr. Meth. Phys. Res. B* **260** pp 685-692
[6] Manam J and Das S 2010 *J. Alloys Compd.* **489** pp 84-90
[7] Mikhail P, Ramseyer K, Frei G, Budde F and Hulliger J 2001 *Opt. Commun.* **188** pp 111-117
[8] Pekgozlu I 2013 *J. Lumin.* **8** pp 8-13
[9] Zhang Y,Lu C, Sun L, Xu Z and Ni Y *Mater. Res. Bull.* **44** pp 179-183

Journal of Physics: Conference Series **619** (2015) 012041 doi:10.1088/1742-6596/619/1/012041

Group velocity equalisation in multimode waveguides using inverse scattering designs

A. R. May and M. N. Zervas
Optoelectronics Research Centre, University of Southampton, SO17 2BJ, UK

E-mail: arm103@orc.soton.ac.uk

Abstract. In this paper, using an inverse scattering approach, we describe how the selection of mode effective indices and thus phase velocities can be used to control group velocity in a waveguide. As such it is shown that differential group delay can be equalised or minimised over a wavelength of choice. A particular feature of the new designs is the development of rings and a peaked core which may split depending upon the number of guided modes. These designs show characteristics comparable with commercially available fibres but with refractive index profiles that differ from typical graded-index designs.

1. Introduction

There is currently great interest in spatial division multiplexing [1] in order to overcome the impending "capacity crunch" [2] of single-core, single-mode optical fibre transmission systems. In particular, mode-division multiplexing is of particular interest because of the large number of modes that can be used in a single fibre. However, in order to use such methods it is important to control the deleterious effect of differential group delay (DGD) as well as mode-coupling. As a result work has been done to investigate the tailoring of optical fibre refractive index (RI) profiles to meet these needs [3–5]. One particular design by Gruner-Nielsen et al. consists of a graded-index core and an outer trench RI. While the performance of this design in terms of minimisation of DGD and low mode coupling between the LP_{01} and LP_{11} modes is very good, it relies upon a parametric design process. We were interested to know whether direct control of mode effective indices could be used to control both of these characteristics.

In this paper we extend our work [6] on the inverse scattering (IS) design of single-mode planar waveguides to the case of multimode planar waveguides through the use of Darboux transformations as have been used previously by Mills[7] in the design of optical interconnects. We show that through manipulation of the phase velocities of the waveguide it is possible to manipulate the group velocities to the extent of controlling the wavelength of equalisation and/or minimising the DGD over large wavelengths. It is clear that we have performed this investigation in planar waveguides rather than optical fibres, but we believe that the qualitative features of the designs will carry over.

2. The Darboux transformations

As has previously been described and utilised by other authors [8,9], there exists a correspondence between a bound state in quantum mechanics and a discrete propagating mode in a waveguide. The IS theory allows for the construction of a potential with a given set of bound states a priori through the use of the Darboux transformations [10] . In this paper we will not describe in detail the process involved other than to say that associated with the set of N propagation constants β_m are a set of bound state eigenvalues k_m defined through the relationship

$$k^2_{\ m} = k_0^{\ 2} n_2^{\ 2} - \beta_m^{\ 2} \tag{1}$$

Where $k_0=2\pi/\lambda_d$ is the free-space wavenumber at the design wavelength λ_d, n_2 is the refractive index of the cladding of the waveguide and β_m is the m'th propagation constant.

Given that there exists Brown's identity [11] connecting the phase velocity and group velocity to a weighed integral of power densities in waveguide regions we proposed that control of dispersive properties would be possible through the above method. With the propagation constants supplied a priori the potential is derived and its dispersive properties obtained through direct scattering. We applied this method to the design of planar waveguides with two, three and four modes. In all cases the design wavelength was chosen to be λ_d=1.55μm and a cladding index of n_2=1.444. In addition, it was assumed in calculating the DGD that material dispersion is the same for all modes and profile dispersion is negligible.

3. Dual-mode designs

We first select the effective index of the TE_0 mode and vary the spacing of the TE_1 mode with respect to this and the cladding. We find that through the variation of the TE_1 mode effective index we may change the wavelength for which group velocity equalisation is achieved. We illustrate this procedure in figure 1 while demonstrating how the refractive index of the design is also altered. It can be seen that as the effective index of the TE_1 mode approaches that of the TE_0 mode we have a deepening of the central dip and a movement towards longer wavelengths of the dispersion equalisation wavelength. This behaviour was originally discussed by Stolen [12] where he considered optical fibres with ring index profiles.

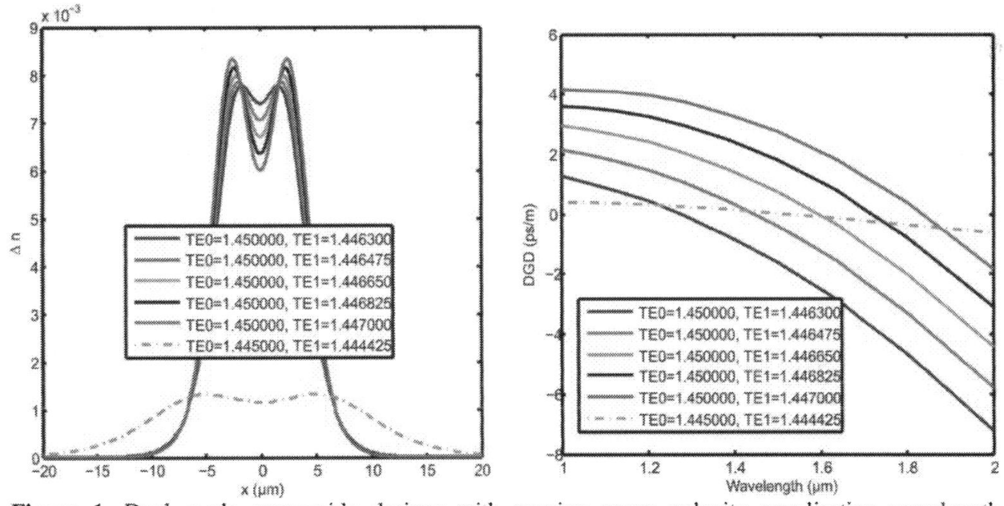

Figure 1. Dual mode waveguide designs with varying group velocity equalisation wavelength specified by changing TE_1 mode effective index for fixed TE_0

It is also interesting to note that reducing the effective index of TE_0 results in a scaling down and across of the refractive index profile as well as a reduction in the slope of the DGD as can be seen in figure 1. As such the DGD can set to zero at a design wavelength and minimised over a large wavelength about this point.

4. Three and four-mode designs

In this section we show that designs can be obtained with minimisation of DGD with respect to TE_0 of all modes. This is achieved by varying the effective indices of the propagating modes. In particular, in figure 2 we show a design for a three mode waveguide, coloured red, for which DGD of all modes is brought to within ~1 ps/m. This DGD is close in range to that associated with graded-index few-mode fibres produced by OFS (±0.4 ps/m) [13] but with a different design.

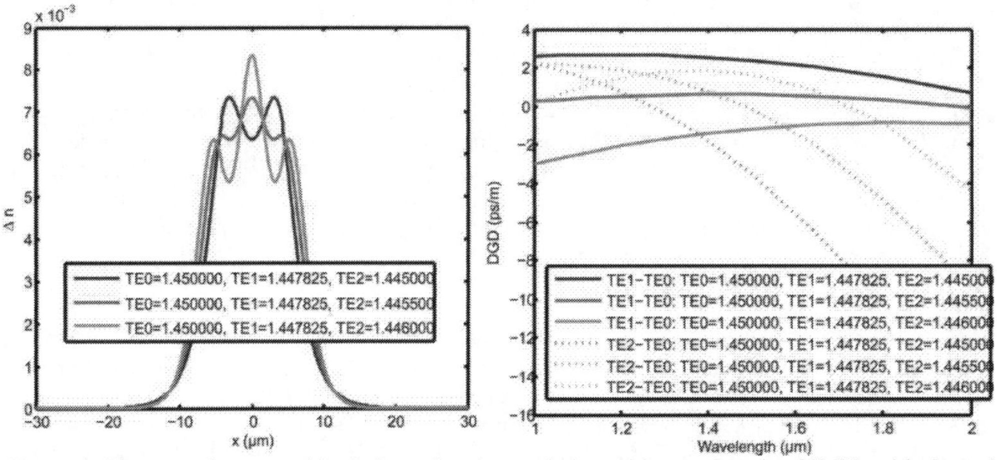

Figure 2. Three mode waveguide designs showing variation with wavelength of DGD with the red curves showing DGD of all modes to within ~1 ps/m

It is shown that equi-distant effective indices (blue curve), which approximate the standard parabolic RI profiles, result in waveguides with the highest DGD. Varying the inter-modal effective-index differences results in a pronounced central peak followed by the development of a ring. Increasing the contrast of the central peak and emphasising the development of the outer ring leads to minimisation of DGD at an ever larger wavelength as shown by the green design.

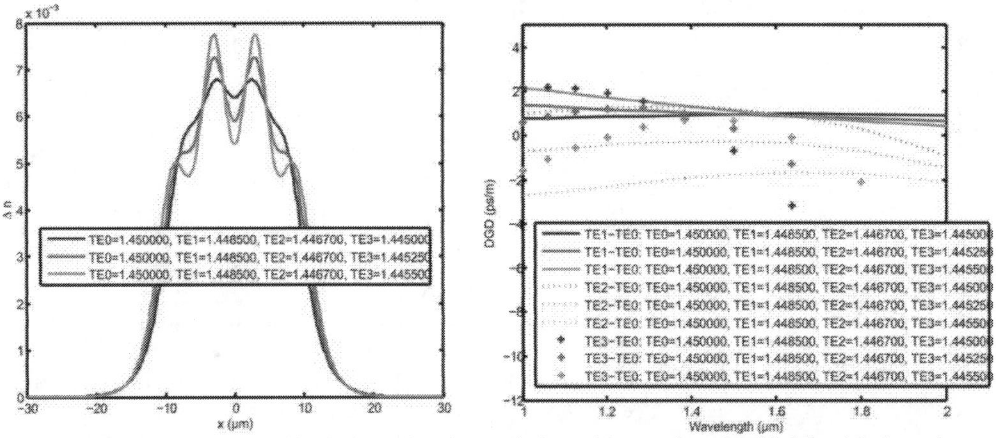

Figure 3. Four mode waveguide designs showing variation with wavelength of DGD with the red curve corresponding to a DGD of all modes of ~1ps/m

In figure 3 we illustrate a four mode waveguide design, once again coloured red, for which DGD of all modes is once again ~1 ps/m. From figure 2 and 3 it can be seen that the overall qualitative shape of the three and four mode designs is the same with the slight development of a ring but in the four mode case an additional splitting of the central peak region. As in the two mode and three mode cases this is to be expected from coupler theory where approximately degenerate modes are associated with separation of the waveguide cores.

5. Conclusions

In this paper we have shown that control of effective indices of modes through the use of the Darboux transformation can be used parametrically to equalise or minimise differential group delay over a range of wavelengths. In addition we have identified that these waveguides can all be described qualitatively as consisting of an outer ring and a peaked core which may in turn split as the number of guided modes increases.

Our designs have shown characteristics comparable with those of commercially available few-mode fibres. While our investigation has been limited to planar waveguides there are similarities in intensity profiles between TE modes and LP modes and it is expected from Brown's identity that characteristics of these planar designs can be adapted for use in future optical fibres.

6. References

[1] Richardson D J, Fini J M and Nelson L E 2013 Space-division multiplexing in optical fibres *Nat. Photonics* **7** 354–62

[2] Essiambre R and Tkach R 2012 Capacity trends and limits of optical communication networks *Proc. IEEE* **100** 1035–55

[3] Grüner-Nielsen L and Sun Y 2012 Few mode transmission fiber with low DGD, low mode coupling, and low loss *J. Light. Technol.* **30** 3693–8

[4] Riesen N and Love J D 2011 Dispersion equalisation in few-mode fibres *Opt. Quantum Electron.* **42** 577–85

[5] Ferreira F, Fonseca D and Silva H 2013 Design of Few-Mode Fibers With Arbitrary and Flattened Differential Mode Delay *IEEE Photonics Technol. Lett.* **25** 438–41

[6] May A R, Poletti F and Zervas M N 2014 Inverse scattering designs of dispersion-engineered single-mode planar waveguides *Proceedings of the SPIE* vol 8988, ed J E Broquin and G Nunzi Conti p 89881S

[7] Mills D W and Tamil L S 1993 Synthesis of Guided Wave Optical Interconnects *IEEE J. Quantum Electron.* **29** 2825–34

[8] Tamil L S and Jordan A K 1991 Spectral Inverse Scattering Theory for Inhomogeneous Dielectric Waveguides and Devices *Proc. IEEE* **79** 1519–28

[9] Jordan A K and Lakshmanasamy S 1989 Inverse scattering theory applied to the design of single-mode planar optical waveguides *J. Opt. Soc. Am. A* **6** 1206–12

[10] Deift P and Trubowitz E 1979 Inverse scattering on the line *Commun. Pure Appl. Math.* **32** 121–251

[11] Chen C-L 2007 *Foundations for Guided-Wave Optics* (Hoboken, New Jersey, USA: John Wiley & Sons)

[12] Stolen R . 1975 Modes in fiber optical waveguides with ring index profiles *J. Appl. Opt.* **14** 1533–7

[13] OFS Speciality Photonics Division 2014 Four Mode Graded-Index, Few Mode Fiber

Effect of laser irradiation on the optical properties of $As_{40}Sb_{15}Se_{45}$ chalcogenide thin films

Ramakanta Naik[1], R. Ganesan[2] and K.S.Sangunni[2]

[1]Department of Physics, Utkal University, Bhubaneswar, 751004, India
[2] Department of Physics, Indian Institute of Science, Bangalore, 560012, India

E-mail: ramakanta.naik@gmail.com

Abstract. The exposure with band gap light of thermally evaporated $As_{40}Sb_{15}Se_{45}$ amorphous film of 800 nm thickness, were found to be accompanied by optical changes. The as-prepared and illuminated thin films were studied by X-ray diffraction, Fourier Transform Infrared Spectroscopy and X-ray Photoelectron Spectroscopy and Raman spectroscopy. The optical band gap was reduced due to photo induced effects along with the increase in disorder. These optical properties changes are due to the change of homopolar bond densities. The core level peak shifting in XPS spectra and Raman shift supports the optical changes happening in the film due to light exposure.

1. Introduction

Chalcogenide glasses are one of the most widely known families of amorphous materials and have been studied for several decades, because of their interesting fundamental properties and wide range of applications [1]. The light illumination of amorphous chalcogenides involves, in most cases, changes in their electronic transport and optical properties, as well as modifications in their local structure [2]. In short, photosensitive chalcogenide glasses are of crucial importance in designing high dense optical holograms and high bit rate optical waveguides. The photo induced effects in amorphous chalcogenides have been extensively studied, partly as an interesting subject for fundamental research in the field of disordered materials and partly due to potential application of these effects in optoelectronics (photo resists, optical memories, optoelectronic circuits, etc.) [3,4]. Among that, photo induced effects such as photo bleaching and photo darkening are among the most explored properties of chalcogenide thin films because of the high potential in a variety of applications in information storage, photonics, opto electronics, opto mechanical transducers, etc [5,6]. Due to the presence of localized states in the gap, photo induced effects are very predominant in such type of materials. The presence of lone pair electrons at the valence band and the structural flexibility associated with the glassy network is mainly responsible for this type of changes. The lone-pair character of the valence tails leads to very rich behavior under the influence of light [7]. The illumination by band gap light of many amorphous chalcogenides changes their internal and/or surface structure and the optical absorption edge will have a red or blue shift. Photo induced changes of optical transmittivity and reflectivity, index of refraction, changes of reactivity, rates of diffusion and inter-diffusion, viscosity and the state (phase) have been observed in many materials. The study of the optical absorption

[1] To whom any correspondence should be addressed.

spectra in chalcogenide glasses provides essential information about the band structure and the energy band gap. The optical band gap of the material plays a major role in the preparation of the device for a particular wavelength. So, the decrease in optical band gap by laser irradiation known as photo darkening is very useful in device making. The present communication reports the effect of laser irradiation on optical properties of amorphous thin films of $As_{40}Se_{45}Sb_{15}$.

2. Experimental

Thin films of $As_{40}Se_{45}Sb_{15}$ were prepared by thermal evaporation method at a base pressure of 1 x10^{-5} Torr from the prepared bulk glass onto the glass substrates. The thicknesses of the films were around 800 nm. The amorphous state of the film was checked by X-ray (Philips, $Cu_{K\alpha}$, λ=1.54 Å) Diffractometer (XRD).To study the photo induced changes, we irradiated the film at room temperature by a diode pumped solid state laser of wavelength 532 nm with a power of 40 mW. The optical absorption spectra of the as-prepared and illuminated films were taken by using the Fourier Transform Infrared (FTIR) spectrometer (Bruker Optics (IFS66v/S) in the visible wavelength range of 400-1200 nm at room temperature. The X-ray Photoelectron Spectroscopy (XPS) core level spectra were obtained with Mg K_α X-rays (1253.6 eV) at a base pressure of ~10^{-9} Torr in Multilab 2000 Thermo Scientific UK instrument. The C 1s line from adventitious carbon on the surface has been widely used for charge referencing [8] and the BE of the reference C 1s line was set as 284.6 eV. The room temperature Raman spectra were recorded in the 180° backscattering geometry, using a 532 nm excitation from a diode pumped frequency doubled Nd–YAG solid state laser and a custom built Raman spectrometer equipped with SPEX TRIAX 550 monochromator.

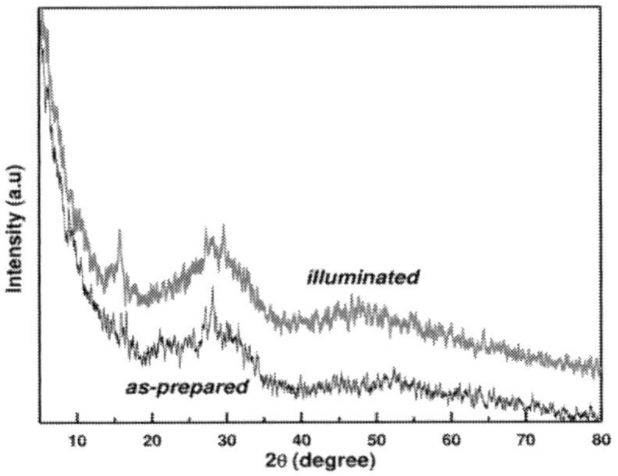

Figure 1. XRD patterns of the $As_{40}Se_{45}Sb_{15}$ thin films

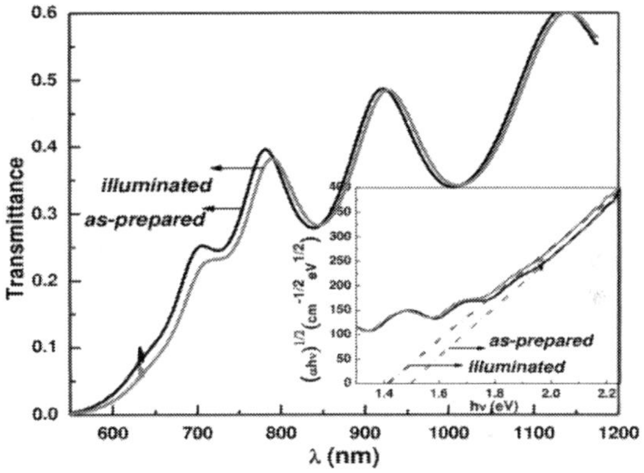

Figure 2. Transmission spectra (inset- E_g change) of the $As_{40}Se_{45}Sb_{15}$thin films.

3. Results and Discussion

Figure 1 shows the X-ray diffraction patterns for the $As_{40}Se_{45}Sb_{15}$ thin films. The absence of the diffraction lines in the X-ray patterns indicates that the films have amorphous structures. Transmission spectra corresponding to the amorphous $As_{40}Se_{45}Sb_{15}$ thin films are plotted in figure 2, which shows a clear difference between the as-prepared and the illuminated films. The absorption coefficient (α) for the studied films were calculated [9] from the transmittance T and reflectance R using the equation

$$\alpha = \frac{1}{d} ln \frac{(1-R)^2}{T} \tag{1}$$

where d is the thickness of the film and T is the transmission and R is the reflectivity of the film. The absorption coefficient increases with illumination. According to Tauc's relation [10] for the allowed non-direct transition, the photon energy dependence of the absorption coefficient can be described by

$$(\alpha h\nu)^{1/2} = B^{1/2}(h\nu - E_g) \tag{2}$$

where B is a parameter that depends on the transition probability and E_g is the optical energy gap. The optical band gaps of the as-prepared and the illuminated films are found to be 1.49±0.001 eV and 1.41±0.002 eV respectively (figure 2, inset). So, photo induced optical band gap change was observed due to light illumination on the as-prepared film. Since, the optical absorption depends on short-range order in the amorphous states and defects associated with it, the decrease in optical band gap may be explained on the basis of "density of state model" proposed by Mott and Davis and on the basis of change in nature of films due to chemical disorderness [11]. The decrease in optical band gap (E_g) with laser irradiation can be attributed to the reduction in the density of tail states adjacent to the band edge. The $B^{1/2}$ values for as-prepared and illuminated films are 510±4 and 468± 2 $cm^{-1/2}eV^{1/2}$ respectively. The $B^{1/2}$ for illuminated film is less than the as-prepared film which indicates the presence of more no of homopolar bonds due to disorder. It is known that the Urbach edge is a useful parameter to evaluate the degree of disorder. The value of E_e for as-prepared and illuminated $As_{40}Se_{45}Sb_{15}$ films is 198 meV and 231 meV respectively. The Urbach energy E_e of the illuminated film is more than that of as-prepared film which indicates the increase in disorder after illumination. The lower values of $B^{1/2}$ and higher values of E_e of the illuminated film over the as-prepared film clearly indicate that the illuminated film is more disordered (chemically) than the as-prepared film, i.e., the creation of homopolar bonds after photo induced process which is well supported in XPS analysis.The increase in homoplar bond because of micro structural disorderness in the films occurs by photo reaction like

$$Sb-Se+As-Se \rightarrow Sb-Sb+As-As+Se-Se \qquad (3)$$

The XPS spectra of the as-prepared and illuminated film for As 3d core level is shown in figure (3a). From the spectra of As 3d peak (figure 3a inset), the position for as-prepared and illuminated films are at 42.81 and 42.59 eVs respectively. This decrease of 0.22 eV binding energy is in accordance with the reduction in optical band gap of the as-prepared film due to creation of more As-As homopolar bonds as a result of photo induced process as observed for other films also [12]. Similarly, the peak position of Sb 4d core level spectra (figure 3a) is shifting towards the lower BE (34.44 eV to 33.72 eV) due to the formation of more Sb-Sb homopolar bonds. The Se 3d spectra of the as-prepared (54.24 eV) and illuminated (54.54 eV) films are shown in figure 3b.The peak shifts towards higher BE shows the formation of Se-Se homopolar bonds due to illumination as the electro negativity of Se (2.55) is more than that of Sb (2.02).

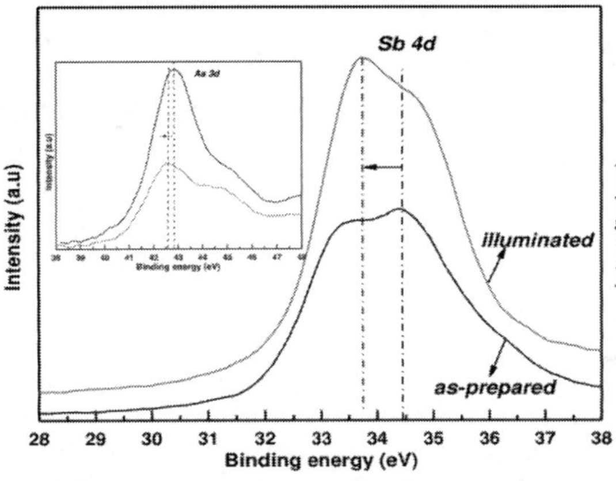

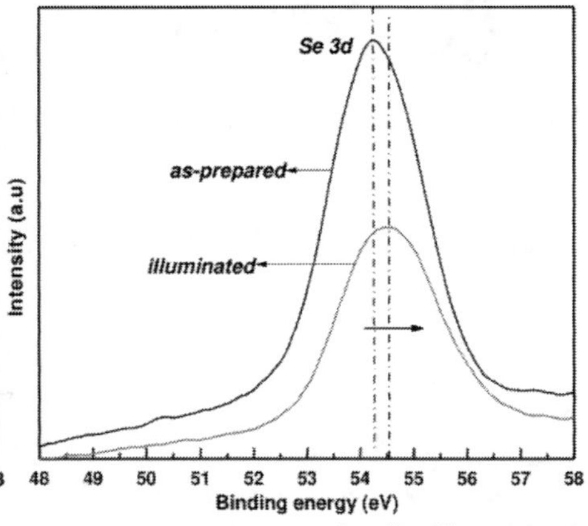

Figure 3a. As 3d and Sb 4d of $As_{40}Se_{45}Sb_{15}$ thin film. **Figure 3b.** Se 3d core level spectra of $As_{40}Se_{45}Sb_{15}$ thin film.

Direct evidence of structural changes in $As_{40}Se_{45}Sb_{15}$ thin films caused by illumination was obtained from Raman Spectra as shown in figure 4. According to the molecular model [13], each Sb atom in the As-Sb-Se ternary glasses is covalently bonded to three Se atoms in a pyramidal unit ($Sb\text{-}Se_3$), and As atoms are covalently bonded to three Se atoms in a pyramidal unit ($As\text{-}Se_3$). The basic structural units $AsSe_3$ and $SbSe_3$ are interconnected through bridging Se atoms. The coupling between the basic structural groups via Se atoms is assumed to be weak, and the vibrational modes are separated into $SbSe_3$ and $AsSe_3$ like vibration in $As\text{-}Se_3$, (Se_2) As-As (Se_2) and the $AsSe_3$ mode is superimposed in that band at 331cm^{-1}.

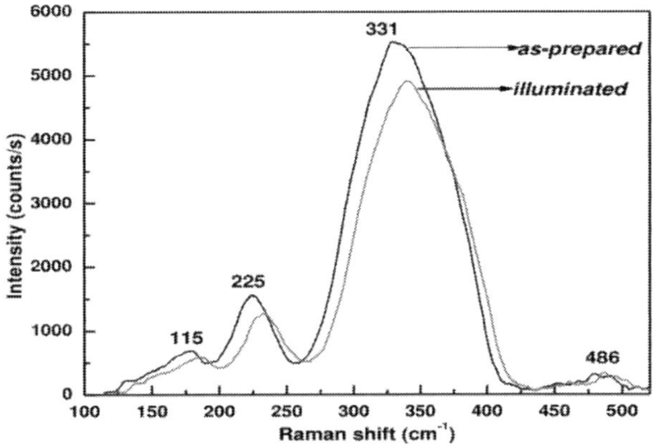

Figure 4. Raman spectra of $As_{40}Se_{45}Sb_{15}$ thin film.

The peak at 115 cm^{-1} and 225 cm^{-1} of the as-prepared film corresponds to Se-Se vibrational modes respectively. The illumination of the as-prepared film causes the increase in homopolar bond density due to which there is shift in the peak position of the various vibrational modes.

4. Conclusions

In conclusion, thermally-evaporated amorphous thin films undergo structural transformations when exposed to band gap illumination (photo-structural transformations). Photo induced changes of optical transmissivity, optical band gap, $B^{1/2}$ value; XPS core level peaks and Raman spectra change are observed. The optical band gap is decreased due to illumination. Photo induced changes can be assigned to chemical bond redistribution.

Acknowledgements

The authors thank Department of Science and Technology (DST), Govt of India for DST-INSPIRE Research grant and using the National Facility for Optical Spectrometry at Department of Physics and Surface Science facility, Indian Institute of Science (IISc) for XPS measurement.

References

[1] Shimakawa K, Kolobov A and Elliot S R 1995 *Adv. Phys.* **44** 475
[2] Marquez E, Jimenez-Garay R and Gonzalez-Leal J M 2009 *Mat.Chem. Phys.* **115** 751
[3] Rosenblum G, Sfez B G, Kotler Z, Lyubin V and Klebanov M 1999 *Appl. Phys. Lett.* **75** 3249
[4] Ramchandran S and Bishop S G 1999 Appl. Phys. Lett. **74** 13
[5] Ganjoo A and Jain H 2006 *Phys. Rev. B* **74** 024201
[6] Krecmer P, Moulin A M, Elliot S R 1997 Science **277** 1799
[7] Zakery A and Elliot S R 2003 *J. Non-Cryst. Solids* **330** 1
[8] Barr T L and Seal S 1995 *J. Vac. Sci. Technol. A* **13** 1239
[9] Naik R, Jain A,Ganesan R and Sangunni K S 2012 *Thin Solid Films* **520** 2510
[10] Tauc J 1979 *Amorphous and Liquid Semiconductors*, Plenum press, New York
[11] Mott N F and Davis E A 1979 *Electronic Procesess in Non-Crystalline Materials*, Oxford, Clarendon p.428
[12] Naik R, Ganesan R and Sangunni K S 2011 *J. Non-Cryst. Solids* **357** 2344
[13] Lucovsky G and Martin R M 1972 *J. Non-Cryst. Solids* **8-10** 185

ICOOPMA2014 IOP Publishing
Journal of Physics: Conference Series **619** (2015) 012043 doi:10.1088/1742-6596/619/1/012043

Sodium yttrium fluoride based upconversion nano phosphors for biosensing

Padmaja Parameswaran Nampi,[1]* Harikrishna Varma,[1] P R Biju,[2] Tarun Kakkar,[3] Gin Jose,[3] Sikha Saha,[4] Paul Millner[5]
[1]Biomedical Technology Wing, Sree Chitra Tirunal Institute for Medical Sciences and Technology, Poojapura, Trivandrum-12, Kerala, INDIA
E-mail: padmavasudev@gmail.com
[2]School of Pure and Applied Physics, Mahatma Gandhi University, Kottayam, India
[3]Institute for Materials Research, Houldsworth Building, University of Leeds, Clarendon Road, Leeds LS2 9JT, United Kingdom
[4]Leeds Institute of Genetics, Health and Therapeutics, University of Leeds, Clarendon Road, Leeds LS2 9JT, United Kingdom
[5]School of Biomedical Sciences, University of Leeds, Clarendon Road, Leeds LS2 9JT, United Kingdom

Keywords: Fluorescence, Upconversion, Nanophosphors, Biosensors

Abstract
In the present study, $NaYF_4$-Yb^{3+}/Er^{3+} having the composition $NaYF_4$-18%Yb^{3+}/2%Er^{3+} and $NaYF_4$-20%Yb^{3+}/2%$Er^{3+}$with and without the addition of PVP (polyvinyl pyrolidone) have been synthesised by a solution method using NaF, yttrium nitrate, ytterbium nitrate and erbium nitrate as precursors. Upconversion spectra of prepared nanomaterial under 980 nm laser excitation have been studied. The variation in upconversion spectra with new born calf serum and myoglobin has been studied. Myoglobin (Mb) may be helpful when used in conjunction with other cardiac markers for rapid determination of acute myocardial ischemia, especially in patients with a typical chest pain or nonspecific ECG changes. The variation of UC fluorescence with addition of Mb indicates the suitability of using $NaYF_4$ based UC nanoparticles in cardiac marker detection. The detailed study is currently under progress.

1. Introduction

Rare earth based materials are strongly fluorescent, low in toxicity and readily synthesised in water which greatly facilitates further biofunctionalization. Hence their biological applications have been reported favourably in recent years. The luminescence of lanthanide doped inorganic materials/nanoparticles is characterized by narrow emission band widths determined by the lanthanide ions, which when used in conjunction with different dopant ions and matrices can generate different emission colours. At the same time, lanthanide doped nanoparticles offer substantial advantage over lanthanide chelates such as high photochemical stability and long fluorescent life time up to several milliseconds. Another unique feature of lanthanide doped phosphors is their ability to emit photons in the visible range after being excited with infrared light, in a process known as upconversion. Upconversion fluorescence labels possess several distinct advantages compared to commonly used down-converting phosphors.

(i) A low optical background is expected due to the absence of autofluorescence of biomolecules upon infrared radiation
(ii) Due to the large wavelength separation between excitation and emission, the optical train is very simple and so there is no need for time resolved detection and
(iii) Simultaneous detection of multiple analytes can be realized since different colours of visible light can be obtained from different upconversion phosphors excited by the same IR laser.

The most efficient UC mechanisms are present in solid-state materials doped with rare-earth ions. Lanthanide-doped upconversion materials, which can convert near-infrared lights to visible lights, have attracted growing interest because of their great potentials in biomedical engineering. However, it remains a grand challenge to manoeuvre the intensity ratio between different emission lines and enable tunable upconversion functions. However many previous upconversion phosphors used for biological labelling are somewhat too large in size and thus cannot be used for the sensitive analysis of moleculs such as DNA, RNA, or protein biomarkers. Accordingly nano sized upconverting phosphors have received considerable attention. Rare earth fluorides, $NaYF_4$ in particular are regarded as ideal host materials of low phonon energy and high photochemical stability. And tuneable upconversion emission is feasible with doping of various lanthanide ions. Among various fluoride host low phonon energy $NaYF_4$ doped either with Yb^{3+}:Er^{3+} or Yb^{3+}:Tm^{3+} trivalent rare earth ions, is recognized as one of the most efficient host for the infrared-to-visible upconversion processes [1-2]. The IR to visible UC particles offer a wide range of technological applications in solid state lasers, three dimensional flate panel displays, energy efficient photovoltaic devices biomedical imaging and photodynamic therapy applications [3-8]. Different synthesis procedures have been adapted to synthesis various upconversion systems and upconversion properties of many such systems for various applications have been studied previously [9-12]. In the present paper an attempt has been made to synthesise $NaYF_4$/Yb-Er system with different Yb^{3+}:Er^{3+} concentration using a simple solution method. PVP has been used as an additive in order to enhance biofuctionalize the material for final applications. The variations in upconversion fluorescence with different Yb^{3+}:Er^{3+} concentration with and without PVP have been dealt with. Initial trials for utilizing the nanophosphors for detection of myoglobin and new born calf serum have been dealt with.

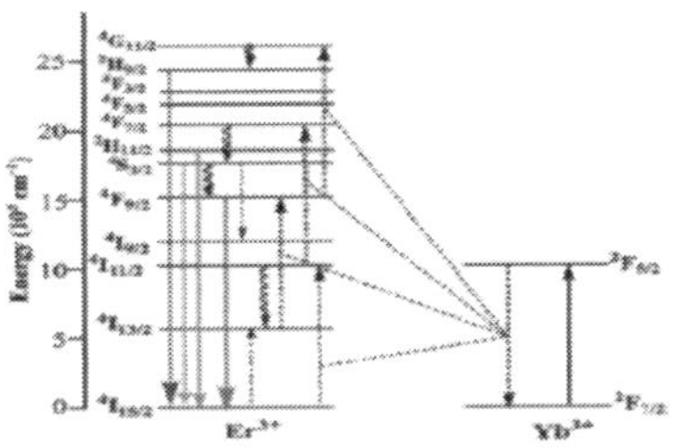

Fig. 1 UC mechanism – Energy level diagram of Yb^{3+}:Er^{3+} system

2. Experimental

For synthesizing the NaYF4/Yb^{3+}/Er^{3+} compositions, NaF (Merck Chemicals, India) in H_2O/EtOH mixture has been added to the rare earth nitrates viz. $Y(NO_3)_3.6H_2O$ (Sigma Chemicals, USA), $Yb(NO_3)_3.5H_2O$ (Sigma Chemicals, USA) and Er $(NO_3)_3.5H_2O$ (Sigma Chemicals, USA). To get 1gm of an ideal composition $NaYF_4$-18%molYb^{3+}/2mol%Er^{3+}, 1.5004 g $Y(NO_3)_3.6H_2O$, 0.4116g $Yb(NO_3)_3.5H_2O$ and 0.0434 g of $Er(NO_3)_3.5H_2O$ is well mixed in 50ml of deionized water and added to a solution of NaF (0.08192 g in 80 ml of 80:20 v/v H_2O/EtOH mixture) and stirred for around 5 hours at RT. The mixture is then centrifuged, washed around 4-5 times with EtOH and then with deionized water and the slurry after decantation is heated to 60 °C by keeping in an air oven. The dried powder is collected and used for further characterisation. For PVP added composition 25 wt% of PVP is added after homogenization of other reagents. The PVP incorporated slurry is stirred for 5-6 hours and centrifugation, washing and drying has been done as for the undoped samples. Following a similar procedure, a composition $NaYF_4$-20mol%Yb^{3+}/2mol%Er^{3+} with and without PVP additive have been synthesised. The change in upconversion fluorescence with the addition of new born calf serum and Mb has been observed, as a part of detection trials.

The fluorescence upconversion has been measured by exciting the sample at 980 nm followed by measuring the emission in the 400 nm-900 nm range at 400 mA using a Spectrofluorimeter (Edinburg Instruments Ltd, UK) equipment. The TEM has been done using a Transmission Electron Microscope (Philips CM 200 High Resolution FEGTEM, UK). The phases of the compounds have been analysed using an X-ray diffractometer (Siemens, Germany) in steps of 0.02° in the 2 theta range 10-70° with a count time of 2 seconds.

3. Results

The nanoclusters are clearly observed in the TEM images of selected samples as shown in Fig.2. It is also seen that more agglomeration is observed in samples with PVP, while nanoparticles are clearly seen in the agglomerated clusters of unmodified samples. The nanocrystallinity is also evident from the diffraction pattern.

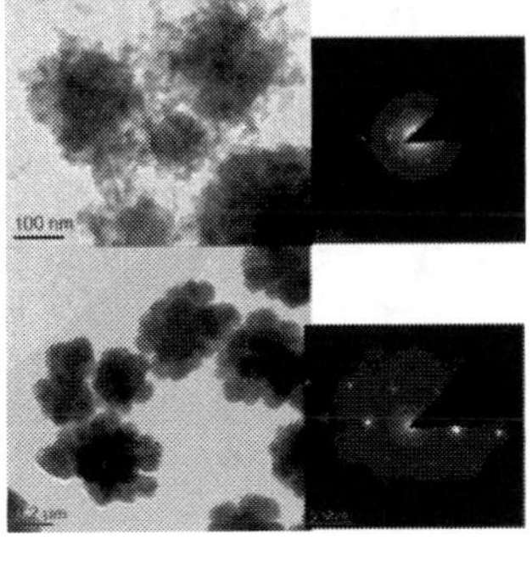

Fig. 2. TEM images of (A) NaYF4-18mol%Yb3+:2mol%Er3+ (B) NaYF4-18mol%Yb3+:2mol%Er3+PVP

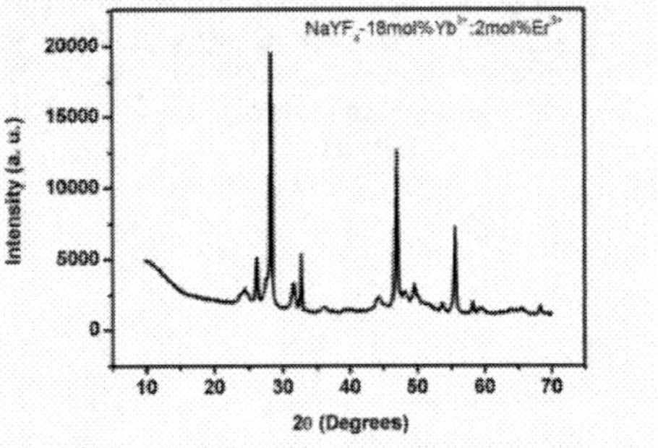

A

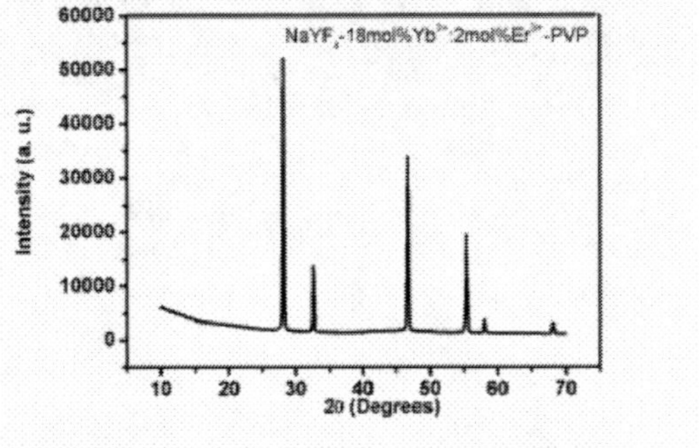

B

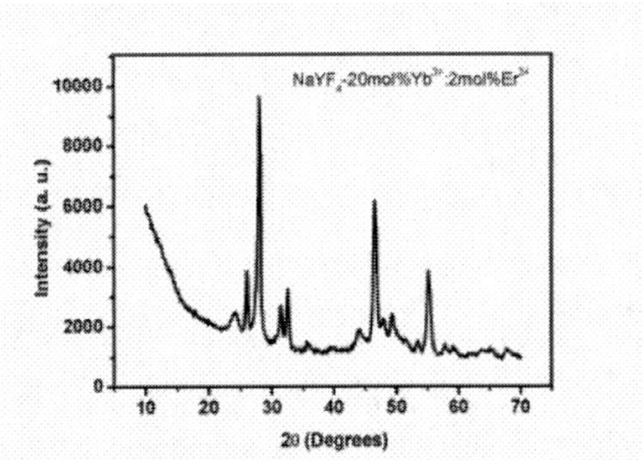

C

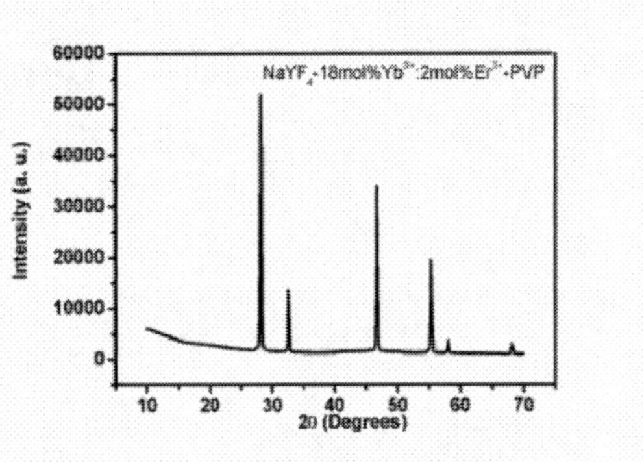

D

Fig.3 XRD patterns of (A) $NaYF_4$-18mol%Yb^{3+}:2mol%Er^{3+} (B) $NaYF_4$-18mol%Yb^{3+}:2mol%Er^{3+}-PVP (C) $NaYF_4$-20mol%Yb^{3+}:2mol%Er^{3+} (D) $NaYF_4$-20mol%Yb^{3+}:2mol%Er^{3+}-PVP

X-ray diffraction patterns (Fig. 3) show phase pure α- $NaYF_4/Yb^{3+}/Er^{3+}$, for PVP modified samples while mixed phases are observed in the unmodified samples.

Fig. 4 shows the fluorescence upconversion spectra of the selected $NaYF_4/Yb^{3+}/Er^{3+}$ based powder samples. It is observed that sharp peaks having very high intensity has been observed for PVP modified samples especially in the 600-700 nm range ($^4F_{9/2} \rightarrow {}^4I_{15/2}$ transition of the Er^{3+} ions). Significant change is seen in the peak at ~900nm. The peaks at ~800 nm also shows significant increase in PVP modified samples while the intensity is almost zero for unmodified samples. The results show PVP considerably enhances the UC of $NaYF_4$/ Yb^{3+}:Er^{3+} system. But for aqueous

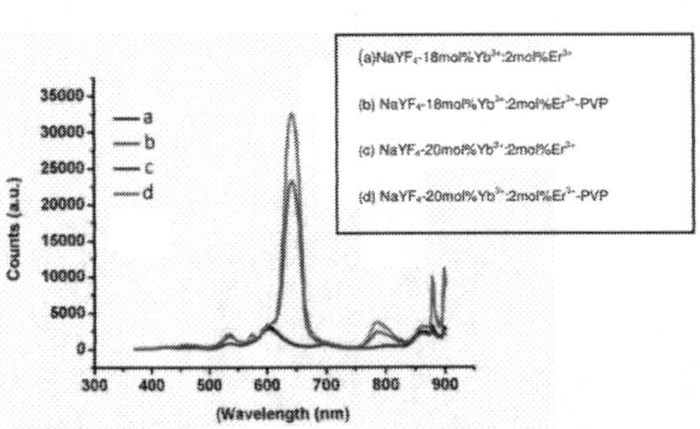

A

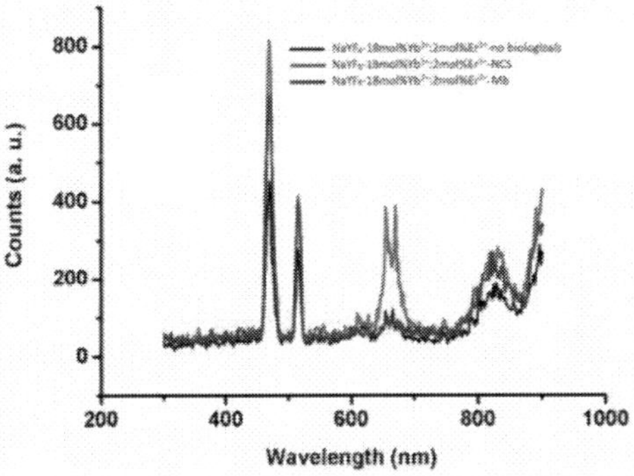

B

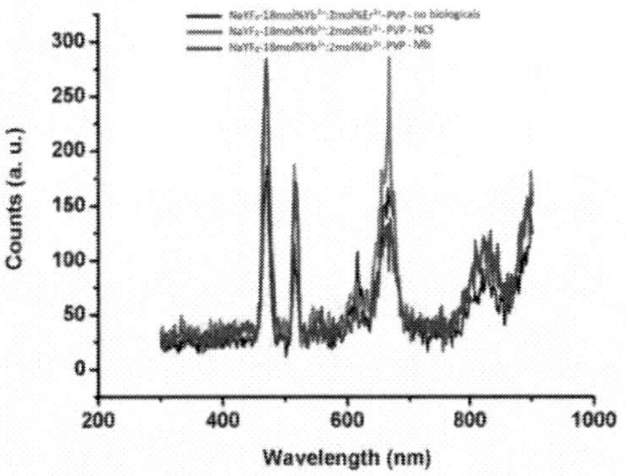

C

Fig. 4: UC fluorescence spectra of (A) Samples with no biologicals (B) &(C) Sample with NCS and Mb and with out biologicals (aqueous solution)

solutions PVP is not giving much enhancement of fluorescence. But when comparing aqueous solutions with and without biologicals, the peaks observed in the range 400-550 nm ($^2H_{11/2} \rightarrow {}^4I_{15/2}$, $^4S_{3/2} \rightarrow {}^4I_{15/2}$) is considerably enhanced when NCS and Mb is added to the system, while the peak in the range 600-800 nm shows more sharpness in presence of Mb. While considering the peaks at 800-900 nm, Mb and NCS added samples show a little higher fluorescence. The variation in fluorescence with addition of biomolecules can be a tool for detection. Further optimization is required for elucidating a standardized procedure for detection.

4. Conclusion

Nano - $NaYF_4$-Yb^{3+}:Er^{3+} having various Yb^{3+}:Er^{3+} concentration have been synthesised by a wet chemical method. TEM picture shows the nano nature of the samples, with more agglomeration for PVP modified samples. It is observed that PVP modified samples show better UC fluorescence compared to unmodified samples. XRD shows pure cubic phase for PVP modified samples. UC spectra of the synthesised samples and possibilities of using nano – $NaYF_4$-Yb^{3+}:Er^{3+} samples in the field of biosensor have been established using new born calf serum and calf myoglobin as examples.

5. References

[1] Tan M C, Al-Baroudi and Riman R E 2011 *ACS Appl. Mater. Inter.* **3** 3910
[2] Yi G S and Chow G M 2006 *Adv. Funct. Mater.* **16** 2324
[3] Philips M L F, Hehlen M P, Nguyen K, Sheldon J M and Cockroft N J 2000 *Proc. Electrochem. Soc.* 2000, **99-40** 123
[4] Scheps R *Prog.* 1996 *Quantum. Electron.* **20** 271
[5] Rapaport A, Milliez J, Bass M, Cassanho A and Jenssen H, 2006 *J. Disply Technol.* **2** 68
[6] Hoeppe H A 2009 *Angew. Chem. Int. Ed.* **48** 3572
[7] Eliseeva S V and Bunzli J-C G 2010 *Chem. Soc. Rev.* **39** 189
[8] Naczynski D J, Andelman T, Pal D, Chen S, Riman R E, Roth C M and Moghe P V 2010 *Small*, **106**, 063118/1-063118/12
[9] Li C and Lin J., 2010 *J. Mater. Chem.* **20** 6831
[10] Tian G, Gu Z, Zhou L, Yin W, Liu X, Yan L, Jin S, Ren W, Xing G, Li S and Zhao Y 2012 *Adv. Mater.* **24** 1226
[11] Wang F and Liu X 2009 *Chem. Soc. Rev.* **38** 976
[12] Nyk M, Kumar R, Ohulchanskyy T Y, Bergey E J and Prasad P N 2008 *Nano Lett.* **8**, 3834

Acknowledgment

One of the authors, PPN acknowledges the Director, Sree Chitra Tirunal Institute for Medical Sciences and Technology for his kind permission to present this work at the ICOOPMA-2014 conference and publish this work in the Journal of Physics: Conference Series (JPCS). PPN also acknowledges the Department of Science and Technology, Government of India for a research fellowship.

Design of erbium doped double clad ZBLAN Fibre laser

A Oladeji[1], A Phillips[1], S Lamrini[2], K. Scholle[2], P Fuhrberg[2], AB Seddon[1], TM Benson[1], S Sujecki[1]

[1]*Division of Electrical Systems and Optics, Faculty of Engineering, University of Nottingham, University Park, Nottingham, NG72RD, England*

[2]*LISA laser products OHG Fuhrberg & Teichmann Max-Planck-Str. 1 D- 37191 Katlenburg-Lindau Germany*

Email: oladeji.ayodele@nottingham.ac.uk

Abstract. A high powered octagonal double clad ZBLAN (33 μm/330 μm, NA=0.13) glass fibre for mid-infrared light generation is studied using a one dimensional rate equation model. The fibre laser design employs the concept of cascade lasing and includes up-conversion phenomena. The results obtained demonstrate that efficient cascade lasing may be achieved in practice without the need for fibre grating fabrication, as a sufficient level of feedback for laser action is provided by Fresnel light reflection at ZBLAN glass fibre air interfaces. Further enhancement of the laser efficiency can be achieved by terminating one of the fibre ends with a mirror. Simulation results show that the laser operation with 20 W of pump power at 0.98 μm wavelength can be achieved at 2.75 μm operating wavelength with Er^{3+} ion concentrations of 60,000 ppm.

1. Introduction

Fibre lasers are the subject of considerable research interest. Erbium-doped silica fibre lasers with output power levels around 297 W [1] have been demonstrated under continuous wave (CW) operation. The natural minimum loss at 1.5 μm possesses significant challenge in extending the emission wavelengths of silica based fibre lasers. At longer wavelengths, the emission efficiency of erbium in silica is limited by the high phonon energy of the glass matrix which introduces competing multi-phonon relaxation paths. ZBLAN on the other hand, is a heavy metal fluoride glass with a wide transmission window of 0.3 – 5 μm and a good rare earth ion solubility. Compared with silicate glasses, ZBLAN glass has a lower maximum phonon energy (of ~565 cm^{-1}[2]) because the bond strength is weaker while atoms forming the glass are heavier [3].

Double clad Fibre provides opportunities for realisation of high power lasers with improved confinement when compared with the traditional step index single clad Fibres. In this work, we focus on two geometries of the double-clad Fibre with different doping levels of the erbium ions. In the model employed we include up-conversion processes that produce the green photoluminescence in ZBLAN fibre laser. Under a continuous wave (CW) pump an investigation of the influence of the fibre length, input pump power and facet reflectivity on the laser operating characteristics are studied. Finally, the modelling results are compared with experimental ones.

2. Fibre laser Modelling

The measured absorption cross section spectra used in this study have been taken from [4] and were fitted using Gaussian functions: ($a\,exp\left(-\ln(2)(\frac{\lambda-\lambda_0}{d\lambda})^2\right)$, where a is the amplitude, λ_0 is the wavelength position of the peak and $d\lambda$ is the half width at half maximum). The corresponding emission cross sections were calculated using McCumber [5] and Judd-Ofelt [6] theory. Figures 1a, 1b, and 1c show cross sections for the pump (~0.98 μm), idler (~1.55 μm and signal (~2.7 μm) waves that were used in the calculations.

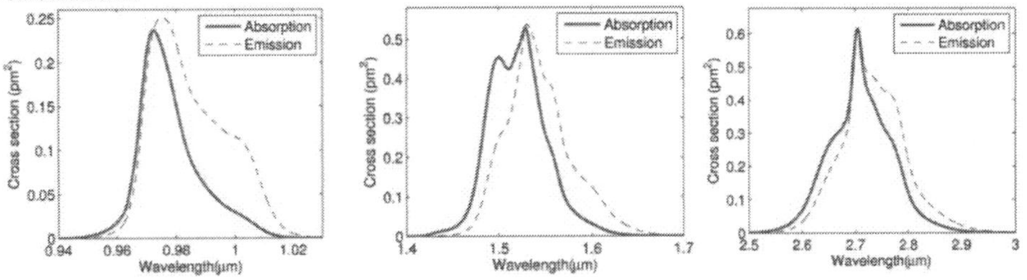

(a) (b) (c)

Figure 1 Emission and absorption cross sections for: (a) – pump, (b) - idler and (c) - signal

Two double-clad fibre geometries that are referred to as 'Fibre 1' and 'Fibre 2' are investigated in this work. Fibre 1 is a 5000 ppm erbium doped ZBLAN fibre with 4.7 μm core diameter and 124 μm circular inner cladding diameter. Fibre 2 is a 60000 ppm erbium doped ZBLAN fibre with a 33 μm core diameter and a 330 μm octagonal inner cladding diameter. We note that both fibres are commercially available. Their schematic diagrams of cross sectional refractive index distributions are presented in figures 2a and 2b.

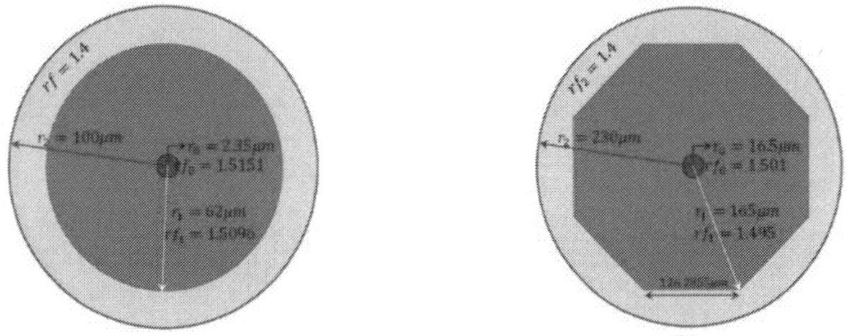

(a) (b)

Figure 2 Refractive index distribution for: (a) – 'Fibre 1', i.e. circular double clad fibre – doped with erbium ions at 5000ppm and (b) – 'Fibre 2', i.e. octagonal double clad fibre - doped with erbium ions at 60,000ppm

The properties of the erbium doped ZBLAN fibre laser are investigated theoretically using the rate equations (1). Figure 3 shows the corresponding energy-level diagram for the erbium ion doped ZBLAN glass. In Fibre 1, the interactions of ion-ion energy transfer up-conversions are labelled as W_{2206} , W_{1103}. The cross relaxation process which in Fig.3 is labelled as C_{0513} was ignored given the relatively low level of erbium doping.

Fibre 2 on the other hand, is highly doped which makes ion-ion interactions highly probable. In our experimental setup, we observed a visible green-glow when using this fibre. This confirms an emission from the thermal levels $^2H_{11/2}/^4S_{3/2}$ which can only be populated by energy transfer from lower lying energy states. Excited ions at the $^4F_{7/2}$ levels are rapidly depleted by multi-phonon transition to the thermal levels $^2H_{11/2}/^4S_{3/2}$ as the energy gap can be bridged by 2 phonons. This effectively makes the 3 levels: $^4F_{7/2}$, $^2H_{11/2}$ and $^4S_{3/2}$ to act as a single band. The energy level $^4F_{9/2}$ is not directly populated by any of the energy transfer processes or indirectly by multi-phonon relaxations. This is because of the relatively large energy gap that separates level $^4F_{9/2}$ from the upper lying level - $^4S_{3/2}$. As a result of this, the contribution of $^4F_{9/2}$ to the system at equilibrium is negligible. The energy diagram is therefore represented by 5 levels: $^4I_{15/2}$, $^4I_{13/2}$ $^4I_{11/2}$, $^4I_{9/2}$ and $^4S_{3/2}/^2H_{11/2}/^4F_{7/2}$. In simulations the coefficients for energy transfer up-conversion and cross relaxation processes were taken from the available literature [7–9] and are listed in Table 1.

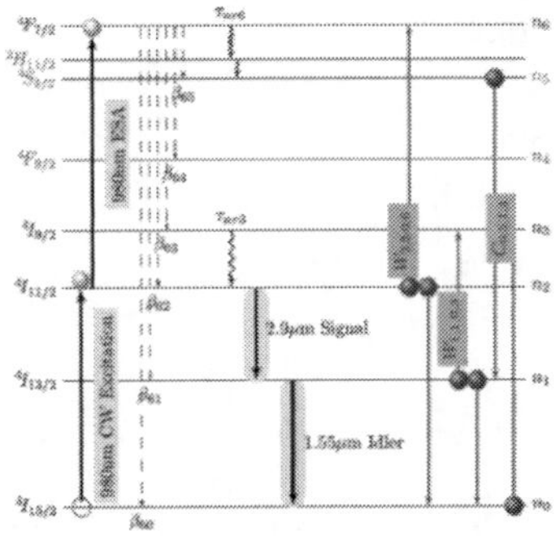

Figure 3. Energy level diagram of Er:ZBLAN showing pump, signal and idler transitions.

Thus we obtain the rate equations (1) which are complemented by the conservation law, equation (2), where $R_\lambda = P_\lambda/h\upsilon_\lambda A_{core}$ is the photon flux rate, P_w denotes the power, h is the Planck's constant, A_{core} is the area of the fibre core. $\sigma_{a\lambda}$ and $\sigma_{e\lambda}$are the absorption and emission cross sections respectively. The subscript λ represents the pump(p), signal(s) and idler(id) wavelengths. τ_i is the total lifetime which includes both radiative (τ_r) and non-radiative (τ_{nr}) processes. β_{ij} is the branching ratio from level i to j. The second order interactions are represented by W_{in} while the number of ions involved in the interactions between levels *l* and *t* are represented by *m*.

$$\frac{dn_k}{dt} = \pm(n_x\sigma_{a\lambda} - n_y\sigma_{e\lambda})R_\lambda \pm mW_{in}n_l n_t + \sum_{i=k+1}^{j}\frac{\beta_{ik}n_l}{\tau_{ri}} - \frac{n_k}{\tau_{nrk}} + \frac{n_k}{\tau_{rk}} + \frac{n_{k-1}}{\tau_{nr(k-1)}} \qquad (1)$$

$$N_{Er} = n_0 + n_1 + n_2 + n_3 + n_4 \qquad (2)$$

The evolution of the pump, signal and idler powers can be calculated by solving a set of ordinary differential equations [10]:

$$\frac{dP_\lambda{}^\pm}{dz} = \pm\Gamma_\lambda P_\lambda{}^\pm(n_x\sigma_{a\lambda} - n_y\sigma_{e\lambda}) \pm \alpha_\lambda P_\lambda{}^\pm \qquad (3)$$

The symbol '±' represents the direction of the travelling waves. The total power P_w at any point along the fibre is the sum of the forward (P_λ^+) and backward (P_λ^-) travelling waves. At the ends of the fibre, (*z=0, z=L*), the forward and backward propagating photon fluxes are partially reflected. If we denote the reflectivity of the input and output mirror by R_{in} and R_{out}, respectively, we obtain the boundary conditions for equations (3): $\phi^+_{(z=0)} = R_{in}\phi^-_{(z=0)} + (1 - R_{in})P_p$ and $\phi^-_{(z=L)} = R_{out}\phi^+_{(z=L)}$. The fibre ends were either terminated by air-glass interface or a mirror. The equations (1)-(3) were solved using coupled solution method [10]. The simulation parameters are τ_{rk} = 0.73-8.52ms[8], τ_{nr2} = 11, τ_{nr3} = 0.58 ms[8][11] W_{2204}=2.8e^{-17}, W_{1103}=1.0e^{-17} , C_{0413}=2.7e^{-17} cm^3/s[7], Pump overlap factor = 0.14% (circular), 1.11% (octagonal).

3. Results

In the simulations we assume that our pump laser module delivers 60 W of power. In Figure 4a,b we show the dependence of the signal and idler power on the fibre length for the pump power equal 60 W. These results show that the optimal length of the fibre for Fibre 1 is 17.3 m for the laser configuration that relies on Fresnel reflections only while for the configuration that uses a mirror, it is 13.2 m. Fibre 2 has an optimum length of 2.8 m for the laser configurations that relies on Fresnel reflections only and 1.8m for the configuration that uses a mirror. Figure 4c,d shows the dependence of the idler and signal powers on the pump power. Fibre 1 produced a slope efficiency of 2.1% (Signal) and 3.7%

(Idler) for the laser configuration that relies on Fresnel reflections only while the one that uses mirror has slope efficiencies of 4.9% (Signal) and 8.4% (Idler). Fibre 2 produced a slope efficiency of 13.0% (Signal) and 22.8% (Idler) for the laser configuration that relies on Fresnel reflections only while the one that uses mirror produced slope efficiencies of 27.8% (Signal) and 48.2% (Idler).

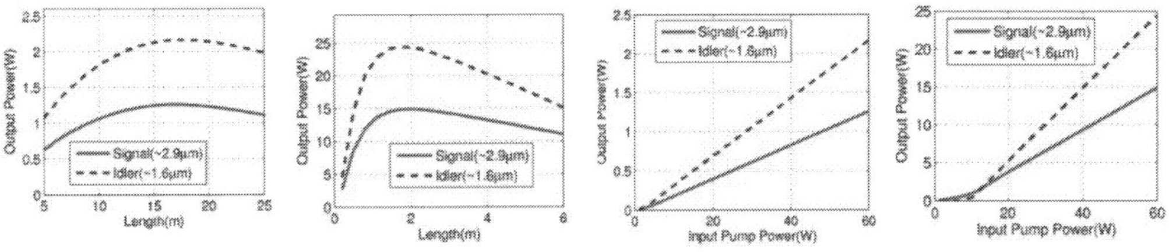

(a) Fibre1: Fresnel, idler and output signal powers versus fibre length

(b) Fibre2: Mirror, idler and output signal powers versus fibre length

(c) Fibre1: Fresnel, idler and output signal versus input pump power

(d) Fibre 2: Mirror, Idler and output signal power versus input pump power

Figure 4. Calculated dependence of idler & output signal powers on length and pump power.

4. Conclusions

In this paper, two double-clad fibres have been considered for the realisation of a mid-infrared fibre laser. The proposed device uses either Fresnel reflections at both ZBLAN glass-air interfaces or the Fresnel reflection on one side and a fully reflective mirror on the other one. The results obtained show that the device that uses a mirror is more efficient. Further, the highly doped, double-clad octagonal geometry with the larger core allows achieving higher optical efficiency than the circular clad fibre with a comparatively lower concentration of erbium.

ACKNOWLEDGEMENTS

This research has been supported by the University of Nottingham, England and by European Commission through the Framework Seven (FP7) Project: MINERVA (317803; www.minerva-project.eu): MId- to NEaR infrared spectroscopy for improVed medical diAgnostics.

References

[1] Jeong Y, Yoo S, Codemard C A, Nilsson J, Sahu J K, Payne D N, Horley R, Turner P W, Hickey L, Harker A, Lovelady M and Piper A 2007 Erbium:ytterbium codoped large-core Fibre laser with 297-W continuous-wave output power *IEEE J. Sel. Top. Quantum Electron.* **13** 573–8

[2] Almeida R M and Mackenzie J D 1981 Vibrational spectra and structure of fluorozirconate glasses *J. Chem. Phys.* **74** 5954–61

[3] Jackson S D 2012 Towards high-power mid-infrared emission from a fibre laser *Nat. Photonics* **6** 423–31

[4] Caspary R 2002 *Applied Rare Earth Spectroscopy for Fibre Laser Optimization* (Shaker Verlag GmbH)

[5] McCumber D 1964 Einstein Relations Connecting Broadband Emission and Absorption Spectra *Phys. Rev.* **136** A954–A957

[6] Walsh B M 2006 Judd-ofelt theory: principles and practices brian m. walsh 403–33

[7] Pollnau M and Jackson S D 2001 Erbium 3- m Fibre Lasers **7** 30–40

[8] Pollnau M and Jackson S D 2002 Energy recycling versus lifetime quenching in erbium-doped 3-μm Fibre lasers *IEEE J. Quantum Electron.* **38** 162–9

[9] Golding P S, Jackson S D, King T A and Pollnau M 2000 Energy transfer processes in Er^{3+} -doped and Er^{3+} , Pr^{3+} -codoped ZBLAN glasses **62** 856–64

[10] Sujecki S, Sójka L, Bereś-Pawlik E, Tang Z, Furniss D, Seddon a. B and Benson T M 2010 Modelling of a simple Dy3+ doped chalcogenide glass fibre laser for mid-infrared light generation *Opt. Quantum Electron.* **42** 69–79

[11]Moos H W 1970 Spectroscopic relaxation processes of rare earth ions in crystals *J. Lumin.* **1–2** 106–21

Non-linear optical response of bulk chalcogenide glasses near the fundamental absorption band edge

E A Romanova[1], A I Konyukhov[1], V I Kochubey[1], T M Benson[2], E Barney[2], N Abdel-Moneim[2], D Furniss[2], A B Seddon[2] and S Guizard[3]
[1] Saratov State University, Astrakhanskaya 83, 410012 Saratov, Russia
[2] University of Nottingham, University Park, NG72RD Nottingham, UK
[3] Ecole Polytechnique, 91128 Palaiseau Cedex, France

E-mail: romanova@optics.sgu.ru

Abstract. Magnitudes of the non-linear coefficients of absorption and refraction have been evaluated near the bandgap wavelengths of chalcogenide glasses of the system As-S-Se by using the interferometric pump-probe method and are compared with literature data. Photo-excited plasma dynamics and long-time scale variation of the dielectric constant have been studied by comparison with the results of numerical modelling of the behaviour of the glasses when heated by the ultra-short laser pulses.

1. Introduction

As other amorphous semiconductors, chalcogenide glasses have a disordered lattice and localised states of electrons in their bandgaps. Due to these specific structural and electronic properties, fundamental absorption band edge of amorphous semiconductors is not sharp. It has a spectral range of exponential decay of the single-photon absorption coefficient α that is usually referred to as the Urbach tail. Chalcogenide glasses have also weak absorption tails with $\alpha < 1$ cm^{-1} [1].

Using of chalcogenide glasses, as highly non-linear materials transparent in the mid-infrared, can enhance non-linear optical effects and reduce the dimensions of integrated and fibre optical devices. However a theory of the non-linear optical response of chalcogenide glasses has not yet been fully developed. For direct-gap crystalline semiconductors, the values of non-linear refractive index n_2 in the low-frequency limit have been obtained by using the bond-orbital approximation [2]. The non-linear Kramers-Kronig relation has been applied to obtain spectral dependencies of n_2 and of the two-photon absorption coefficient β_2 near the bandgap frequencies of the direct-gap and indirect-gap crystalline semiconductors in [3] and [4], respectively, by using two-band models. For amorphous semiconductors, only some experimental results of measurements of n_2 and β_2 are currently available. In particular, for chalcogenide glasses, these parameters have been measured mostly in the spectral range above their two-photon bandgap wavelengths [5-8].

In this paper, we study the non-linear optical response of chalcogenide glasses of the compositions $As_{40}Se_{60-x}S_x$ (x = 0,10,15,20,30,40,45,60) (atomic %) by using the time-resolved interferometric pump-probe method [9]: evaluate n_2 and β_2 near their bandgap wavelengths and characterise plasma dynamics at the Urbach tail. Different compositions fit the pump pulse peak wavelength of 790 nm (the photon energy is shown in figure 1 as a dashed line) at different points of their Urbach tails.

[1] To whom any correspondence should be addressed.

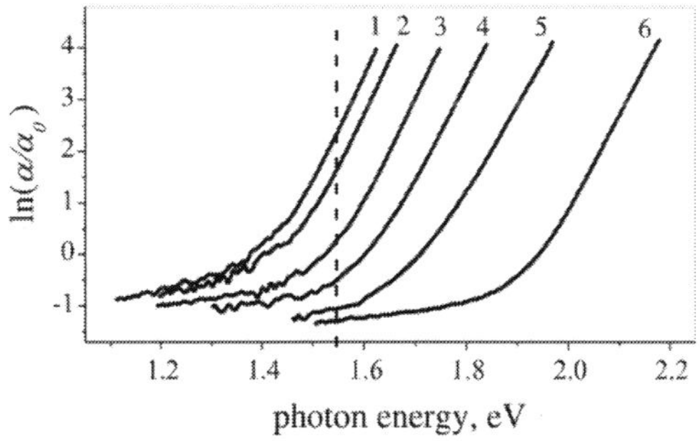

Figure 1. Logarithm of the absorption coefficient of the compositions $As_{40}Se_{60-x}S_x$, x = 0 (1), 10 (2), 20 (3), 30 (4), 40 (5), 60 (6). The linear parts of the curves correspond to the Urbach tails, α_0=1 cm^{-1}.

2. Non-linear optical response at low energies of the pump pulse

In the experiment, a glass sample (hot pressed between two precision aligned tungsten carbide plates with flatness of 0.08 µm and a surface finish of 0.009 µm at ~25 °C above its glass transition temperature in an in-house-built vacuum rig) was probed with two collinear 50 fs pulses separated by a fixed time delay. When a high intensity pump pulse was focused on the sample (by a lens with a focal distance of 300 mm after a 3 mm aperture) at some time between the two probe pulses, the second probe pulse experienced a disturbed dielectric constant. The induced phase shift φ and absorbance were measured by a spectrometer as a function of time by changing the delay Δt between the pump pulse and the second probe pulse. At relatively low pump energies E, the phase shift was induced by non-linear refraction and two-photon absorption (figure 2a) or single- and two-photon absorption (figure 2b) of the pump pulse. The effect of cross-phase modulation is described by the system of differential equations for the pump pulse intensity I_{pu} and the probe pulse intensity I_p:

$$\frac{\partial I_{pu}}{\partial z} = -(\alpha + \beta_2 I_{pu})I_{pu}\,; \quad \frac{\partial I_p}{\partial z} = -(\alpha + 2\beta_2 I_{pu})I_p\,; \quad \frac{\partial \varphi}{\partial z} = 2kn_2 I_{pu} \qquad (1)$$

Parameters n_2 and β_2 evaluated by solution of Eqs.(1) are shown in figure 3 for some compositions (filled circles at 790 nm). For comparison, the wavelength dependencies of n_2 and β_2 have been plotted by using the spectral function obtained in [3] for direct-gap crystalline semiconductors.

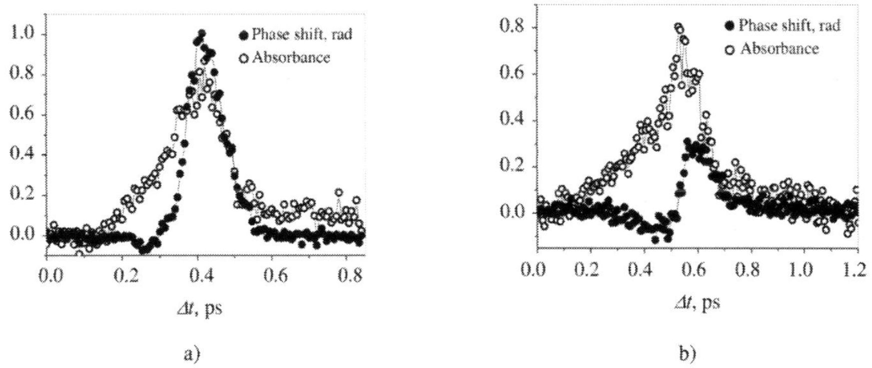

Figure 2. Time-resolved non-linear optical response of the compositions: (a) $As_{40}S_{60}$, E = 1.0 µJ and (b) $As_{40}Se_{50}S_{10}$, E = 0.7 µJ.

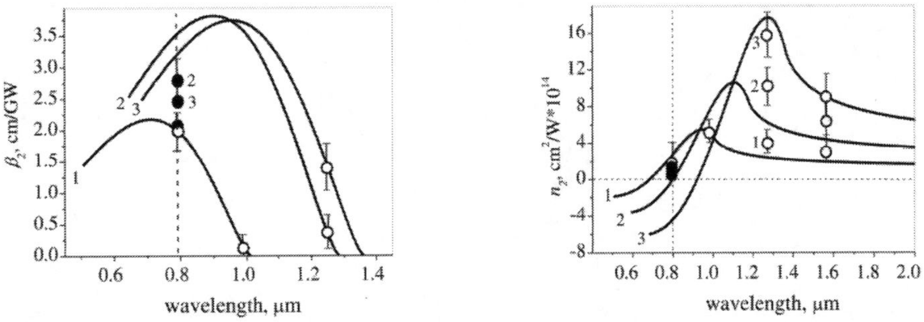

a) b)

Figure 3. Measured magnitudes of (a) β_2 and (b) n_2 for the compositions: $As_{40}S_{60}$ (1), $As_{40}Se_{30}S_{30}$ (2), $As_{40}Se_{50}S_{10}$ (3) (filled circles) and literature data [5-8] (hollow circles).

The curves in figure 3b were obtained by fitting the values of n_2 calculated in the low-frequency limit [2]. Each curve ends at the bandgap wavelength λ_g of a particular composition. In some range of wavelengths near λ_g, these curves exhibit negative values of n_2. In our measurements, only positively-valued n_2 have been obtained for all the glass compositions. In estimations of n_2 and β_2, maximum magnitudes of the measured phase shift and absorbance at the probe beam axis have been used.

3. Non-linear optical response at high energies of the pump pulse

At higher energies, the effect of plasma formation due to the single- and two-photon absorption is to be taken into account. In [10], three basic scenarios of the plasma dynamics have been distinguished depending on the ratio $P=h\nu/E_g$ of the irradiating photon energy to the bandgap energy. In figure 4, the positive-valued maximum corresponding to the self-induced refraction of the pump pulse is followed by the fast phase shift decrease in $As_{40}S_{60}$ sample (P=0.65) due to plasma formation and a subsequent slow phase shift increase due to the recombination of carriers. The effect of plasma formation is weakening when partially replacing S by Se. For the $As_{40}S_{20}Se_{40}$ sample, P=0.74 and for the $As_{40}S_{15}Se_{45}$ sample, P=0.77. In the latter sample, there were no free electrons after the pump pulse.

Permanent positive variation Δn_p of the refractive index observed at the long-time scale and associated with photodarkening or the formation of excitons, has been compared with the results of numerical simulations of the glass heating [11]. A correlation with a sample surface heating above the glass transition temperature (T_g) and energy threshold of Δn_p observation has been revealed (figure 5).

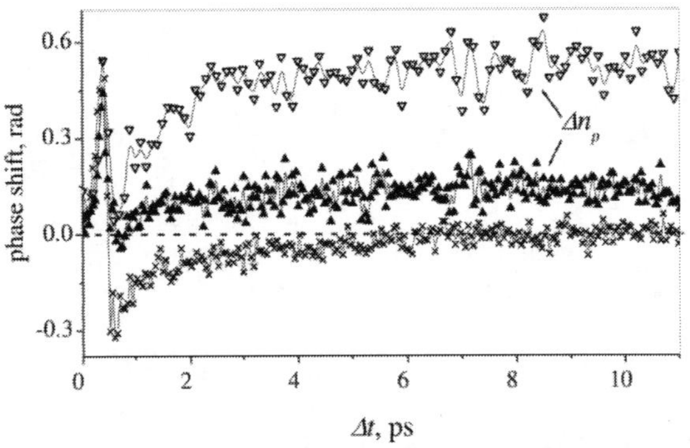

Figure 4. Time-resolved non-linear optical response of the compositions: $As_{40}S_{60}$ (**x**), E = 3μJ; $As_{40}S_{20}Se_{40}$ (▲), E = 4 μJ; $As_{40}S_{15}Se_{45}$ (▼), E = 9 μJ.

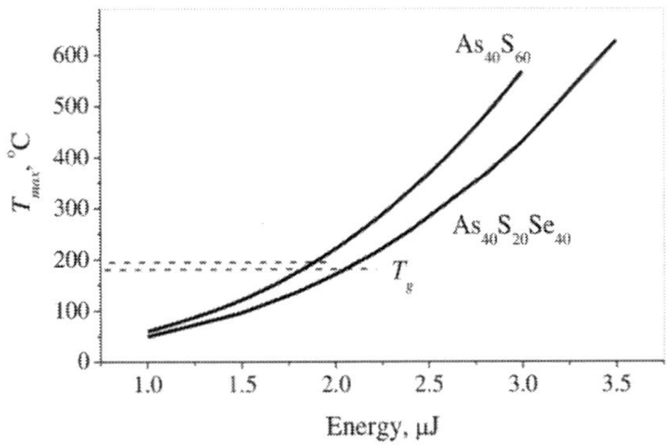

Figure 5. Maximum temperature in the irradiated zone at different pump pulse energies.

4. Conclusions

Study of the non-linear optical response of chalcogenide glass samples of the system As-S-Se has demonstrated that all the samples have the positive-valued n_2 unlike the direct-gap crystalline semiconductors having the negative-valued n_2 near their bandgap wavelengths. By comparison with the available literature data we have shown that the spectral curve obtained in [3] for the direct-gap crystalline semiconductors, agrees with the spectral dependence of the non-linear coefficients of chalcogenide glasses at wavelengths more than the two-photon bandgap wavelength.

A scenario of the photo-excited plasma dynamics depends on the ratio of the photon energy to the bandgap energy. If $P > 0.75$, fast trapping of carriers results in the lack of free electrons after the pump pulse. By numerical simulation of the glass sample heating in the irradiated zone, we have found a correlation between the permanent change of the dielectric constant at the long-time scale and glass heating above the glass transition temperature.

Acknowledgments

The research leading to these results has received funding from LASERLAB-EUROPE (grant agreement no. 284464, EC's Seventh Framework Programme) and The Royal Society (IES-2013/R2). E.Romanova and S.Guizard are grateful to S.Klimentov for taking part in the measurements.

References

[1] Zakery A, Elliot S R 2007 Optical nonlinearities in chalcogenide glasses and their applications (Springer-Verlag, Berlin, Heidelberg, New York)
[2] Lines M E 1990-II *Phys.Rev.B.* **41** 3383
[3] Sheik-Bahae M, Van Stryland E W, Hagan D J, Hutchings D C 1991 *IEEE J. of Quantum Electron.* **27** 1296
[4] Dinu M 2003 *IEEE J. of Quantum Electron.* **39** 1498
[5] Sanghera J S, Shaw L B, Pureza P, Nguyen V Q, Gibson D, Busse L, and Aggarwal I D 2010 *Int. Journ. of Applied Glass Science* **1** 296
[6] Boudebs G, Sanchez F, Troles J, Smektala F 2001 *Opt.Commun.* **199** 425
[7] Qian Z G, Shen W Z, Ogawa H and Guo Q X 2002 *J. Appl. Phys.* **92** 3683
[8] Blonskyi I, Kadan V, Shpotyuk O, Iovu M, Pavlov I 2010 *Opt. Materials* **32** 1553
[9] Mao S S, Quere F, Guizard S, Mao X, Russo R E, Petite G, Martin P 2004 *Appl.Phys.A* **79** 1695
[10] Romanova E, Chumakov K, Mouskeftaras A, Guizard S, Abdel-Moneim N, Furniss D, Seddon A B, Benson T M 2013 *15th International Conference on Transparent Optical Networks* 1-4
[11] Romanova E A, Kuzyutkina Yu S, Konyukhov A I, Abdel-Moneim N, Seddon A B, Benson T M, Guizard S and Mouskeftaras A 2014 *Opt. Engineering* **53** 071812

Journal of Physics: Conference Series **619** (2015) 012046 doi:10.1088/1742-6596/619/1/012046

Dynamics of silver photo-diffusion into Ge-chalcogenide films: time-resolved neutron reflectometry

Y Sakaguchi[1], H Asaoka[2], Y Uozumi[2], Y Kawakita[2], T Ito, M Kubota[2], D Yamazaki[2], K Soyama[2], M Ailavajhala[3], M R Latif[3], K Wolf[3], M Mitkova[3] and M W A Skoda[3]

[1] Comprehensive Research Organization for Science and Society (CROSS), Tokai, Ibaraki 319-1106 Japan
[2] Japan Atomic Energy Agency, Tokai, Ibaraki 319-1195 Japan
[3] Boise State University, Boise, ID U.S.A
[4] ISIS, STFC, Rutherford Appleton Laboratory, Didcot, OX110QX, UK

E-mail: y_sakaguchi@cross.or.jp

Abstract. Silver diffuses into an amorphous (a-) chalcogenide layer while visible light illuminates Ag/a-chalcogenide films and neutron reflectometry is a suitable technique probing time evolution of the depth profiles without damaging the sample by the probe beam itself. In this paper, we report the results of time-resolved neutron reflectivity measurements of a-$Ge_{40}Se_{60}$/Ag/ Si films taken while the films are exposed to visible light. From the measurements, we found enormous changes in the neutron reflectivity profile, including a loss of total reflection region, with continuous illumination even after forming one homogeneous layer, which occurred about 50 min after starting illumination. At this stage, a clear off-specular scattering was observed by a linear detector and a surface roughness was observed with naked eyes.

1. Introduction

Silver photo-diffusion in amorphous (a-) chalcogenide films such as a-As_2S_3 and a-$GeSe_2$ has attracted much interest because of the potential applications in, for example, photo-lithography [1, 2] and non-volatile memory devices [3]. Investigation of the dynamics of the silver photo-diffusion is considered to be quite important in order to get good performance in devices, and in fact, previous numerous studies have shown that this diffusion exhibits unique behaviour in the dynamical process. According to Rutherford Backscattering (RBS) studies [4, 5, 6], there is a step-like silver concentration in the depth profile after the silver diffusion takes place. This feature is different from typical diffusion processes in which the concentration gradually decreases or fades out. Although RBS is a powerful non-destructive technique for determining depth profiles, the experiments in this case were very challenging because a strong He^+ beam can also induce silver diffusion in the films. For this reason, a lower He^+ beam intensity was used in the reported experiments and the time-evolution measurement was not successful. Considering such a situation, Wagner et al. performed a combinational method using RBS and modified optical reflectivity measurements [6]. In the modified optical reflectivity measurement, they measured the reflected beam from the film irradiated from the chalcogenide side as a function of time. It resulted in an oscillatory change due to interference with the optical path in the film and the wavelength of the probe laser beam. From the result, they estimated the thickness of the chalcogenide layer as a function of time. After that, they selected several illumination times in order to see all the features of the diffusion process, and prepared several films with different illumination times. For these films, they measured RBS with an appropriate He^+ beam intensity and found the dynamics of the depth profile. According to their conclusion, there is a step-like silver concentration profile in the silver reaction layer, and the reaction layer proceeds to the chalcogenide layer, resulting in the disappearance of the chalcogenide layer in the final stage. Of course, there is no doubt that their carefully treated measurements provide a conclusive picture on silver diffusion dynamics. However, there is one unanswered question in their treatment: Is the diffusion process really frozen by stopping the illumination? In this sense, it is desirable to measure the depth profile *in situ* during exposure of the sample to light. X-ray and Neutron reflectivity are powerful techniques that can be used to reveal such transient depth profiles. Of these, synchrotron radiation provides brilliant photon fluxes for time-resolved X-ray reflectivity measurements and can readily discern small changes in surface or interface structure. However, it is known that X-rays can induce silver diffusion in chalcogenide materials [7] and the strong possibility of rapid sample decomposition due to the synchrotron radiation beam – even before any data can be collected – cannot be overlooked. Neutrons, on the other hand, offer a safer approach by excluding the possibility of changes induced by the probe beam and, the use of an intense pulsed neutron source is well-suited to realize time-resolved measurements. So far, we have applied time-resolved neutron reflectivity technique for the investigation on the silver diffusion dynamics into germanium sulphide films and obtained successful results to reveal the diffusion process [8]. In this paper, we report our recent results of time-resolved neutron reflectivity measurement for Ag/Ge-Se films while being exposed to visible light, and discuss the silver photo-diffusion process into the films.

2. Experimental

The neutron reflectivity measurements of Ag/Ge-Se films were carried out on INTER [9, 10] at ISIS (UK). Pulsed neutrons are provided at a 10Hz repetition rate. The specular reflectivity was measured with a ^{3}He detector with the time-of-flight technique. The available wavelength range was 1.5 to 15 Å. The kinetic data were measured at an incident angle of 0.55°, resulting in a Q range from 0.008 to 0.05 Å^{-1}. The full profile before and after illumination was measured at two angles, 0.55° and 1.6°, resulting in a Q range from 0.008 to 0.23 Å^{-1}. The data were acquired within a 30 second interval and they were averaged to improve their statistics. In addition, a linear detector was used to measure off-specular scattering for a specific case. White light from a 300 W xenon lamp (MAX-303, ASAHI Spectra, Co., Ltd.) was used as an excitation light source, and the exposure of the sample was under computer-control. The samples were prepared using thermal evaporation. The thicknesses were estimated using the output from a quartz crystal microbalance.

3. Results and Discussion

Figure 1 shows the time evolution of the neutron reflectivity profiles of a-$Ge_{40}Se_{60}$ 1500Å / Ag 500 Å / Si substrate film. At the initial stage, before light exposure, we confirmed that silver diffusion had already taken place forming a reaction layer with about 250Å thickness, and leaving a thin Ag layer with 30Å thickness (see, Table 1). After starting light exposure, silver diffusion proceeds further, and one homogenous layer was formed at 20.5-25.5 min as shown in Table 1. Usually, the diffusion dynamics are completed at this stage. However, further changes in the neutron reflectivity profile have been observed. Firstly, the intensity at total reflection region decreased around 50 min, and the flat total reflectivity region was totally lost at 76 min. Secondly, the fringes in the neutron reflectivity profile also changed after 50 min and the osculated pattern was totally lost at 76 min. Such a situation cannot be followed as long as only a typical specular scattering is assumed. Therefore, we used a linear detector instead of a point ^{3}He detector and measured the intensities as a function of neutron wavelength and scattering angle. Figure 2 shows a scattering angle-wavelength map for the film after having the unusual changes, together with a "usual" map obtained from Ag 500Å film. In the map of Ag 500Å film, scattered neutrons are observed at the same angle as the incident angle, 0.55°, suggesting specular scattering. At the angle, a bright yellow line (6-15 Å), indicating a total reflection, and repeated light blue-bright blue pattern (1.75-6 Å), indicating fringes on the reflectivity profiles, are observed. On the other hand, in the map of a-$Ge_{40}S_{60}$ 1500Å / Ag 500Å /Si substrate film, the scattered neutrons are diffused from 0.1 ° to 1.0 ° around 0.55°. It is apparent that the losses of the total reflection region and the fringes, observed in Fig.1, result from the off-specular scattering, and this is attributed to a large (e.g. micrometer-size) surface roughness. In fact, the surface roughness is visible even with the naked eyes as shown in Figure 3(a). Also, the image of an optical microscope in Figure 3 (b) shows that there are inhomogeneous patterns on micrometer-scales on the surface.

Considering the change in the depth profiles (Table 1), the enormous surface roughness is caused by a photo-induced structural change on a homogeneous silver-diffused film. The further light illumination must trigger a structural transformation from the homogeneous Ag-Ge-Se layer to more stable state. Also, it is noted that the change occurred only when the Ge composition is 40% and the $Ge_{40}Se_{60}$ layer was on top. Such change was not observed for other films; Ag/$Ge_{20}Se_{80}$/Si, Ag/$Ge_{40}Se_{60}$/Si, and $Ge_{20}Se_{80}$/Ag/Si. Therefore, the occurrence of this phenomenon seems to be composition-sensitive and the compositional stability must be related to the change. At present, we cannot conclude what the more stable state is. However, there are some candidates to explain the results. The first possibility is a photo-induced surface deposition, in which silver particles appear [11]. This is the opposite process to photo-diffusion. The second possibility is a photo-induced phase separation, in which Ag-rich stable domains and Ag-poor stable domains co-exist. The third possibility is particle-formation, in which self-aggregation takes place. In any case, further investigations are required to understand the physical origin of the surface roughness.

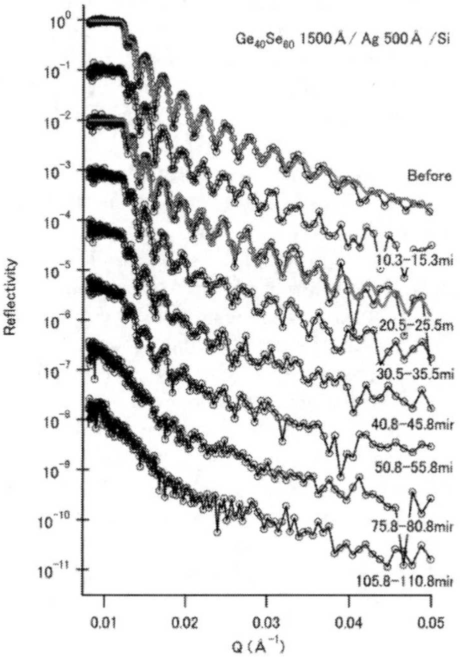

Figure 1 Time evolution of the neutron reflectivity profile of a-$Ge_{40}Se_{60}$ 1500Å / Ag 500Å / Si substrate film. Thick red curves indicate the fitted curves using the parameters in Table 1.

	thickness	SLD	roughness
$Ge_{40}Se_{60}$	1594	3.01	15.3
Ag-$Ge_{40}Se_{60}$	252	3.24	2.1
Ag	30	3.47	0.5

Before light illumination

	thickness	SLD	roughness
Ag-$Ge_{40}Se_{60}$	1943	3.13	18.0

At 20.5-25.5 min

Table 1 Parameters used for curve fitting; thickness (Å), scattering length density (SLD) (x 10^{-6} $Å^{-2}$), and roughness (Å).

4. Conclusion

We performed time-resolved neutron reflectivity measurements for Ag/$Ge_{40}Se_{60}$ films at ISIS to clarify the silver photo-diffusion into amorphous chalcogenide. From the measurement of the $Ge_{40}Se_{60}$/Ag film, we found that enormous surface roughness appeared with continuous light illumination even after completing the silver diffusion. The structural origin of this change is not clear. However, other techniques such as X-ray diffraction, scanning electron microscope, and atomic force microscopy would be useful to clarify the origin.

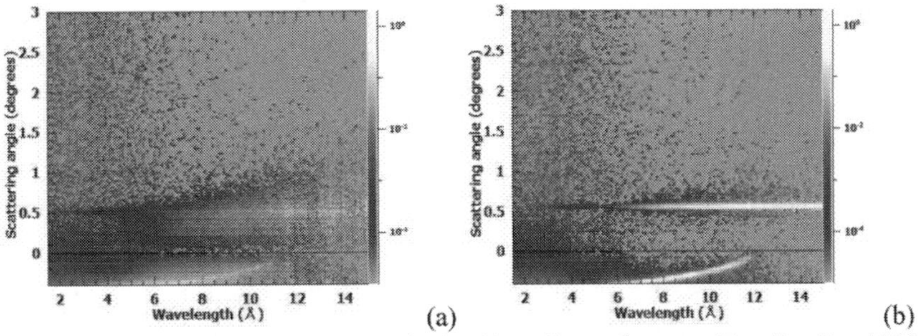

Figure 2 Scattering angle-wavelength maps obtained by a linear detector. (a) a-$Ge_{40}Se_{60}$ 1500Å /Ag 500Å /Si substrate , (b) Ag 500Å / Si substrate. The incident angle was 0.55°.

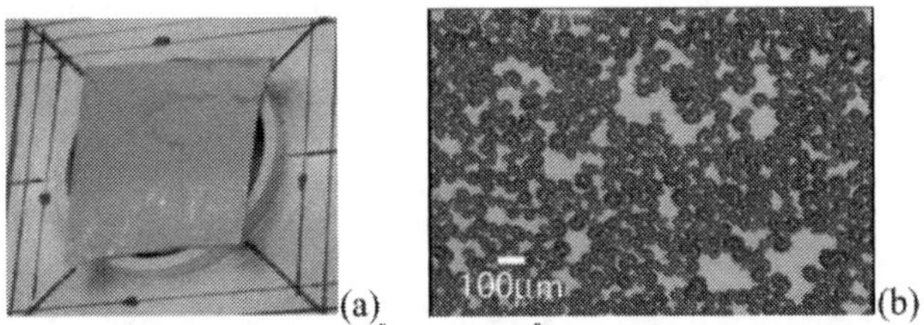

Figure 3 (a) Photograph of a-$Ge_{40}Se_{60}$ 1500Å /Ag 500Å /Si substrate, taken after the whole neutron reflectivity measurements, in which a large off-specular scattering was observed by a linear detector. The film on a Si wafer is about 30mm x 30mm. (b) Image of an optical microscope.

Acknowledgements

This work was supported by JSPS Grant-in-Aid for Scientific Research (C) Grant Number 25400435 and by Defense Threat Reduction Agency under Grant HDTRA 1-11-1-0055R. The neutron reflectivity measurements were performed on INTER in ISIS under Project No. RB1320307. We would like to thank Dr. G. Foran (CROSS) for valuable discussions

References

[1] Yoshikawa A, Nagai H and Mizushima Y 1976 *Jpn. J. Appl. Phys.* **16** Suppl .16-1 67
[2] Leung W, Cheung N W and Neureuther A R 1985 *Appl. Phys. Lett.* **46** 481
[3] Mitkova M and Kozicki M N 2002 *J. Non-Cryst. Solids* **299-302** 1023
[4] Yamamoto Y, Itoh T, Hirose Y and Hirose H 1976 *J. Appl. Phys.* **47** 3603
[5] Rennie J, Elliott S R and Jeynes C 1986 *Appl. Phys. Lett.* **48** 1430
[6] Wagner T, Peřina V, Vlček M, *et al.* 1997 *J. Non-Cryst. Solids* **212** 157
[7] Kovalskiy A, Miller A C and Jain H 2008 *J. Am. Ceram. Soc.* **91** 760
[8] Sakaguchi Y, Asaoka H, Uozumi Y, et al. 2014 *Can. J. Phys.* **92** 654
[9] Webster J, Holt S and Dalgliesh R 2006 *Physica* B **385-386** 1164
[10] Webster J R P, Langridge S, Dalgliesh R M, and Charlton T R 2011 *Eur. Phys. J. Plus* **126** 112
[11] Kawaguchi T, Maruno S, and Elliott S R 1996 *J. Appl. Phys.* **79** 9096

X-ray excited optical luminescence of CaF_2: A candidate for UV water treatment

Chen W[1], Ma L[1], Schaeffer R[1], Hoffmeyer R[2], Sham T[3], Belev G[4], Kasap S[5] and Sammynaiken R[2,*]

[1] Department of Physics, University of Texas at Arlington, Texas, USA
[2] Saskatchewan Structural Sciences Centre, University of Saskatchewan, Saskatoon, SK S7N 5C9, Canada
[3] Department of Chemistry, University of Western Ontario, London, ON N6A 5B7
[4] Canadian Light Source Inc., 101 Perimeter Road, Saskatoon, SK S7N 0X4, Canada
[5] Department of Electrical Engineering, University of Saskatchewan, Saskatoon, SK S7N 5A9

r.sammynaiken@usask.ca

Abstract

Secondary optical processes are becoming more and more important in health and environmental applications. Ultraviolet produced from secondary emission or scintillation can damage DNA by direct photoexcitation or by the creation of reactive oxygen species. X-ray Excited Optical Luminescence (XEOL) and Time Resolved XEOL (TRXEOL) results for the fast emitter, CaF_2:ZnO, that have been treated by heating in air and in vacuum, show that the scintillation from the Self Trapped Exciton (STE) emission of CaF_2 at 282 nm is dominated by a slow process (>100 ns). A faster but weaker 10 ns component is also present. The ZnO and CaF2 show independent emission. The ZnO bandgap emission at 390 nm has dominant lifetimes of less than 1 ns.

1. Introduction

Use of optical processes in health and environment is well established. Deep ultraviolet light is known to directly destroy DNA and cells to the extent that it is used in environment technology to purify and clean water [1,2]. Nanomaterials such as CaF2 and ZnO are good UV emitters but before they are used in any applications, the scintillations characteristics need to be investigated for intensity, frequency and persistence of luminescence. The persistence has implications for the stability and efficiency of the particles and the type of high energy radiation that can be used i.e. pulsed versus continuous wave excitation. The optical emission of both ZnO and CaF_2 are well investigated but not fully understood. The main emission from CaF_2 (282 nm) is from a self- trapped exciton whereas the ZnO emission at 380 nm is a bandgap emission [3-6].

For environmental application, such as clean water technology, it is important to disperse the emitters so that they are as close as possible to the target molecule or microbes so that the emitted UV photons will not be attenuated by the matrix (water). In addition to an increase in effectiveness, the entire process becomes very energy efficient to the extent that it could compete with UV lamps used in water treatment.

An understanding of the scintillation process would enable optimization of the nano emitters. A dual particle [CaF2: ZnO] that independently emits in two UV ranges would increase the range of activity.

2. Experimental

2.1 Synthesis : Zn(acac) powder was dissolved in 70% ethanol and was ultrasonicated and refluxed. Sodium hydroxide (1.5 equivalent) kept at 0°C was added and the mixture refluxed for 45 minutes. The white precipitate was filtered and washed with acetone then with DI water and left to dry in air for 2 days[7]. The CaF_2:ZnO mixture was synthesized from a solution of ZnO, $CaCl_2$ and NH_4F in ethanol and then washed with DI and dried in air for 2 days. Samples were subjected to several treatments by heating in a vacuum oven or by heating in air. The mixture was then characterized with XRD to determine the sizes of particles and percentage composition. Pure CaF_2 particles were carefully precipitated from an ethanol solution containing $CaCl_2$ and NH_4F.
2.2 X-ray Absorption : X-ray Absorption Spectroscopy measurements were conducted at the Ca $L_{3,2}$ edge, the F K edge, Zn $L_{3,2}$ edge and at the O K edge for powder samples. The powders were embedded in conductive carbon tape and excited with a beam size of 25 μm x 25 μm at the SGM beamline at the Canadian Light Source. Absorption was measured by both total electron yield (TEY) and fluorescence yield (FLY). The FLY is relatively weak and has the same features as the TEY.

2.2.1 XEOL and TRXEOL : XEOL measurements were made by exciting at several wavelengths below and above the absorption edge of all the constituent elements. UV-Vis photons were collected by using an Ocean Optics QE65000 spectrometer which was connected by fibre optic and a collimating lens to the sample chamber. Several excitation energies that yielded high intensity of a characteristic emission were used and the monochromator was set at the appropriate emission (282 nm and 532 nm for CaF_2, and 390 nm and 510 nm for ZnO). Lifetimes were fitted by using a multi-exponential function [8]. In most cases the best fit was obtained with two exponential. However, in some cases three exponentials were required for a proper fit.

3. Results and discussion

XANES measurements shown in figure 1a-1d indicate that the materials were relatively pure. Within the level of detection and as expected, a small amount of calcium oxide was formed at the CaF2:ZnO interface. A sample that is heated in air would have more surface oxide and evidence for the oxide of calcium is shown in figures 1a and 1c (labelled a,b,c,d) by the broadening of the Ca $L_{3,2}$ edge of the ZnO: CaF_2 heated in air. XANES of the Zn $L_{3,2}$ edge and the O K edges of the mixed samples were relatively weak compared to pure ZnO, but overall the results indicated that ZnO remains in the pure nanocrystallite form. The small change observed at the O K edge is consistent with some surface reaction to form an oxide of calcium.

XEOL data are shown in figures 2a-2c. Figure 2a shows spectra obtained by scanning over the edges (shown by arrows in figures 1a and 1c) of Ca and F in CaF_2. Scanning across the Ca $L_{3,2}$ edge (330-360 eV) resulted in a primary STE emission band at 282 nm. Also, an extremely weak band at 532 nm suggests that the CaF_2 nano crystals have few additional defects. Excitation of CaF_2 below and above the fluorine edge (670-710 eV) results in higher XEOL intensity compared to the Ca $L_{3,2}$ edge. The stoichiometry of CaF_2 suggests that more photons will be absorbed across the edge of F which would result in increasing the numbers of F_2^- STE formed.
The XEOL of Caf2:ZnO blended materials are shown figures 2b and 2c. The emission intensity is significantly increased when the Zn or the O edges of ZnO: CaF_2 are excited compared to the Ca and F edges being excited. Heating the sample in air suggests that surface and passivation and not interfacial oxide formation affects the number of STE generated. Because the d orbitals of Zn^{2+} is effectively full, the absorption cross-section of the Zn $L_{3,2}$ edge is low which results in the emission intensity being lower than the O K edge excitation. This increased intensity at the O K edge is due to more secondary processes and radiation tracks are created. The weak O K edge and Zn $L_{3,2}$ XANES is due to reduced x-ray and electron emission from the sample. This correlates very well with the increased STE emission when the O K edge and the Zn L edge are excited. Secondary electrons and x-ray emitted from the ZnO are of sufficient energies to create large numbers of STE in the CaF_2. Effectively, ZnO collects the high energy photons and the final state is from a STE emission from CaF_2.

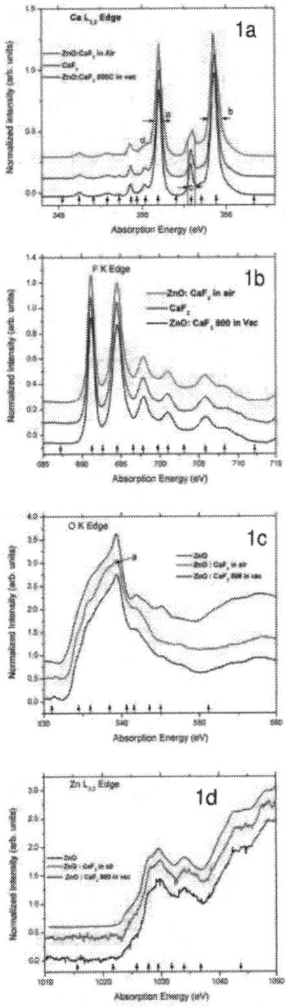

Figure 1: XANES at the Ca $L_{3,2}$ edge, F K edge, O K edge and Zn $L_{3,2}$ edge of ZnO and CaF_2 compared with ZnO:CaF_2 heated in air and in vacuum

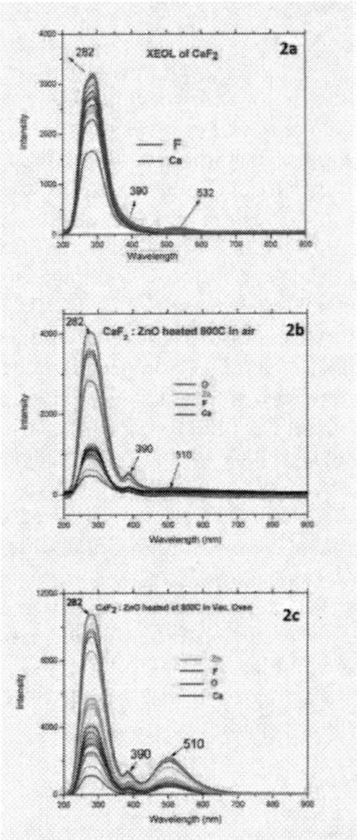

Figure 2: (a) XEOL CaF_2 excited at the F K edge region (670-710 eV) and at the Ca $L_{3,2}$ edge region (330- 360 eV), (b and c) XEOL of ZnO:CaF_2 heat treated in air excited at the Zn $L_{3,2}$ region (1010-1050), O K edge region (520-560 eV) and at the regions for CaF_2

Emission (nm)	CaF_2	ZnO: CaF_2 (Air)	ZnO: CaF_2 (Vac)
282 nm	10 ns, 20% >100 ns, 80%	10 ns, 33% 100 ns, 67%	10 ns, 33% 100 ns, 67
390 nm		<1 ns	<1 ns
510 nm		6 ns, 32% 86 ns,31% >100 ns, 37%	6 ns, 32% 86 ns,31% >100 ns, 37%

Table 1: Lifetimes determined from TRXEOL for the three CaF2 samples.

Annealing of the ZnO:CaF2 in vacuum results in two processes. First, the CaF2 nanocrystallites become more ordered and the F_2^- anion become more aligned in particular crystal orientations e.g. <111> or <110> resulting in greater overlap with the hole centres. This results in an increased transition probability of the STE and hence higher intensity of emission. Second, the ZnO emission increases because upon annealing, more ZnO is available near the surface to observe the bandgap emission. The results of our TRXEOL investigations are shown in table 1. The lifetime of ZnO emission for both ZnO:CaF2 samples are the same. The CaF_2 STE emission is different in annealed samples compared to the unannealed CaF_2 which has two main lifetimes (10 ns and >100 ns). For both materials containing ZnO, there was 20% at 10 ns and 80% with lifetime greater than 100 ns. Lifetimes for CaF_2 were determined to be 33% at 10 ns and 67% at >100 ns. Annealing in air resulted in a decrease in intensity of STE emission due to surface passivation but the lifetime relative to vacuum annealed sample does not change, suggesting that the STE emission from the CaF_2 nanocrystals are not from just surface sites. The shorter lifetimes of the annealed samples are a result of an increase in order in the nanocrystals.

4. Summary

XEOL of CaF_2 is an efficient method to produce UV photons for biological applications if the CaF_2 nanomaterials are close to the biomaterial of interest. By mixing the CaF_2 with x-ray collectors such as ZnO, the intensity of CaF_2 emission is significantly increased. Annealing the samples in air and in vacuum shows that surface oxidation attenuates the CaF_2 STE emission intensity but increases the percentage of fast emission. Overall, the CaF_2 STE emission can be increased by mixing with a collector material and annealing at 800C in vacuum.

5. Acknowledgements

Research described in this paper was performed at the Canadian Light Source, which is supported by the Natural Sciences and Engineering Research Council of Canada, the National Research Council Canada (NSERC), the Canadian Institutes of Health Research, the Province of Saskatchewan, Western Economic Diversification Canada, and the University of Saskatchewan (USask). Additional support was provided by the Agriculture Development Fund of Saskatchewan. Drs. T. Regier, S. Brunet and J. Voigt are thanked for providing assistance in setting up the TRXEOL experiments.

6. References

[1]Moan J and Peak M J, (1989)J. Photochem. Photobiol. B **4**, 21
[2]Sommer R, Cabaj A, Hirschmann G, and Haider T, (2008) *Ozone: Sci. Eng.* , **30**, 43
[3]Naftel S, Yiu Y M, Sham T K and Yates, B W, (2001) *J. Electron Spectrosc. Relat. Phenom.* **119**, 215
[4]Tanimura K, (2001*) Phys. Rev. B*, 63, 184303
[5]Sils J, Hausfeld S, Clauß W, Pahl U, Lindner R and Richling M, (2009) J. Appl. Phys. **106**, 063109
[6]Williams R. T, Kaber M N, Hayes W, and Scott J P, (1976) *Phys. Rev. B*, **14**, 725
[7]Woo B K, Chen W, Joly A J, and R Sammynaiken (2008) *J. Phys. Chem.*, 112, 14292
[8] James D R, Demmer D R M, Verrall R E, and Steer R P, (1983) *Rev. Sci. Instrum.* **54**, 1121

Effects of host glasses on luminescence properties of Sm^{3+}, Pr^{3+} co-doped glass

Y. Shimizu*[1], K. Watanabe[1], H. Uemura[2], S. Fuchi[1] and Y. Takeda[2]

[1]Aoyama Gakuin Univ., Fuchinobe, 5-10-1 Chuo-ku, Sagamihara-shi, Kanagawa 252-5258, Japan
[2]Aichi Science & Technology Foundation, 250-3 Yamaguchi-cho, Seto-shi, Aichi 489-0965, Japan

*E-mail: simizu356@gmail.com

Abstract. We calculated "density"/"molecular weight" ratio of glass former oxides and modifier oxides to increase a number density of rare-earth ions in glass phosphor. Based on the calculated results, GeO_2 and ZnO were chosen for the glass former oxide and the modifier oxide, respectively. The effects of substituting GeO_2 for B_2O_3, ZnO for Sb_2O_3, or ZnO for Bi_2O_3 on luminescence spectra in Sm^{3+}-doped glass were not observed. On the other hand, a drastically luminescence spectrum changing of Pr^{3+}-doped glass was observed by substituting GeO_2 for B_2O_3. The output power of Sm^{3+} -doped glass and that of Pr^{3+} -doped glass increased with substituting ZnO for Bi_2O_3. We successfully achieved an ultra-wideband luminescence from 760 nm to 1100 nm with the output power of 2.5 mW by combining a blue LED with $0.15Sm_2O_3$-$0.12Pr_6O_{11}$-10ZnO-$45Sb_2O_3$-$45GeO_2$ glass in one package.

1. Introduction

Wideband light sources are used in fiber gyroscopes [1, 2] and absorption spectrometry for agricultural applications [3, 4]. A central wavelength in a near-infrared (NIR) region is suitable for absorption spectrometry, leading to penetration depths in biological samples larger than that at other wavelength [5]. The wideband light-source is desirable for the absorption spectrometry in order to measure a broader region of an absorption spectrum.

Light emitting diodes (LEDs) in the NIR region and halogen lamps are usually used as the NIR wideband light-sources. LEDs are small and have longer lifetime. However, a spectral width of LEDs is 50 nm at maximum and it is insufficient for most spectroscopic applications. On the other hand, halogen lamps have a very wide spectral width. However, a size of halogen lamps is much bigger than that of LEDs. Moreover, a lifetime of halogen lamps is much shorter than that of LEDs.

Thus, we proposed a novel wideband NIR light-source by combining a wideband NIR phosphor and the LED as an excitation light-source in one package. To realize this light-source, we synthesized a phosphor that emits light at approximately 1000 nm using Bi_2O_3-B_2O_3 based glasses doped with Yb^{3+} and Nd^{3+} [6-8]. We successfully fabricated a high power (over 1 mW) NIR light-source with a Gaussian-shape spectrum (spectral width ranging from 950 nm to 1100 nm) using this glass phosphor and a high power amber LED [9]. We successfully achieved the maximum output power of 1.15mW with luminescence from 760 nm to 1100 nm by a Sm^{3+}, Pr^{3+} co-doped glass phosphor combined with a blue LED [10]. In this paper, we investigated the effects of host glasses on luminescence properties of Sm^{3+}, Pr^{3+}-doped glass to increase the output power keeping wideband luminescence.

2. Experimental

The samples used in this study were synthesized by a melt-quenching method. Powders of Sm_2O_3, Pr_6O_{11}, Bi_2O_3, Sb_2O_3, ZnO, H_3BO_3, and GeO_2 were used. The mixed powders in an alumina crucible were melted at 1250 °C in an electric furnace. After keeping for 10 min, a molten liquid was poured between two stainless-steel mould plates kept at room temperature for a formation of the glass.

Output power (P_{out}) measurements and photoluminescence (PL) measurements were carried out at room temperature. A LED with a central wavelength of 470 nm was used as a photo-excitation light-source. A Si power meter (ML9001A, Anritsu, Japan) was used for P_{out} measurements. Si photodiode mounted on a 0.20 m grating monochromator was adopted for a lock-in detection in the NIR region.

3. Results and discussions

3.1 Calculation of the number density of rare-earth ions in the glass

The number density of rare-earth (RE) ions is an important parameter, since the P_{out} should be depended on the number density of RE ions in the same volume of glasses. It is well known that the number density of RE ions is proportional to an average density of samples and is inversely proportional to an average molecular weight of samples [11]. Therefore, we calculated the "density"/"molecular weight" ratio. For candidates of modifier oxides, we selected small environmental load, transparent, and not expensive oxides. From calculated results, GeO_2 (glass former oxide) and ZnO (modifier oxide) were chosen for host glasses.

However, increase of the number density of RE ions leads decreasing an average RE ion - RE ion distance. It is well known that a concentration quenching occurs by short RE ion - RE ion distance. Therefore, the effects of number density of RE ions on the concentration quenching are investigated.

3.2 Dependence of luminescence properties on the host glass (Sm^{3+}-doped glass)

Figure 1 shows the dependence of the P_{out} on the number density of RE ions in the $10Bi_2O_3$-$45Sb_2O_3$-$45B_2O_3$ glass, in the $10Bi_2O_3$-$45Sb_2O_3$-$45GeO_2$ glass, in the $10Bi_2O_3$-$10Sb_2O_3$-35ZnO-$45GeO_2$ glass, and in the 10ZnO-$45Sb_2O_3$-$45GeO_2$ glass. The P_{out} of 10ZnO-$45Sb_2O_3$-$45GeO_2$ is the highest in these glasses. This is due to decreasing the absorption of excitation light by the host glass. Moreover, the number density of concentration quenching depends on the glass composition. Especially, $10Bi_2O_3$-$45Sb_2O_3$-$45GeO_2$ glass shows the highest number density ($13\times10^{19}cm^{-3}$) in these glasses. This result suggests that the concentration quenching do not simply depend on the number density of RE ions. Since luminescence properties of RE ions depend on local structures around RE ions, an atomic level structure analysis is now in progress. Figure 2 shows the PL spectra of four samples. In Fig. 2, luminescence peaks at approximately 910 nm (from $^4G_{5/2}$ to $^6F_{3/2}$), 960 nm (from $^4G_{5/2}$ to $^6F_{5/2}$), and 1030 nm (from $^4G_{5/2}$ to $^6F_{7/2}$) were observed.

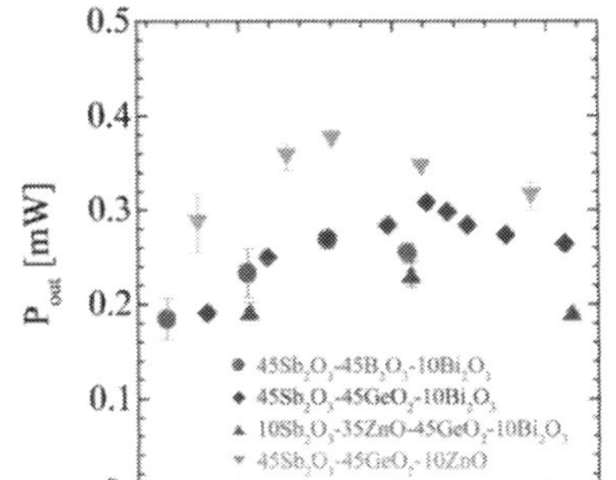

Figure 1. Dependence of the P_{out} on the host glass (Sm^{3+}-doped glass).

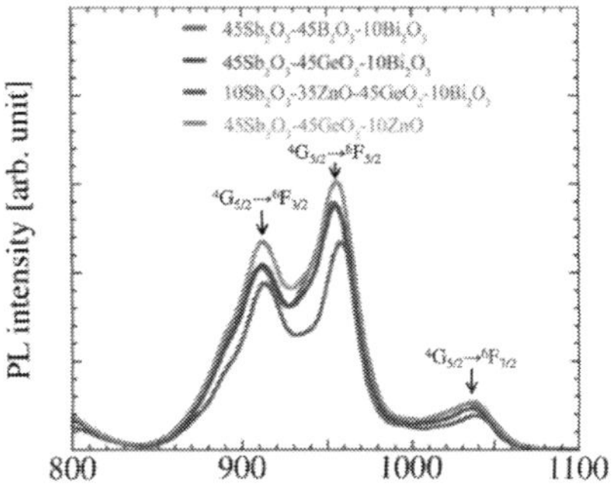

Figure 2. Dependence of the PL spectra on the host glass (Sm^{3+}-doped glass).

From these results, it is revealed that the substituting GeO_2 for B_2O_3, ZnO for Sb_2O_3, or ZnO for Bi_2O_3 have no effects on luminescence spectra. On the contrary, substituting GeO_2 for B_2O_3, ZnO for Bi_2O_3 increase the P_{out}. Therefore, it is concluded that a choosing GeO_2 as the glass former oxide and ZnO as the modifier oxide is a good approach to increase the P_{out}.

3.3 Dependence of luminescence properties on the host glass (Pr^{3+}-doped glass)

Figure 3 shows the dependence of the P_{out} on the number density of RE ions in the $10Bi_2O_3$-$45Sb_2O_3$-$45B_2O_3$ glass, in the $10Bi_2O_3$-$45Sb_2O_3$-$45GeO_2$ glass, in the $10Bi_2O_3$-$10Sb_2O_3$-35ZnO-$45GeO_2$ glass, and in the 10ZnO-$45Sb_2O_3$-$45GeO_2$ glass. The P_{out} of 10ZnO-$45Sb_2O_3$-$45GeO_2$ and that of $10Bi_2O_3$-$45Sb_2O_3$-$45GeO_2$ are higher than that of other glasses below the number density of $2.8\times10^{19}cm^{-3}$. The number density of concentration quenching in $10Bi_2O_3$-$45Sb_2O_3$-$45GeO_2$ is the highest ($2.8\times10^{19}cm^{-3}$) in these glasses. Figure 4 shows the PL spectra of four samples. In Fig. 4, luminescence peaks at approximately 810 nm (from $^1D_2$ to $^3H_6$), 870 nm (from $^1D_2$ to $^3F_2$), 1000 nm (from $^1D_2$ to $^3F_3$), and 1030 nm (from $^1D_2$ to $^3F_4$) were observed in $10Bi_2O_3$-$45Sb_2O_3$-$45B_2O_3$. On the contrary, in $10Bi_2O_3$-$45Sb_2O_3$-GeO_2, in $10Bi_2O_3$-$10Sb_2O_3$-35ZnO-$45GeO_2$, and in 10ZnO-$45Sb_2O_3$-$45GeO_2$, luminescence peaks at approximately 750 nm (from $^3P_0$ to $^3F_4$), 810 nm (from $^1D_2$ to $^3H_6$), 890 nm (from $^1D_2$ to $^3F_2$), 930 nm (from $^3P_0$ to $^1G_4$), 1000 nm (from $^1D_2$ to $^3F_3$), and 1050 nm (from $^1D_2$ to $^3F_4$) were observed. From assignment of luminescence peaks, it is clear that the luminescence related to $^3P_0$ increases in $10Bi_2O_3$-$45Sb_2O_3$-$45GeO_2$, in $10Bi_2O_3$-$10Sb_2O_3$-35ZnO-$45GeO_2$, and in 10ZnO-$45Sb_2O_3$-$45GeO_2$. In general, non-radiative multi-phonon relaxation rate depend on vibration energy of the host glass. And it is well known that the vibration energy of GeO_2 is lower than that of B_2O_3 [12, 13]. Therefore, lower vibration energy of GeO_2 should decrease the relaxation from $^3P_0$ to $^1D_2$, which leads the spectrum changing and the increasing the output power by generating radiative relaxation paths related to $^3P_0$.

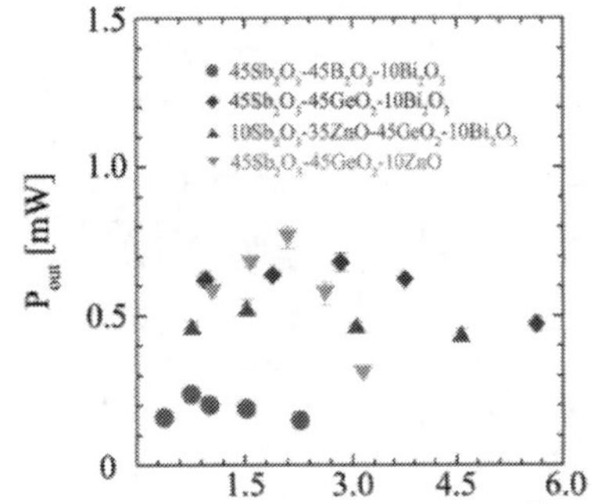

Figure 3. Dependence of the P_{out} on the host glass (Pr^{3+}-doped glass).

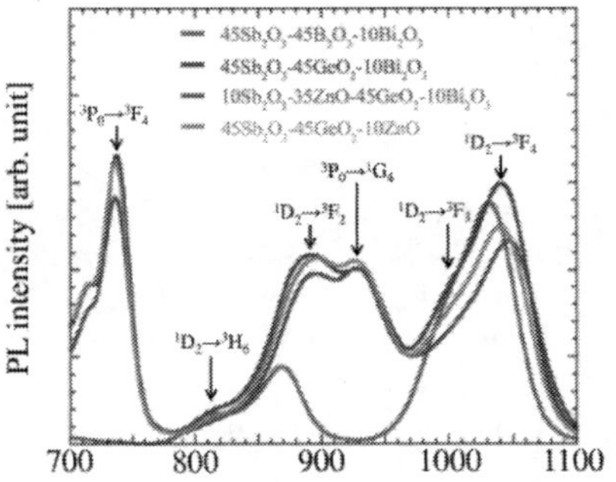

Figure 4. Dependence of the PL spectra on the host glass (Pr^{3+}-doped glass).

From these results, it is revealed that the substituting GeO_2 for B_2O_3 has strong effects on the luminescence spectra and the P_{out}. Therefore, it is concluded that a choosing GeO_2 as the glass former oxide is a good approach to increase the P_{out}.

3.4 Sm^{3+}, Pr^{3+} co-doped 10ZnO-$45Sb_2O_3$-$45GeO_2$ glass

From the above results and discussions, Sm^{3+}, Pr^{3+} co-doped 10ZnO-$45Sb_2O_3$-$45GeO_2$ glass was synthesized. In order to increase the P_{out} with wideband luminescence, we optimized the Sm_2O_3 and Pr_6O_{11} concentration as Sm_2O_3 of 0.15 mol% and Pr_6O_{11} of 0.12 mol%. Detailed optimization method will report elsewhere.

Figure 5 shows the PL spectrum of $0.15Sm_2O_3$-$0.12Pr_6O_{11}$-10ZnO-$45Sb_2O_3$-$45GeO_2$ glass phosphor combined with a blue LED. The ultra-wideband luminescence from 760 nm to 1100 nm was observed.

Figure 6 shows the dependence of the P_{out} on the injection current of the blue LED. Based on the technical datasheet for the blue LED, the maximum injection current was set to 1000 mA. The P_{out} increased with increasing injection current, and a maximum P_{out} of 2.5 mW was achieved. Maximum P_{out} of 2.5 mW is twice higher than that of Sm^{3+}, Pr^{3+} co-doped $10Bi_2O_3$-$45Sb_2O_3$-$45B_2O_3$ glass combined with a blue LED (1.15 mW).

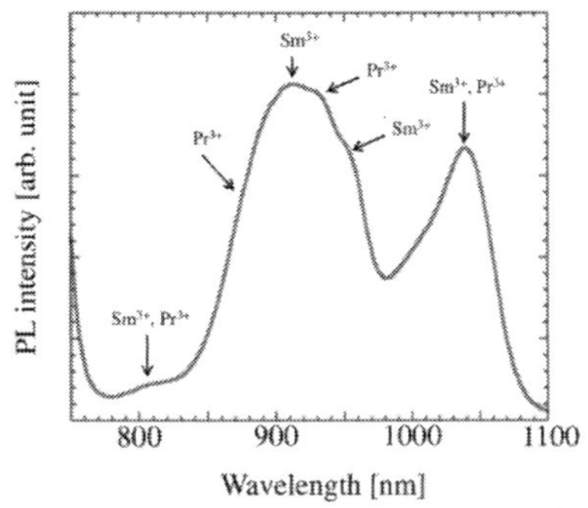

Figure 5. Luminescence spectrum of Sm^{3+}, Pr^{3+} co-doped glass phosphor combined with the blue LED.

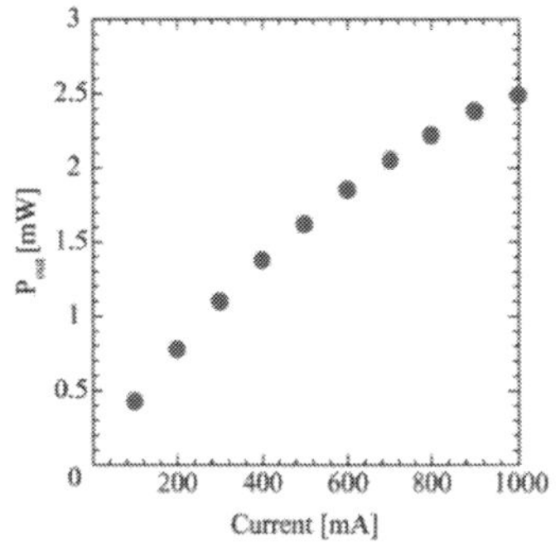

Figure 6. Dependence of the P_{out} on the injection current of the blue LED.

4. Summary

We have investigated effects of host glasses on luminescence properties of Sm^{3+}, Pr^{3+}-doped glasses. Substituting GeO_2 for B_2O_3 and ZnO for Bi_2O_3 increased the P_{out}, while luminescence spectra were same in Sm^{3+}-doped glasses. On the contrary, substituting GeO_2 for B_2O_3 changed the luminescence spectra and increased the P_{out} in Pr^{3+}-doped glasses. These phenomena should be due to the suppression of the multi-phonon relaxation process form $^3P_0$ to $^1D_2$ of Pr^{3+} ions. We synthesized Sm^{3+}, Pr^{3+} co-doped $10ZnO$-$45Sb_2O_3$-$45GeO_2$ glass and combined with the blue LED. The wideband luminescence from 760 nm to 1100 nm with maximum output power of 2.5 mW was obtained. These luminescence properties are suitable for absorption spectrometry of agricultural products.

Acknowledgements

This work was supported in part by JSPS KAKENHI Grant Number 24760008 and by Knowledge Hub Aichi.

References

[1] C. C. Cutler, S. A. Newton, and H. J. Shaw, Opt. Lett. **5**, 488 (1980).
[2] S. Blin, H. K. Kim, M. J. F. Digonnet, and G. S. Kino, J. Lightwave Technol. **25**, 861 (2007).
[3] H. W. Siesler, Y. Ozaki, S. Kawano, and H. M. Heise: Near-Infrared Spectroscopy-Principles, Instruments, Applications (Wiley-VCH, Weinheim, Germany, 2002).
[4] Y. Ozaki, Anal. Sci. **28**, 545 (2012).
[5] A. F. Fercher, W. Drexler, C. K. Hitzenberger, and T. Lasser, Rep. Progr. Phys. **66**, 239 (2003).
[6] S. Fuchi, A. Sakano, and Y. Takeda: Jpn. J. Appl. Phys. **47** 7932 (2008).
[7] S. Fuchi, A. Sakano, R. Mizutani, and Y. Takeda: Eur. J. Glass Technol. A **50** 319 (2009).
[8] S. Fuchi, A. Sakano, R. Mizutani, and Y. Takeda: Appl. Phys. **B 105** 877 (2011).
[9] S. Fuchi, A. Sakano, R. Mizutani, and Y. Takeda: Appl. Phys. Express **2** 32102 (2009).
[10] S. Fuchi, Y. Shimizu, K. Watanabe, H. Uemura, and Y. Takeda: Appl. Phys. Express **7** 072601 (2014).
[11] Zahra Ashur Said Mahraz, *et al.*, Journal of luminescence, **144** 139 (2013).
[12] I. Simmon, "Modern Aspects of Vitreous State" vol. **1**, 138 (1960).
[13] D.E.C. Corbridge et al., J. Chem. Soc., 4555 (1954).

Journal of Physics: Conference Series **619** (2015) 012049 doi:10.1088/1742-6596/619/1/012049

Cathodoluminescent images and spectra of single crystals of $Y_2O_2S:Tb^{3+}$ and $Gd_2O_2S:Tb^{3+}$ nanometer sized phosphor crystals excited in a field emission scanning transmission electron microscope

J. Silver,[1] X. Yan,[1] G.R. Fern[1] and N. Wilkinson[2]
1) Centre for Phosphors and Display Materials, Wolfson Centre for Materials Processing, Brunel University, Uxbridge, Middlesex, UB8 3PH, UK
E-mail: jack.silver@brunel.ac.uk; george.fern@brunel.ac.uk

2) Gatan UK, 25 Nuffield Way, Abingdon, Oxon. OX14 1RL, UK

Abstract. Cathodoluminescence (CL) spectra have been collected from single nanometer-sized crystals of $Y_{1.98}Tb_{0.02}O_2S$ and $Gd_{1.98}Tb_{0.02}O_2S$ using a Gatan Vulcan cathodoluminescence imaging spectrometer. Slight variations observed in the CL spectra taken from the crystals are explained, and discussed in relation to bulk samples.

1. Introduction

Y_2O_2S and Gd_2O_2S crystals are both wide band-gap (4.6 – 4.8 eV) semiconductors that have been considered ideal host matrixes for trivalent rare earth cations [1] where the Ln and the O atoms have C_{3v} site symmetry and the S atoms have D_{3d} symmetry [2,3]. Tb^{3+} activated Y_2O_2S and Gd_2O_2S have been used as bulk phosphor scintillator materials for medical diagnostics because of their inherent properties such as hard radiation stability, high X-ray absorption, high conversion efficiency from X-ray to visible light, short decay times and low afterglow and is used in X-ray intensifying screens [4-8]. In a paper whose subject was "contrasting behaviour of the co-activators in the luminescence spectra of $Y_2O_2S:Tb^{3+}$, Er^{3+} nanometre sized particles under UV and red light excitation" [9] we reported the properties of nanometre sized particles of $Y_2O_2S:Tb^{3+}$. Therein we discussed the synthesis and characterisation of the nanometre sized crystals of $Y_2O_2S:Tb^{3+}$.

A question that phosphor chemists are often asked is "how homogeneous is the phosphor sample?" This question is not easy to answer as PL or CL spectra collected from phosphor samples are the average spectra of the samples making them up. Now that it is possible to collect CL spectra from individual nanometre sized particles the question of homogeneity can begin to be addressed.

The first objective was to compare the spectra of nanometre sized crystals of $Y_{2-x}Tb_x O_2S$ and $Gd_{2-x}Tb_x O_2S$ (where x has the value of 0.02) to determine the variability of the emission properties in such crystals made in the same reaction.

2. Experimental

Tb^{3+} activated Y_2O_2S and Gd_2O_2S nanometre sized phosphor particles were synthesized by the two-step method described in our earlier work,[6, 9, 10]. Scanning TEM (STEM) (and CL STEM) was

[1] To whom any correspondence should be addressed.

carried out at a range of temperatures in a JEOL 2100F analytical TEM operated at 200 kV. STEM was carried out using a JEOL HAADF detector. CL images and the in microscope CL emission spectra were collected using a Gatan Vulcan™ system.

3. Results and Disscussion

Figure 1 shows the HAADF STEM images of the nanometre-sized single crystals from which four of the spectra were recorded for the nanometre sized crystals of $Y_{1.98}Tb_{0.02}O_2S$ studied in this work.

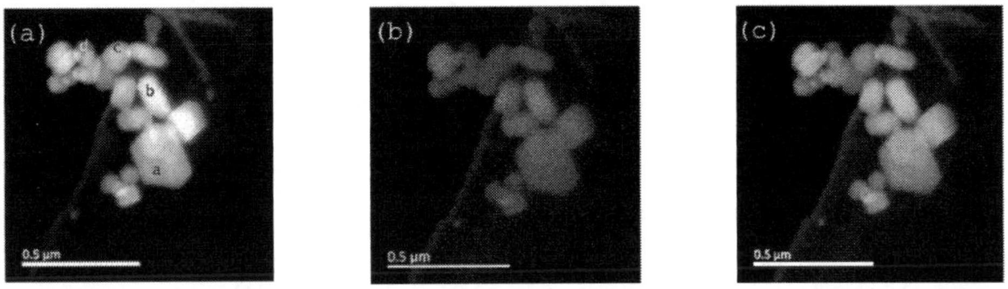

Fig. 1 STEM micrographs of (a) the nanometer sized single crystals of $Y_{1.98}Tb_{0.02}O_2S$; (b) Total visible light (CL) image collected simultaneously with (a) using the VULCAN detector; (c) Overlay of (a) and (b).

In Figure 1a all the crystals can be seen to be discrete with apparently smooth surfaces. In Figure 1b the light distribution image of the cluster is seen, the distribution of the light (generated by CL of the sample) emerging from the crystals is very even. This light originates from the activator sites within the crystals; therefore this possibly indicates an even distribution of the Tb^{3+} cations in the group of crystals. It appears that there is no reason to assume the distribution of the Tb^{3+} cations is not even from this picture. An overlay of Figures 1a and 1b is presented in Figure 1c; additional information is immediately apparent from the overlay which was less obvious on the separate images. For example at the upper left of Figure 1c the small crystal below the larger shows discrete parallel lines. These lines of light may be emitted between lattice stacking planes or fault planes; hence although the crystals appear smooth in Figure 1a the overlay gives more information.

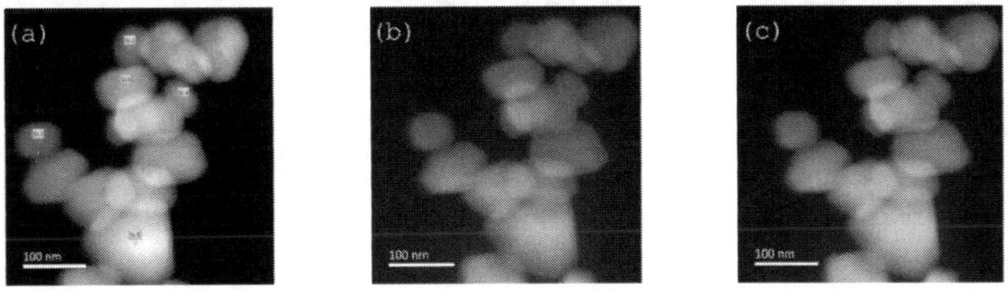

Fig. 2 STEM micrographs of (a) HAADF STEM of single crystals of $Y_{1.98}Tb_{0.02}O_2S$; (b) Total visible light (CL) image collected using the VULCAN detector; (c) overlay of images shown in (a) and (b).

As in Figure 1a all the crystals seen in Figure 2a are discrete with apparently smooth surfaces. In Figure 2b the light distribution image of the cluster is seen, unlike that of Figure 1b although the distribution of the light (generated by CL) emerging from some of the crystals is very even, some areas are brighter. These brighter areas originate from areas of either contact between crystals or from where crystals are overlapping; much less evidence for such effects was seen in Figure 1. From Figure 2b it may be thought that in this cluster there is not such an even distribution of the Tb^{3+} cations in the group of crystals as in Figure 1. To address this point the normalised emission spectra for the crystals shown in Figures 1 and 2 are presented in Figures 3 and 4.

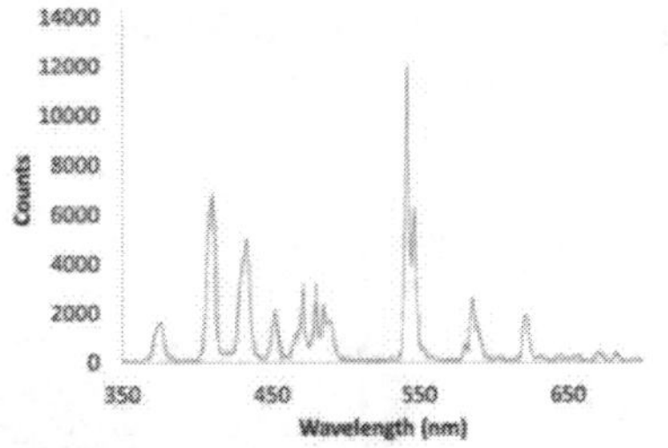

Fig. 3 CL emission spectra of $Y_{1.98}Tb_{0.02}O_2S$ at 20°C showing the variation in CL spectra from the 4 separate nanometer sized crystals shown in Figure 1a.

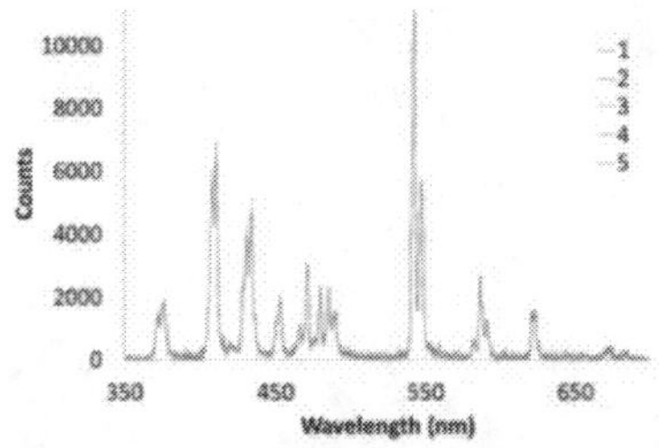

Fig. 4 CL emission spectra of $Y_{1.98}Tb_{0.02}O_2S$ at 20°C showing the variation in CL spectra from the 5 positions marked in Figure 2a.

Each CL emission spectrum shown in Figure 3 of $Y_{1.98}Tb_{0.02}O_2S$ was recorded from a different nanometre sized crystal in the group shown in Figure 1a (crystals a-d). Similarly each of the five CL emission spectra presented in Figure 4 were recorded from Figure 2a (marked as crystals 1-5). In both figures 3 and 4 the spectra are all very similar as would be expected with crystals prepared in the same batch. There are no significant differences observed in the spectra in either the $^5D_3 \rightarrow ^7F_J$ or the $^5D_4 \rightarrow ^7F_J$ transitions. [6, 9-11]. These transitions usually loose intensity by a cross relaxation process. In fact for the 1 mol% Tb^{3+}activated Y_2O_2S phosphor the green emission band at 545 nm is the most intense one, the intensity ratio between the $^5D_4 \rightarrow ^7F_J$ and $^5D_3 \rightarrow ^7F_J$ transitions is dependent on the doping concentration of Tb^{3+} cations. Thus the emission colour and intensity of Tb^{3+} emission bands are also strongly affected by the doping concentration [9] and the former can be blue to green. [11]
Evidence that the concentration of the Tb^{3+} cations is very similar if not exactly the same in these nanometer sized crystals can be taken from the fact that the spectra all show the same intensity ratio between the $^5D_3 \rightarrow ^7F_J$ transitions and the $^5D_4 \rightarrow ^7F_J$ transitions. So even though there appeared to be some areas of brighter emission in Figure 2b it appears that the activator concentration is the same.

Fig. 5 STEM micrographs of (a) single crystals of $Y_{1.98}Tb_{0.02}O_2S$ used in the CL spectral comparison (spectrum imaging) studies (inset image square shows the area where the 4 different survey positions were taken); (b) Total visible light (CL) image collected using the VULCAN detector; (c) overlay of images shown in (a) and (b).

In Figure 5a all the crystals can be seen to be discrete with smooth surfaces. In Figure 5b the distribution of the light emerging from the crystals is very even with features in common to Figure 2b.

In this case we chose to survey four positions close together across a boundary so we covered different apparent bright areas in close proximity. The four resulting CL emission spectra are presented in Figure 6. These high resolution spectra manifests all the emission bands present in the range 350 nm to 700 nm and the relative intensities of the emission bands in the spectra presented in Figures 3 and 4 are in good agreement with these spectra. The normalised spectra in Figure 6 are all identical which is evidence for an even distribution of the Tb^{3+} activators and shows the even distribution of activator cations. At low Tb^{3+} concentrations the blue emissions from the $^5D_3 \rightarrow ^7F_J$ transitions are predominant in the $Y_{1.98}Tb_{0.02}O_2S$ spectrum while green emissions from $^5D_4 \rightarrow ^7F_J$ transitions are dominant in the equivalent Gd compound's spectrum (Figures 4 and 5). This is due to the positions of the CTSs [11].

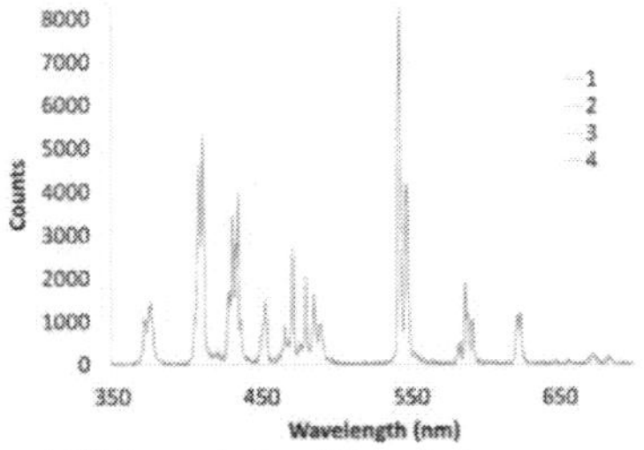

Fig. 6 High resolution CL emission spectra of nanometre-sized single crystals of $Y_{1.98}Tb_{0.02}O_2S$ at 20°C showing the variation in CL spectra from the 4 different survey positions in Figure 5a.

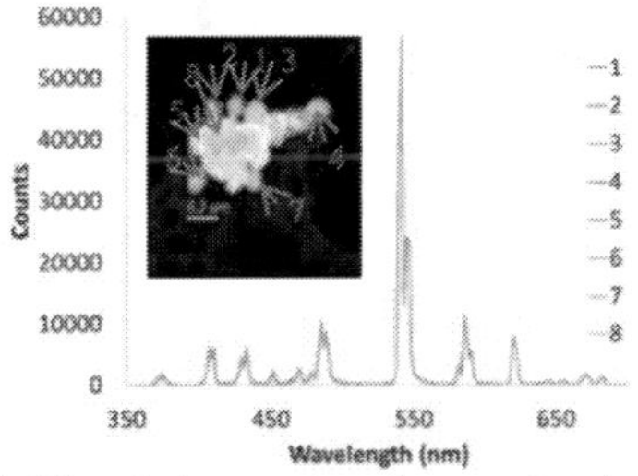

Fig. 7. CL emission spectra of nanometre-sized single crystals $Gd_{1.98}Tb_{0.02}O_2S$ at -172°C showing the variation in CL spectra from 8 separate crystals (Inset image shows the position of the 8 survey positions).

The CL spectra of eight different single nanometre-sized crystals of $Gd_{1.98}Tb_{0.02}O_2S$ are presented in Figure 7. The spectra are all very similar as would be expected with crystals prepared in the same batch. The main differences are observed in the $^5D_3 \rightarrow {}^7F_J$ transitions, particularly the $^5D_3 \rightarrow {}^7F_5$ and the $^5D_3 \rightarrow {}^7F_4$. These transitions usually loose intensity in two ways as discussed above for the 1 mol% Tb^{3+}activated Y_2O_2S phosphor, the first is by the cross relaxation process, but in this case unlike the Y^{3+} compounds the $Gd_{2-x}Tb_xO_2S$ samples are all green as not only does the above relaxation process take place but in addition the second mechanism that loses intensity in the blue bands is more dominant. The CTS in Gd_2O_2S lie closer to the $^5D_3$ excited state of Tb^{3+} than it does in the Y_2O_2S lattice. Therefore most of the electrons on the $^5D_3$ excited state of Tb^{3+} in Gd_2O_2S are fed to $^5D_4$ state at room temperature, yielding a noticeable reduction on the intensities of the $^5D_3 \rightarrow {}^7F_J$ transitions [11].

4. Conclusions

The work presented herein has led to a number of important findings:- CL spectra have been easily collected from two clusters of single nanometer-sized crystals (one from four and one from five separate crystals) of $Y_{1.98}Tb_{0.02}O_2S$. In addition CL spectra have been collected from eight single nanometer-sized crystals of $Gd_{1.98}Tb_{0.02}O_2S$; High quality light images demonstrate uniform emission of light from individual phosphor crystals; Evidence the concentration of the Tb^{3+} cations is very similar in these nanometer sized crystals of $Y_{1.98}Tb_{0.02}O_2S$ and $Gd_{1.98}Tb_{0.02}O_2S$ can be taken from the fact that the spectra all show the same intensity ratio between the $^5D_3 \rightarrow {}^7F_J$ and $^5D_4 \rightarrow {}^7F_J$ transitions.

5. Acknowledgements

The authors are grateful to the UK Technology Strategy Board (TSB) for funding NoveLELs, ACTIVEL & BEDs and TSB and EPSRC for funding PURPOSE, HTRaD & CONVERTED projects.

6. References

1. Blasse G and Grabmaier BC, Luminescent Materials, Springer-Verlag, Berlin, 1994
2. Zachariassen WH 1949 *Acta Crystallogr.* 2 57
3. Sovers OJ and Yoshikoka T, J. 1968 *Chem. Phys.* 49 4945
4. Phosphor Handbook, ed. Shionoya S, Yen WM and Yamamoto H, CRC press, Boca Rotan, 2nd edn, 2006
5. van Eijk CWE 2002 *Phys. Med. Biol.* 47 R85
6. Fern GR, Ireland T, Silver J, Withnall R, Michette A, McFaul C and Pfauntsch S 2009 *Nucl. Instrum. Methods Phys.Res., Sect. A* 600 434
7. Kalivas N, Costaridou L, Kandarakis I, Cavouras D, Nomicos CD and Panayiotakis G 2002 *Nucl. Instrum. Methods Phys. Res., Sect. A* 490 614
8. Jung ID, Cho MK, Lee SM, Bae KM, Jung PG, Lee CH, Lee JM, Yun S, Kim HK, Kim SS and Ko JS 2009 *J. Micromech. Microeng.* 19 015014
9. Yan X, Fern GR, Withnall R and Silver J 2013 *Nanoscale* 5 1091
10. Silver J, Withnall R, Marsh PJ, Lipman A, Ireland TG and Fern GR 2005 *Dig. Tech. Pap. - Soc. Inf. Disp. Int. Symp.* 36 594
11. Yan X, Fern GR, Withnall R and Silver J 2013 *Nanoscale*, 5 8640

Materials Suitable for preparing Inorganic Nanocasts of butterflies and other insects

J. Silver, G R Fern and T G Ireland

Centre for Phosphors and Display Materials, Wolfson Centre for Materials Processing, Brunel University, Kingston Lane, Uxbridge, Middlesex, UK

Jack.silver@brunel.ac.uk

Abstract. Replication of 3D-structures, in particular those that have a periodic modulation of a dielectric material at optical wavelengths and below have proven very difficult to fabricate. The majority of such replication techniques are complex or use moisture sensitive precursors requiring the use of for example a glove box. Here we demonstrate how an air stable supersaturated europium-doped yttrium nitrate phosphor precursor solution has the ability to easily impregnate a structure or produce a cast yielding faithful replicas composed of $Y_2O:Eu^{3+}$ after a final short annealing step. New replicas of Lepidoptera (moth) wing scales using field emission scanning electron microscopy, structures down to 10 nm have been imaged. Moreover as these replicas are made of phosphors, their luminescence in some cases may be modulated by the internal periodic modulation built into their structures. In this work we will discuss more recent results on the use of the phosphors for making nanocasts of moth wing scales and show a range of beautiful pictures to show what the method can achieve.

1. Introduction

The class of insects belonging to the order Lepidoptera is a large, varied range of butterflies and moths containing at present approximately 175,000 known species, with upward estimates including unclassified species numbering 250,000 (more than any other type of insects except beetles). Various estimates have put a figure of approximately seven percent of all life forms for the members of this order. There are around 15,000 butterfly species, the rest of the order are moths, (the number of species increases as new species are discovered) [1,2]. The inexact nature of the science of taxonomy in conjunction with the introduction of various new techniques such as DNA sequencing [3,4], microbiology [5] and larval morphology [6] has resulted in reclassification of the relationships between species.

Butterflies and moths have different wings than those of other insects [7]. They are formed from a transparent double epithelial chitin membrane which is crisscrossed with tubular veins which nourish and hold their structure in place, in addition to transporting insect blood (haemolymph), functioning as part of the insect's respiratory system and providing ducts for their nervous system [8-11]. The wings have a covering of overlapping scales attached by a single stem on both their upper and lower sides. The size of butterfly and moth wing scales differ from species to species, and individual scales are approximately 50 to 100 μm in width by 150 to 200 μm in length. There are a small number of butterflies that have areas of their wing that are not covered by scales such as *Danais tytia, Greta-oto (glasswinged butterfly)* and *Graphium ridleyanus*. There is a great diversity in wing scale form yet they all share a common structure. The upper layer of a wing scale consists of longitudinal ridges connected by transverse crossribs creating an open rectangular netlike structure. These openings may contain various structures or hollow air spaces depending on the butterfly species. Columnar pillars called trabeculae occur within the upper and lower layers of the scale. Lattices of microribs and ridge-lamellae, in these structures are found along the longitudinal ridges, in these structures the iridescent colours of the Lepidoptera are created by preferential reflection/diffraction of certain wavelengths of light [12].

The soft interior of Lepidoptera members are supported by a hard exoskeleton of chitin and a matrix of fats and proteins, chitin is found in the wing scales of these insects in the form of fibrils approximately three nanometers in diameter [13]. The chitin fibrils within the Lepidopteran wing scales are composed of molecular chains crosslinked together by hydrogen bonding endowing the material with exceptional toughness [14]. The Lepidoptera wing usually has two types of scale; (a) cover scales that contain the structural colour (iridescent blues, greens and metallic lustres) and pattern elements of their individual design and (b) ground scales that are generally pigmentary forming the background of the wing scales (producing black, brown, red and yellow colouring) [9]. Lepidopteran wing scales in common with other macrochaetes are grown from specialised epidermal hair producing cells (trichogen) that have a socket enclosed in a tormogen cell [15]. The pedicel of each individual scale is attached to these sockets. These wings have been evolved by many species to suit their many survival strategies and pre-reproductive approaches.

The majority of butterfly wings display vibrant, highly saturated colours (produced at the sub-micrometer level by the intrinsic structure of these wing scales) which are used as communication signals for attraction, repulsion, or camouflage [9]. Some butterfly species such as *Morphidae* and *Papillo* have developed exceptionally striking iridescent colours (within the blue to green spectral colour range); this phenomenon has attracted extensive studies resulting in both an accurate classification and a deep understanding of the colour generation by the hierarchical micro-structuring of the wing scales. Whereas *Lycenid* butterlies produce iridescent colours from photonic bandgap (PBG) crystal-like structures situated in lower part of the wing scale [16,17], other species exhibit varying shades of black, brown or a range of spectral colours (in some cases going beyond the visible into the ultraviolet), as well as displaying optical polarization effects. The different effects of structural colour and colour as perceived from pigments and dyes are obvious in butterfly and some moth wings. Structural colours are vibrant and do not fade over time, in contrast to many pigments and dyes which fade at rates that depend on their colour fastness. This is because the structural colour of such wings arises from light interference that is caused by its periodic structure, in contrast to the colour of a pigment or dye molecule that is due to the selective absorption of light.

The thrust of our work was to explore the feasibility of producing casts of nanostructures that had known photonic properties using inorganic metal oxides and find ways to eventually mass fabricate these three-dimensional PBG structures. As introduced above Lepidoptera wing scales and hairs are highly structured natural photonic materials and were chosen herein and previously [18-21] as templates to fabricate replica structures out of phosphor materials. A $Y_2O_3:Eu^{3+}$ precursor solution was used previously to prepare Lepidopteran wing scale [18-22] and hair replicas successfully and further studies will be presented in this work. There are two reasons for the choice of yttrium oxide as a lattice for these structures: (a) it has been synthesised into a range of structures; and (b) it is an important technological material.

Herein we report biomorphic mineralisation replication studies on a moth wing scale template replicas using optical, field emission gun scanning electron microscopy (FEGSEM).

Materials and Methods

2.1. Specimens

The Lepidoptera specimen used for these biomorphic mineralisation replication studies was the east African sunset moth *Urania Croesus* from Mafia Island, Tanzania.

2.2. Chemicals
Yttrium and europium oxides (Neomaterials, UK, 99.99%), nitric acid (Merck, UK, AnalaR), ethanol (Merck UK, AnalaR 99.5%) and titanium ethoxide (Sigma Aldrich, UK). All chemicals were used without further purification.

2.3. Preparation of Y_2O_3:Eu^{3+} Precursor Infilling Solutions, infilling and annealing of Moth Wing Scales
An in-depth description of the preparation, infilling and annealing of the moth wing scales (and butterflies) is described elsewhere [46].

2.4. Electron microscopy
The original *Urania croesus* moth wing scale samples (before any treatment) and the post-annealed Y_2O_3:Eu^{3+} (moth) replicas were imaged using a field emission gun scanning electron microscope (FEGSEM), (Zeiss, Germany). The samples were mounted on carbon tabs fixed to aluminium pin stubs and coated with a sputter coated layer of gold to eliminate charging of the specimens.

3. Results and Discussion

FEGSEM images from a topside forewing section of the natural *Urania Croesus* specimen's green wing scales before replication treatment are presented in figure 1. The iridescent curved cover scales formed at their base by air-cuticle multilayers are apparent in figure 1 (a) (to the left) to the right of the image the flatter dentated terminated ground scales that in this species produce the black or darker scales are apparent. In figure 1 (b) a higher magnification study illustrating the extreme curvature of the iridescent cover scales.

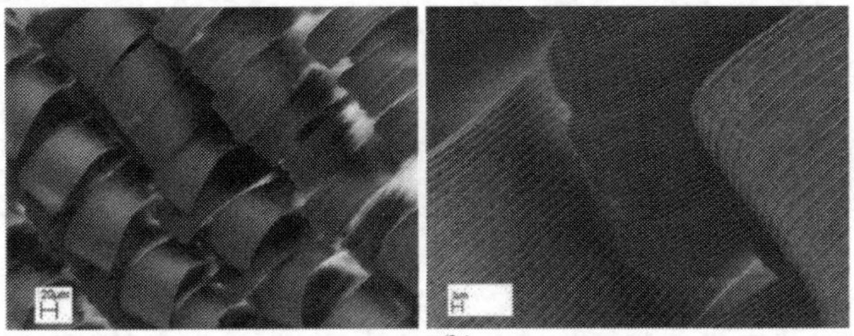
(a) (b)

Figure 1: FEGSEM images of *Urania Croesus* topside forewing scales (a) scale bar 20 μm, (b) scale bar 2 μm.

In figure 2 FEGSEM studies showing faithful Y_2O_3:Eu^{3+} *Urania Croesus* topside forewing scale replicas that are overlaying one another are shown. In figure 2 (a) are cover scales on top of ground scales, that have had their curvature flattened due to the pressure applied during infilling the scales with Y_2O_3:Eu^{3+} precursor solutions. The upper scale to the left of the image has a complete pedicel replica which had become detached from the epithelial wing membrane before infilling, the uppermost scale to the right has a small tear on its surface. In the higher magnification study in Figure 2 (b) it is possible to see through the tear to the scale below which has been replicated indicating that the Y_2O_3:Eu^{3+} precursor solution has permeated through the entire wing scale section. From these initial studies it can be seen that there is some shrinkage between groups of scales whilst the scales themselves appear to be intact, there is minor wrinkling upon their surfaces. The curling at some edges observed in figure 2 (a) and (c) may be due either to the separating of the quartz slides after infilling with the precursor solutions or the flattening of the curved cover scales, if it was due to the heat treatment more scales would display this feature. Also in figure 2 (c) is another Y_2O_3:Eu^{3+} replication of a complete scale pedicel of which a higher magnification is presented in figure 2 (d), fine radiating vein-like structures and flat bottomed furrows can be seen as the pedicel starts to transform into the scales longitudinal ridges (it is normal that they are not fully formed at this position). Also the crossribs are not mature at this transformation zone, they form circular openings that lead into the scale interior instead of rectangular voids [12].

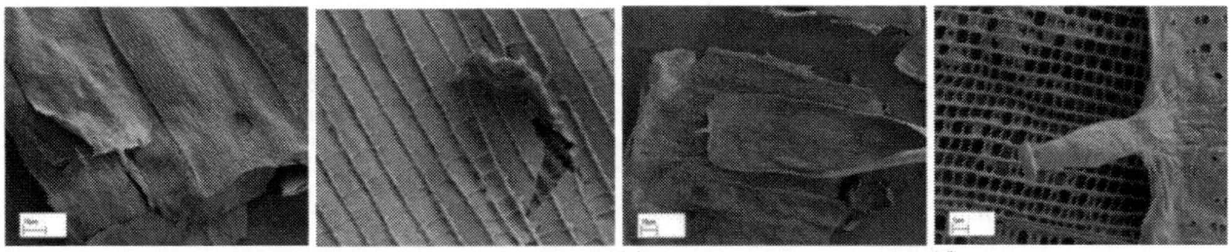

(a) (b) (c) (d)

Figure 2: FEGSEM images showing Y_2O_3:Eu^{3+} replicas of *Urania Croesus* topside forewing scales (a) stacks of ground and cover scales, scale bar 10 μm, (b) expansion of previous image showing faithful replication of ground scale below tear in cover scale, scale bar 1 μm (c) is another group of wing scales, scale bar 10 μm, and (d) an intact pedicel, scale bar 1 μm.

4. Conclusions

The work described here has shown that the nanostructures of Lepidoptera can be replicated using precursor solutions of europium-doped yttrium nitrate solution, by a faithful replacement of the chitin structures forming the wing scale. We have previously shown that it is possible to look at a range of spectral properties of such casts [18-22].

5. References

[1] Michael G P 2009 Biodiversity of Lepidoptera – in *Insect biodiversity: Science and society* (Foottit R G and Adler P H, Wiley-Blackwell, Oxford, UK)
[2] MacRae I J 2007 *Can. J. Environ. Ed.* **13** 11-29
[3] Herbert P D N and Gregory T R 2005 *Syst. Biol.* **54** 852-59
[4] Wiemers M and Fiedler K 2007 *Front. Zool.* **4** (8) 1-16
[5] Hammer T J, McMillan W O and Fierer N 2014 *PLoS ONE* **9** (1) 1-8
[6] Goldstein PZ and DeSalle R 2011 **33** 135-47
[7] Scoble M J 1992 *The Lepidoptera: Form, function and diversity* (Oxford University Press, Oxford, UK)
[8] Boggs C L, Watt W B and Ehrlich P R 2003 *Butterflies: Ecology and evolution taking flight* (University of Chicago Press, Chicago, USA)
[9] Ghiradella H and Radigan W 1976 *J. Morph.* **150** 279-98
[10] Ghiradella H 1991 *Appl. Opt.* **30** 3492-500
[11] Ghiradella H 1998 Hairs, bristles and scales – in *Microscopic anatomy of invertebrates* **11A:** Insecta (ed. Locke M Wiley-Liss New York) pp 257-87
[12] Vukusic P, Sambles J R and Ghiradella H 2000 *Photonic Sci, News* **6** 61-6
[13] Merzendorfer H and Zimoch L 2003 *J. Exper. Biol.* **206** 4393-412
[14] Dutta P K, Dutta J and Tripathi V S 2004 *J. Sci. Ind. Res.* **63** 20-31
[15] Ghiradella H and Butler M W 2009 *J. R. Soc. Interface* **6** S243-51
[16] Morris R B 1975 *J. Ent. (A)* **49** 149-54
[17] Biró L P, Bálint Z, Kertész K, Vértesy Z, Márk G I, Horváth Z E, Balázs J, Méhn D, Lousse V and Vigneron J P 2003 *Phys. Rev. E* **67** 021907
[18] Silver J, Ireland T G and Withnall R 2005 *J. Modern Opt.* **7** 999-07
[19] Withnall R, Ireland T G, Martinez-Rubio M I, Fern G R and Silver J 2002 *J. Mod. Opt.* **49** 965-76
[20] Withnall R, Martinez-Rubio M I, Fern G R, Ireland T G and Fern G R 2003 *J. Opt. A-Pure Appl. Opt.* **5** S81-5
[21] Silver J, Withnall R, Ireland T G, Fern G R and Zhang S 2008 *Nanotechnology* **19** 095302
[22] Withnall R, Silver J, Ireland T G, Zhang S and Fern G R 2011 *Opt. Laser Technol.* **43** 401-9

Cathodoluminescence studies of phosphors in a scanning electron microscope

Paul Harris, Daniel den Engelsen, Terry Ireland, George Fern and Jack Silver
Centre for Phosphor and Display Materials, Wolfson Centre for Materials Processing, Brunel University, Uxbridge, Middlesex, UB8 3PH, UK
E-mail: jack.silver@brunel.ac.uk; terry.ireland@brunel.ac.uk

Abstract. Cathodoluminescence studies are reported of phosphors in a field emission scanning electron microscope (FESEM). A number of phosphor materials have been studied and exhibited a pronounced comet-like structure at high scan rates, because the particle continued to emit light after the beam had moved onto subsequent pixels. Image analysis has been used to study the loss of brightness along the tail and hence to determine the decay time of the materials. This technique provides a simple and convenient way to study the decay times of individual particles.

1. *Introduction*

The detection of cathodoluminescence (CL) is a standard technique in many scanning electron microscopes (SEM). Important applications of CL microscopy are in the field of mineralogy, studies of defects in luminescent crystalline materials and characterization of nanostructures [1-4]. CL studies of phosphor particles, however, have received surprisingly little attention in the literature. Poelman and Smet have recently published results on time-resolved CL spectroscopy of phosphor particles by blanking a static electron beam [5].
The objective of this investigation was to study the decay behaviour of phosphor particles in a conventional SEM.

2. *Microscope and materials*

A Zeiss Supra 35VP was used in this work. The system was equipped with four detector systems. The first is an Everhart-Thornley (ET) secondary electron (SE) detector, which detects SEs plus a contribution from backscattered electrons (BSE). There is also an in-lens SE detector, for use when very short working distances are required, and this detects only SEs. An annular (retractable) Robinson™ solid state BSE detector is mounted immediately above the sample.
The microscope has the facility to operate in high nitrogen pressures (<133Pa) and an additional detector is fitted to facilitate this, which operates by using a photomultiplier to detect the fluorescence generated when low-energy SEs, emitted from the surface (under bombardment from the primary electron beam), excite the gas (nitrogen) in the chamber. If this detector is used under high vacuum conditions, then these signals are absent, and it is capable of generating high quality CL images from suitable phosphor materials. Image analysis of the panchromatic CL-micrographs was performed using ImageJ (Public Domain) software.
ZnO:Zn phosphor was obtained from Kasei Optonix (Japan), with particle sizes of 0.2-1.5μm. Monosized $Y_2O_3:Eu^{3+}$ has been synthesized in this laboratory with an average size was 330nm. The synthesis has been described in detail in our previous work [6]. Commercial $Y_2O_3:Eu^{3+}$ and $Gd_2O_2S:Tb^{3+}$ samples (micrometer sized) were obtained from Nichia, Japan. $Zn_2SiO_2:Mn^{2+}$ was supplied by Sylvania, USA. Finally, $InBO_3:Tb^{3+}$ and $SrGa_2S_4:Eu^{2+}$ were obtained from Phosphor Technology Ltd., UK. All commercial samples were analyzed without further treatment.

Journal of Physics: Conference Series **619** (2015) 012051 doi:10.1088/1742-6596/619/1/012051

Two types of substrates were used in this work, conductive graphite loaded pads and aluminium stubs. The graphite loaded pads were obtained from Agar Scientific and were slightly sticky. Phosphor powder was deposited either by a dusting technique or by dispensing a diluted phosphor suspension in ethanol. In the first case layers with 1-5 particles on top of each other were obtained, in the second case isolated particles could be deposited. In the case of aluminium substrates phosphor was applied by the alcohol dispensing technique.

3. Decay

Figure 1 shows micrographs of a cluster of Y_2O_3:Eu^{3+} particles and a single Gd_2O_2S:Tb^{3+} particle on a graphite-loaded pad. The SE-micrograph is a clear picture and shows details of the particles, whereas the CL-micrograph is smeared out. The reason for this blurring is the rather long decay times of Y_2O_3:Eu^{3+} and Gd_2O_2S:Tb^{3+}. In other words, smearing of CL-images offers the opportunity to measure decay times of CL-phosphors.

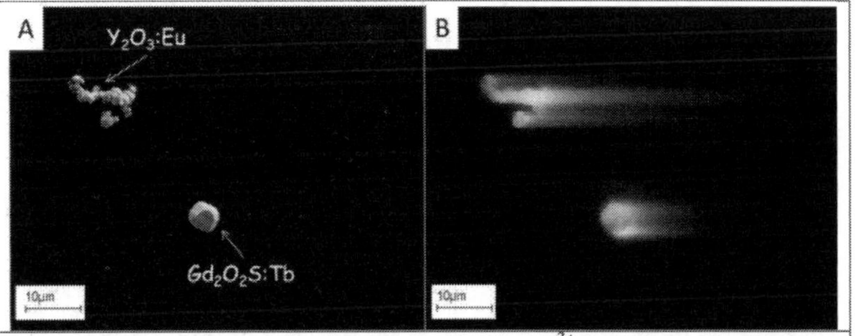

Figure 1. A: SE-micrograph of a cluster of Y_2O_3:Eu^{3+} particles and a single Gd_2O_2S:Tb^{3+} particle recorded at 10keV. B: CL-micrograph of same area as shown in A at 10keV and scanning rate of 10.1s/frame.

The decay times can be determined by analyzing the greyscales in comet-like structures as shown in figure 1B. We made these analyses with the ImageJ software; an example is shown in figure 2.
The comets shown in figure 1B are represented in spatial dimensions. From the known scanning speed and the number of pixels (1024x768) the length can be converted into time, as shown in figure 2.
Figure 2 shows two curves: the noisy curve has been determined with ImageJ from the CL-micrograph, while the exponential is the greyscale representation (GS) of the experimental data in terms of one exponential function:

$$GS = GS_{max} e^{-t/\tau} + BG \qquad (1)$$

where GS_{max} is the maximum greyscale, t indicates the time, being 0 when $GS=GS_{max}$, τ is the time constant, being the 1/e-value of the decay time ($\tau_{1/e}$), and BG is the background correction. This latter correction depends on the gain setting of the photomultiplier tube that amplifies the weak CL from individual phosphor particles in the SEM chamber.

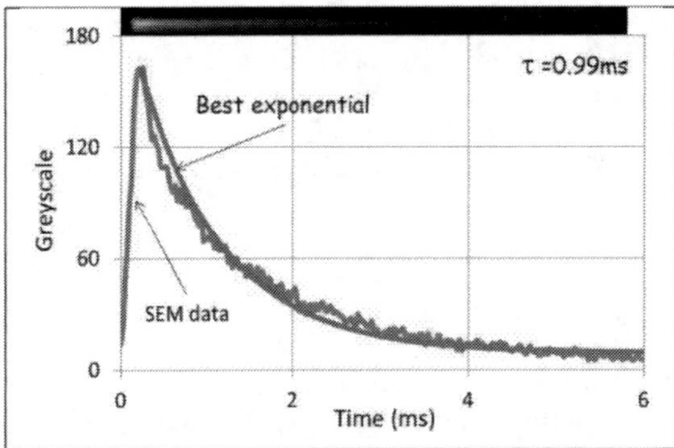

Figure 2. Greyscale of Cl-micrograph (10keV) versus time for $Y_2O_3:Eu^{3+}$ particles made in house.

The comet, from which the GS curve is derived, is also indicated in figure 2. It can be seen in figure 1B that in the case of a cluster of particles it will not be possible to determine GS_{max} unambiguously. In that case the magnification can be lowered to enable an accurate determination of the start time. We have measured the decay times of various commercial phosphors to get an impression of the range of decay times that can be measured by this method. The result is shown in table 1.

Table 1. Decay times (1/e) determined from CL-micrographs

Material source	$\tau_{1/e}$ (ms) This work	$\tau_{1/e}$ (ms) Literature	Ref.
$Y_2O_3:Eu^{3+}$ (Nichia)	1.2	1.2, 1.12, 1.1	7, 8, 9
$Y_2O_3:Eu^{3+}$ (Brunel)	0.99	n.a.	
$Zn_2SiO_4:Mn^{2+}$ (Sylvania)	3.3	8*)	10
$InBO_3:Tb^{3+}$ (PT)	3.2	2.1	11
$SrGa_2S_4:Eu^{2+}$ (PT)	0.0023	~0.0001	12
$Gd_2O_2S:Tb^{3+}$ (Nichia)	0.56	0.558	13

*) This refers to $\tau_{10\%}$. The $\tau_{10\%}$ of the $Zn_2SiO_4:Mn^{2+}$ sample in our work is 7.6ms, which indicates that the Mn-doping is ~5% according to the information in [12].

The shortest decay time determined was 2.3μs for $SrGa_2S_4:Eu^{2+}$. In this case, the fastest scan rate of the SEM (90ms per frame) was used. At this high scan rate it was necessary to use frame averaging to achieve good signal to noise ratios. It appears that the minimum decay time that can be determined with the SEM used in this work is ~1μs. The longest decay time that could be determined is about 0.1s; so, the range of decay times that can be determined with this method is about 5 decades.

It should be noted that the decay time of the nano $Y_2O_3:Eu^{3+}$ made in-house is shorter than that of the coarser commercial sample and shorter than values reported in the literature [7-9]. We also tried to measure decay times of phosphors deposited on Al substrates. Figure 3 is a CL-micrograph of monosized $Y_2O_3:Eu^{3+}$ (2%Eu) spheres with a diameter of 300nm on an Al substrate.

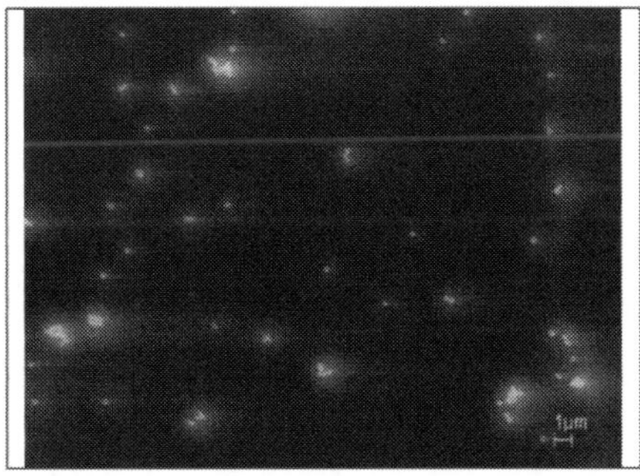

Figure 3. CL micrograph of monosized $Y_2O_3:Eu^{3+}$ spheres on Al substrate. Beam energy 10keV, scan rate 5.1s/frame.

Unlike the CL micrographs of phosphors on the graphite-loaded pads, the CL micrograph shown in figure 3 has a noticeable halo effect surrounding each particle that is of similar intensity to the comet tails. What appears to be happening is that a fraction of the electrons striking the substrate immediately around each phosphor particle are backscattered onto the particle, causing it to emit light. The reason that we can see this more clearly from an aluminium substrate rather than the graphite-loaded polymer substrate is the much larger backscatter coefficient of aluminium ~13% at 10keV, as compared to that on graphite-loaded polymers of ~ 2.2% at 10keV. Although it is possible to evaluate decay times from CL micrographs as presented in figure 3, the accuracy is reduced and the preferred measuring method is to use substrates with low backscatter coefficients, such as graphite-loaded pads or Formvar-coated TEM grids.

4. Conclusions

CL-microscopy is found to be well suited for measuring decay times in the range between 1μs and 0.1s for individual phosphor particles that may have micrometer or nanometer dimensions. No additional investments are needed to make these analyses with a SEM that is equipped with a CL-sensor. For this measuring method substrates with low backscatter coefficients are recommended. We believe that our new measuring method of analyzing the temporal information in CL-micrographs can be applied simply in studies of individual micro and nano luminescent particles.

5. References

[1] Yacobi B G and Holt D B 1986 *J. Appl. Phys.* **59** R1.
[2] Götze J 2002 *Anal. Bioanal. Chem* **374** 703
[3] Maestre D, Cremades A and Piqueras J 2004 *J. Appl. Phys.* **95** 3027
[4] Dierre B, Yuan X and Sekiguchi T 2010 *Sci. Technol. Adv. Mater.* **11** 043001
[5] Poelman D and Smet P F 2014 *Physica B* **439** 35
[6] den Engelsen D, Harris P G, Ireland T G, Withnall R and Silver J 2013 *ECS J. Solid State Sci. Technol.* **2** R201
[7] Chung Y, Park S, Jang K, Kim I, Lee Y I and Seo H J 2003 *J. Korean Phys. Soc.* **42** 543
[8] Tallant D R, Seager C H and Simpson R L 1999 *Extended Abstracts of The fifth International conference on the science and technology of display phosphors*, November 8-10, San Diego, CA, 325-328
[9] Thomes J, Seager C, Tallant D and Holloway P 2000 *Extended Abstracts of The sixth International conference on the science and technology of display phosphors*, November 6-8, San Diego, CA, 249-251
[10] Taghavinia N, Lerondel G, Makino H, Yamamoto A, Yao T, Kawazoe Y and Goto T 2001 *Nanotech.* **12** 547-551
[11] Chiu Z W, Fang T H and Hsiao Y J 2012 *J. Lumin.* **132** 2608
[12] Talin A, Dean K A and Jaskie J E 2001 *Solid-State Elect.* **45** 963
[13] Glendinning A G, Hunt S G and Bonnett D E 2001 *Phys. Med. Biol.* **46** 517

Journal of Physics: Conference Series **619** (2015) 012052 doi:10.1088/1742-6596/619/1/012052

Optical parametric amplification in dual-pumped tellurite hybrid microstructured optical fiber with engineered chromatic dispersion

Tong Hoang Tuan[1], **Edmund Samuel**[1], **Tonglei Cheng**[1], **Koji Asano**[1], **Takenobu Suzuki**[1] **and Yasutake Ohishi**[1]
[1]Research Center for Advanced Photon Technology,
Toyota Technological Institute.
2-12-1 Hisakata, Tempaku-ku, Nagoya, Japan (468-8511)
E-mail: tonghoangtuan@gmail.com

Abstract. We numerically demonstrated broad and flattened gain spectra of optical parametric amplification employing highly nonlinear tellurite optical fiber and dual-pump configuration. The chromatic dispersion profile of the tellurite hybrid microstructured optical fiber was engineered to be near-zero and flattened, having four zero-dispersion wavelengths at 1421, 1643, 1928 and 2169 nm. The effect of pump wavelength was investigated when the two pump powers were kept at 1 W, the fiber length was 25 cm and the fiber nonlinearity was as high as 6642 $W^{-1}km^{-1}$. It is shown that OPA gain bandwidth with gain ripples could be as broad as 1393 nm at 10-dB signal gain. When the central pump wavelength approached the third zero-dispersion wavelength and the pump powers were 1.25 W, an ultra-flat (±0.01-dB gain fluctuation) and broad gain bandwidth (658 nm at 30-dB signal gain) could be achieved.

1. Introduction

Fiber optical parametric amplifiers (FOPAs) with flattened gain spectra and broad bandwidth are very promising for all-optical signal processing applications such as signal generation, broadband wavelength conversion, optical sampling, switching, wavelength division multiplexing (WDM), optical time division multiplexing (OTDM), etc [1-4]. Although a degenerate FWM configuration (single-pump) has been widely used for FOPA experiments, dual-pump FOPAs are now of great interest because they can present much broad and flattened gain spectra when the two pumps were tuned symmetrically to the zero-dispersion frequency (ω_0) of the fiber as first reported by Marhic et al [5]. There have been several efforts on dual-pump FOPA, both experimentally [6, 7] and numerically [2, 8, 9]. However, the performances are limited due to the uses of silica fibers. The low nonlinearity of silica fiber requires the fiber length of several kilometers to obtain practical FOPA gain and it is pointed that the fluctuation of zero-dispersion wavelength (ZDW) in such long fibers strongly affects FWM process [10-12].

Recently, much attention has been given to tellurite fibers due to their high nonlinearity, wide transmission windows and high glass stability [13]. But FOPAs in tellurite optical fibers are facing a great challenge because their ZDWs typically locate at long wavelengths far from the wavelengths of common and commercial pump sources. In order to control the dispersion of tellurite fibers with high

[1] Tong Hoang Tuan (tonghoangtuan@gmail.com)

freedom in telecommunication window, our group has developed hybrid microstructured optical fibers (HMOFs) whose refractive index difference (Δn) between core and cladding materials could be as large as 0.5. By using tellurite HMOF, a near-zero and flattened chromatic dispersion with four ZDWs was reported in [14] and a single-pump FOPA with broad gain bandwidth was demonstrated in [3].

In this paper, we present the simulated performance of dual-pump FOPA for tellurite HMOFs which have near-zero and flattened dispersion with four ZDWs. The effects of pump wavelengths and pump powers were investigated to achieve an ultra-flat and broad signal gain bandwidth.

2. Theory of dual-pump FOPA

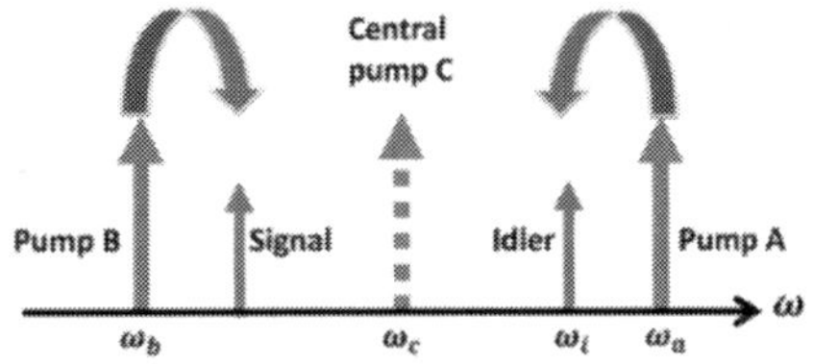

Figure 1. Schematic diagram of dual-pump FOPA configuration.

The parametric signal gain (G) in dual-pump FOPA configuration is given by equation (1) [2].

$$G = 1 + \left(\frac{2\gamma\sqrt{P_1 P_2}}{g} \right)^2 \sinh^2(gL) \tag{1}$$

where g is the gain coefficient shown in equation (2), γ is the nonlinear coefficient and L is the fiber length.

$$g = \left[(\gamma P_p r)^2 - \frac{1}{4}(\Delta\beta + \gamma P_p)^2 \right]^{1/2} \tag{2}$$

In equation (2), $P_p = P_1 + P_2$, $r = 2\sqrt{P_1 P_2} / P_p$, P_1 and P_2 are pump powers at pump frequencies ω_a and ω_b, respectively. As shown in Fig. 1, ω_s and ω_i are the signal and idler frequencies, respectively. They locate at the positions that the condition of $\omega_a + \omega_b = \omega_s + \omega_i$ is satisfied. The linear phase-mismatch $\Delta\beta$ in equation (2) is expressed by Eq. (3)

$$\Delta\beta = \beta_2\left[(\Delta\omega_s)^2 - (\Delta\omega_p)^2\right] + \frac{1}{12}\beta_4\left[(\Delta\omega_s)^4 - (\Delta\omega_p)^4\right] + \frac{1}{360}\beta_6\left[(\Delta\omega_s)^6 - (\Delta\omega_p)^6\right] \tag{3}$$

where β_2, β_4 and β_6 are the second, the fourth and the sixth-order fiber dispersion coefficients. The frequency deviation between the signal frequency ω_s and the central frequency $\omega_c = (\omega_a + \omega_b)/2$ is defined as $\Delta\omega_s = \omega_s - \omega_c$ while $\Delta\omega_p = (\omega_a - \omega_b)/2$ is the half-frequency difference between the two pumps. Since the value of the gain coefficient g depends on $\Delta\beta$, the signal gain G depends on the phase-matching condition which is the key factor for efficient performance of FOPA. Broad gain bandwidth can be generated if the phase-matching condition given in equation (4) is satisfied [15].

$$\Delta\beta + \gamma(P_1 + P_2) = 0 \tag{4}$$

3. Results and discussions

3.1. Tellurite HMOF and chromatic dispersion control

Figure 2 shows the microstructure of the proposed tellurite HMOF with a ring of six air holes in the cladding. It was designed with the core diameter (D=0.8946 μm), air hole diameter (d=2.26 μm) and the pitch which is the distance between the center of two adjacent air holes (Λ=1.999 μm). At 1550 nm, the refractive indices of core and cladding were n_{core}=2.058 and n_{clad}=1.568 (Δn=0.49). The calculated nonlinear coefficient was as large as γ=6642 $W^{-1}km^{-1}$. The full-vector finite element method (FEM) was applied to calculate the chromatic dispersion. In Fig. 3, a near-zero and flattened chromatic dispersion profile from 1.3 to 2.3 μm is shown with four ZDWs at 1421, 1643, 1928 and 2169 nm.

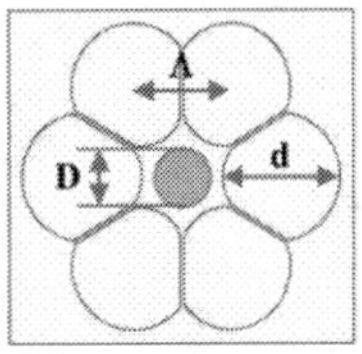

Figure 2. The design of tellurite HMOFs.

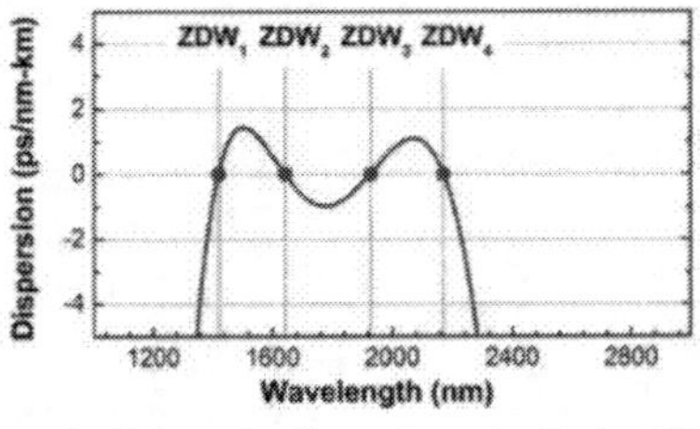

Figure 3. Chromatic dispersion of tellurite HMOFs

3.2. Numerical calculation of dual-pump FOPA in tellurite HMOF

All of the results in Figs. 4 and 5 were calculated when the two pump powers were the same $P_1 = P_2$ =1 W and the fiber length was L=25 cm. The signal gains were plotted as a function of signal wavelength. The symbols of square, circle and triangle corresponded to the pump wavelength A, B and the central C which are obtained from the pump frequencies ω_a, ω_b and the central angular frequency ω_c. The black, red and blue colors show different pump conditions. The gain bandwidths of each pump condition were determined in the shadowed range of the signal gain spectra. The details of pump wavelength conditions and gain bandwidths have been tabulated in Table 1.

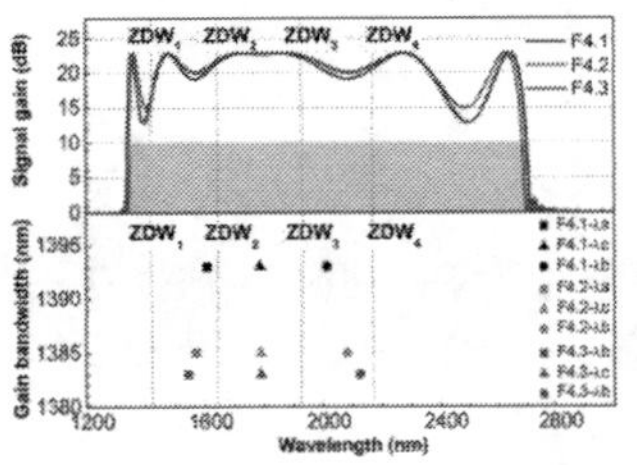

Figure 4. The effects of pump wavelength on signal gain spectra and bandwidths for tellurite HMOF

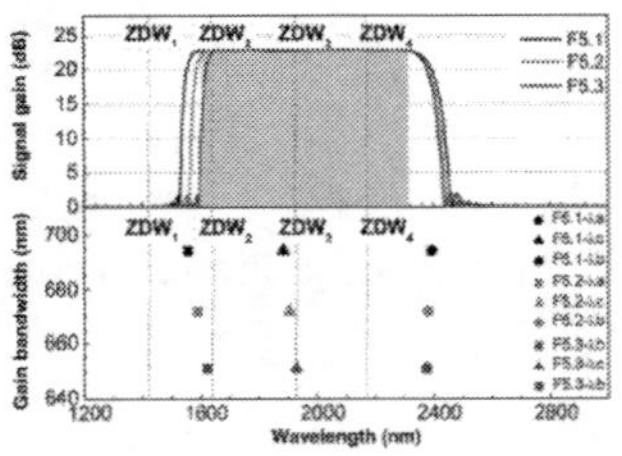

Figure 5. The effects of pump wavelength on signal gain spectra and bandwidths for tellurite HMOF

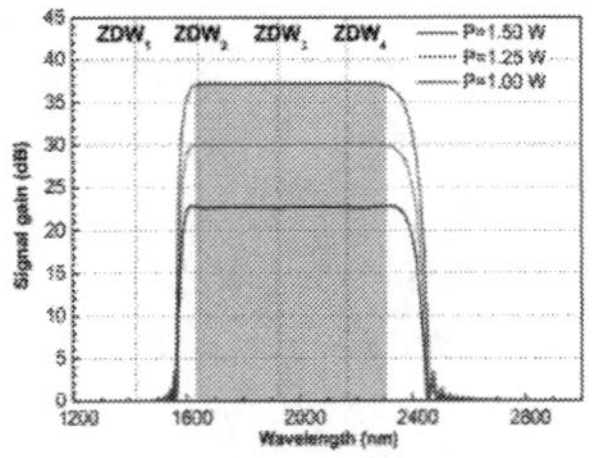

Figure 6. The effects of pump power on signal gain spectra and bandwidths for tellurite HMOF

Table 1. The pump wavelength conditions and corresponding gain bandwidths when $P_1 = P_2$ =1 W, L=25 cm and γ=6642 $W^{-1}km^{-1}$

Name	λ_a (nm)	λ_c (nm)	λ_b (nm)	Bandwidth (nm)	Name	λ_a (nm)	λ_c (nm)	λ_b (nm)	Bandwidth (nm)
F4.1	1607	1787	2013	1393	**F5.1**	1555	1884	2392	694
F4.2	1569	1788	2081	1385	**F5.2**	1589	1906	2381	647
F4.3	1544	1789	2126	1383	**F5.3**	1623	1928	2376	645

In Fig. 4, the calculated gain bandwidths of dual-pump FOPA which signal gains fluctuated from 13 to 23 dB were shown. At 10-dB signal gain, those bandwidths were very broad and could be 1393 nm for the pump condition F4.1. When the position of the pump λ_a and λ_b varied, the magnitude of gain fluctuation slightly changed but the shape of the gain spectra were kept. In contrast, flattened signal gain spectra could be found in Fig. 5. It is observed that when λ_c approached the third ZDW, the parametric gain bandwidth reduced but the spectrum became flatter. The black, red and blue spectra in Fig. 5 had ±0.1, ±0.01 and ±0.03-dB gain fluctuations, respectively. The bandwidth of the flat-gain region of the black (F5.1) was as broad as 694 nm which is much broader than those reported for most of highly nonlinear fibers so far [8, 9].

In Fig. 6, the effect of pump powers was studied for the black gain spectrum (F5.1) that was obtained in Fig. 5. The signal gain increased with the value of pump power. When $P_1 = P_2 = 1.25$ W, the gain flatness improved. An ultra-flat signal gain at 30 dB with ±0.01-dB gain fluctuation was obtained having 658-nm gain bandwidth from 1634 to 2292 nm. When $P_1 = P_2 = 1.5$ W, the signal gain could reach 37 dB, however, the gain spectrum was obtained with ±0.1-dB gain fluctuation. Another significant feature that could be noticed from Figs. 4, 5 and 6 is that very low values of the pump power were used for those calculations. It is attributed to the great contribution of the high nonlinearity and the near-zero and flattened dispersion profile of the proposed tellurite HMOF.

4. Conclusions

Highly nonlinear tellurite HMOFs were employed to study the performance of dual-pump FOPA by simulation. Due to the near-zero and flattened chromatic dispersion control from 1.3 to 2.3 μm with four ZDWs at 1421, 1643, 1928 and 2169 nm, broad and ultra-flat signal gain spectra could be realized for dual-pump FOPA. The results showed that when the central pump wavelength approached the third ZDW, an ultra-flat signal gain at 30 dB (±0.01-dB gain fluctuation) having broad gain bandwidth of 658 nm could be obtained. Moreover, the gain spectra with gain ripples could be as broad as 1390 nm at 10-dB gain. To the best of our knowledge, the potential, which is the ultra-flat and broad gain bandwidth of dual-pump FOPA using tellurite HMOF, has been reported for the first time with very low values of pump power.

Acknowledgments
This work was supported by MEXT, the Support Program for Forming Strategic Research Infrastructure (2011-2015).

5. References

[1] S. Radic, and C. McKinstrie 2003 Opt. Fiber. Technol **9** (1) 7-23
[2] Y. Tian, X. Xiao, S. Gao, S. Lu, and C. Yang 2006 Opt. Commun **263** (1) 116-119
[3] T. H. Tuan, T. Cheng, K. Asano, Z. Duan, W. Gao, D. Deng, T. Suzuki, and Y. Ohishi 2013 Opt. Express **21** (17) 20303-20312
[4] J. Hansryd, P. A. Andrekson, M. Westlund, J. Li, and P.-O. Hedekvist 2002 IEEE. J. Sel. Top. Quant **8** (3) 506-520
[5] M. Marhic, F. Yang, L. Kazovsky, and Y. Park 1996 Opt. Lett **21** (17) 1354-1356
[6] J. M. Chavez Boggio, J. D. Marconi, and H. L. Fragnito 2005 IEEE. Photonic. Tech. L **17** (9) 1842-1844
[7] J. D. Marconi, J. M. Chavez Boggio, H. L. Fragnito, and S. R. Bickham 2007 Nearly 100 nm bandwidth of flat gain with a double-pumped fiber optic parametric amplifier *Optical Society of America* OWB1
[8] X. Zhang, and C. Jiang 2007 Opt. Laser. Technol **39** (3) 638-643
[9] M. Gao, C. Jiang, W. Hu, and J. Wang 2006 Opt. Laser. Technol **38** (3) 186-191
[10] N. Wong, and K. K.-Y. Wong 2007 Opt. Commun **272** (2) 514-520
[11] J. C. Boggio, P. Dainese, and H. Fragnito 2003 Opt. Commun **218** (4-6) 303-310
[12] P. Velanas, A. Bogris, and D. Syvridis 2006 J. Lightwave. Technol. **24** (5) 2171
[13] X. Feng, A. K. Mairaj, D. W. Hewak, and T. M. Monro 2005 J. Lightwave. Technol. **23** (6) 2046
[14] Z. Duan, H. T. Tong, M. Liao, T. Cheng, M. Erwan, T. Suzuki, and Y. Ohishi 2013 Opt. Matter **35** (12) 2473-2479
[15] M. Gao, J. Wang, C. Jiang, W. Hu, and H. Ren 2006 Appl. Phys, B **84** (3) 433-438

Development of near-infrared absorption spectrometry system by using NIR wideband glass phosphor LED

H Uemura[1], S Fuchi[2], R Kato[3], K Amano[4], K Hiraizumi[5], H Hayase[4], and Y Takeda[1]

[1]Aichi Science & Technology Foundation, 1267-1 Akiai, Yakusa-cho, Toyota, Aichi 470-0356, Japan

[2]Aoyama Gakuin Univ., Fuchinobe, 5-10-1 Chuo-ku, Sagamihara-shi, Kanagawa 252-5258, Japan

[3]Toyohashi University of Technology, 1-1 Hibarigaoka Tenpaku-cho, Toyohashi, Aichi 441-8580, Japan

[4]Mitsui Kinzoku Instrumentations Technology Co., 2-88 Kokihigashi, Komaki, Aichi 485-0059, Japan

[5]Mitsui Mining & Smelting Co., 1-11-1 Osaki, Shinagawa, Tokyo 141-8584, Japan

E-mail: JP20000017@chinokyoten.pref.aichi.jp

Abstract. We developed a NIR absorption spectrometry system for detection of toxic substances by using a glass phosphor based LED. Using this NIR absorption spectrometry system, phosphoric acid solution samples were measured by molybdenum-blue method. Absorption band around 900 nm and that around 960 nm were observed. The absorption band around 900 nm increased with increasing of the phosphoric acid concentration. Partial least squares (PLS) analysis was revealed that a lower phosphoric acid concentration limit of 0.01 ppm. Furthermore, Cu dilute solutions were measured. Although there was no clear absorption band related to Cu, PLS analysis was revealed that a lower Cu concentration limit of 0.1 ppm. These results indicated that this NIR absorption spectrometry system is useful for practical applications.

1. Introduction

In Japan, a ''Positive list system'' was introduced for the regulation of residual agricultural chemicals, pesticides, feed additives and veterinary drugs in 2006. This regulation prohibits the distribution of food containing agricultural chemicals above their maximum residue limits (MRLs). Furthermore, if the MRL has not been set, maximum upper limit of 0.01 ppm was applied. Additionally, agrochemicals are regulated by many countries and Codex Alimentarius Commission [1,2]. Therefore, rapid and reliable measurement method is required for the detection of hundreds of agrochemicals. In general, a liquid or gas chromatography and a mass-spectroscopy combination system (LC/MS or GC/MS) are used [3-9]. However, it is necessary to comminute samples. In addition to it, a long measurement time (typically, > 60 minutes) is required.

A Near-infrared (NIR) absorption spectroscopy is used in applications of agricultural field [10-12].Since the NIR absorption spectroscopy is known as rapid and non-destructive method [10], it is suitable for the total inspection. In general, a wide-band light source is desirable for absorption spectrometry for measuring a broad region of the absorption spectrum. Thus, halogen lamps or light-emitting diodes (LEDs) are usually used as NIR light sources. Halogen lamps have a wide spectral width, but are big and have a short lifetime. On the other hand, LEDs are small and have a long lifetime, but have a narrow spectral width. Therefore, we have proposed a new-type light sources that emits NIR region with a wideband spectrum, and have the long-lifetime. To realize this light source, we synthesized a wideband NIR phosphor and combined with an LED in one package. The detail has reported in our previous papers [13-17]. Therefore, we tried the development the prototype of NIR absorption spectrometry system by using this light source for analysis of agrochemicals.

2. Prototype of NIR absorption spectrometry system

Figures 1(a), (b), and (c) show photos of the NIR absorption spectrometry system, an inside this system and wideband NIR glass phosphor combined LED, respectively. The size of this system is W500 ×H170 ×D300 mm. This system has enough space to change optical configurations. The basic configuration is consisted from our wideband NIR light source, a sharp-cut filter (SURUGASEIKI Co., LTD, IR-62), a sample holder (Ocean Optics, Inc., CUV-VAR), an integrating sphere (Opto Sirius Co., FOIS-1), and a multi-channel spectrometer (Hamamatsu Photonics K.K., C9405B). The temperature of above optical components is able to keep at constant by a Peltier device and a temperature controller. The light source is a Pr^{3+}-doped glass phosphor stacked on a Sm^{3+}-doped glass phosphor combined with a high-power blue LED in one package, and it has the luminescence from 760 to 1100 nm and has the maximum output power of 1.1 mW. Figure 2 shows the schematic diagram of this system. The multi-channel spectrometer is connected to the PC by USB cable, and spectra are displayed on the monitor in real time. After correcting spectra, a partial least squares (PLS) analysis was carried out on the PC. In general, a detection of sub-ppm level of toxic substances is difficult due to low light absorption. Thus, we used the PLS method for the analysis of a small spectrum changing.

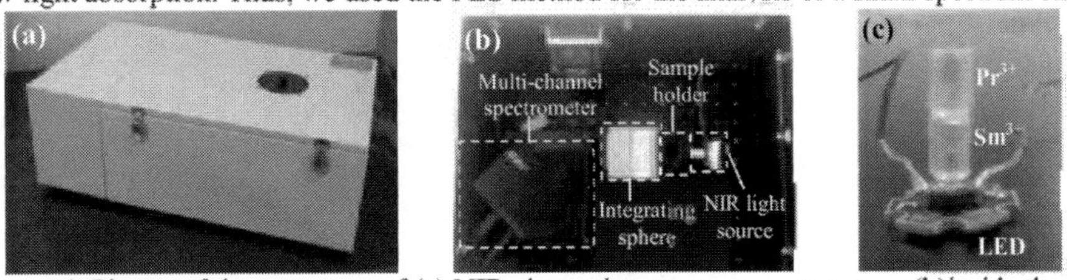

Figure 1. Photos of the prototype of (a) NIR absorption measurement system, (b)inside the system, and (c) glass phosphor combined LED as the light source.

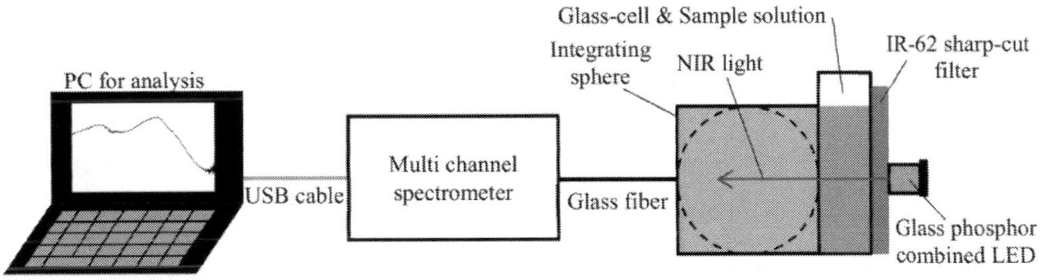

Figure 2. Schematic diagram of the prototype of the NIR absorption spectrometry system.

3. Experimental

Phosphoric acid solution samples were measured by a molybdenum blue method, since phosphorus-based agrochemical is used for a growth of agro-products. The molybdenum blue method is well known as a practical technique for measuring phosphoric acid in a few ppm levels [18]. Samples were prepared by diluting the 1000 ppm phosphoric acid standard solution (Wako Pure Chemical Industries, Ltd.) with ultrapure water. Phosphoric acid concentration was set to 0.01 to 0.1 ppm, and ultrapure water was used as a zero ppm sample. NIR spectra of these samples were measured by using a ten mm quarts cell. An integration time of multi-channel spectrometer was set to

50 msec, and absorption spectra were averaged by 50 times. Therefore, total measurement time for one spectrum is 2500 msec. The absorption spectrum (A_λ) was calculated from the transmission spectrum (S_λ), spectrum of light-source (L_λ) and dark spectrum (D_λ) by using following equation,

$$A_\lambda = -ln\left[\frac{(S_\lambda - D_\lambda)}{(L_\lambda - D_\lambda)}\right]. \quad (1)$$

We also measured the Cu diluted solutions as the preliminary experiment of Cu based agrochemicals. Samples were prepared by diluting the 1000 ppm Cu standard solution (Wako Pure Chemical Industries, Ltd.) with methanol. Cu concentration was set to 0.1 to 1.0 ppm, and methanol was used as a zero ppm sample.

4. Results and Discussion

4.1 Phosphoric acid

Absorption spectra of samples are shown in Figure 3 (indicated in every 0.02 ppm). These spectra were smoothed by the Savitzky-Golay (SG) method (5 points). In Fig. 3, a large absorption band corresponding to the second overtone of the O-H stretching of water is observed around 960 nm. Moreover, it is observed that a small absorption band around 900 nm increases with increasing the phosphoric acid concentration. This result is in agreement with Ref 18.

PLS method is applied for quantitative phosphoric acid concentration analysis. The second derivative spectra with SG method (23 points) were used for PLS analysis to remove both baseline and linear trend. In this study, Unscrambler X (version 10.2, CAMO) was used for PLS analysis. Figure 4 shows the relationship between the nominal and predicted phosphoric acid concentrations by PLS analysis. Nominal and predicted phosphoric acid concentration show good agreement. In this analysis, the values of R^2 (determination coefficient) and RMSEP (root-mean-square-error of prediction) are 0.981 and 0.004 ppm, respectively. These values indicate the high correlativity and accuracy. From these results, it is estimated that the lower detection limit of phosphoric acid is 0.01 ppm.

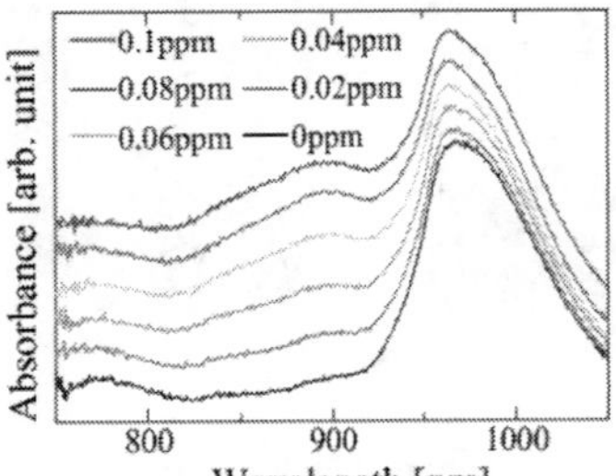

Figure 3. Absorption spectra of phosphoric acid aqueous solution.

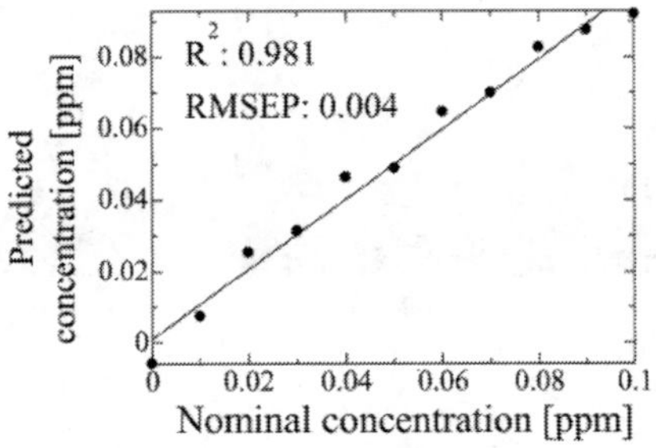

Figure 4. The relationship between nominal and predicted concentration of phosphoric acid.

4.2 Cu solutions

Absorption spectra and regression coefficient spectrum of second derivative spectra are shown in Figures 5 (a) and (b), respectively. In Fig. 5 (a), spectra indicated in every 0.2 ppm. These spectra were smoothed by the SG method (5 points). The strong absorption around 900 and 1000 nm corresponding to the third overtone of C-H stretching and the second overtone of O-H stretching are observed, respectively. Therefore, clear absorption related to Cu was not observed. The second derivative with SG method (23 points) was carried out and used for the PLS analysis. The regression coefficient spectrum is one of the PLS analysis result, and it shows the correlation between Cu concentration and spectrum change. In Fig. 5 (b), strong correlation is observed around 820 nm, which should be due to the d-d transition of Cu [19].

Figure 6 shows the relationship between the nominal and predicted Cu concentrations by PLS analysis. Nominal and predicted Cu concentration show good agreement. In this analysis, the values of $R^2$and RMSEP are 0.986 and 0.037 ppm, respectively. These values indicate the high correlativity and accuracy. From these results, it is estimated that the lower detection limit of Cu is 0.1 ppm.

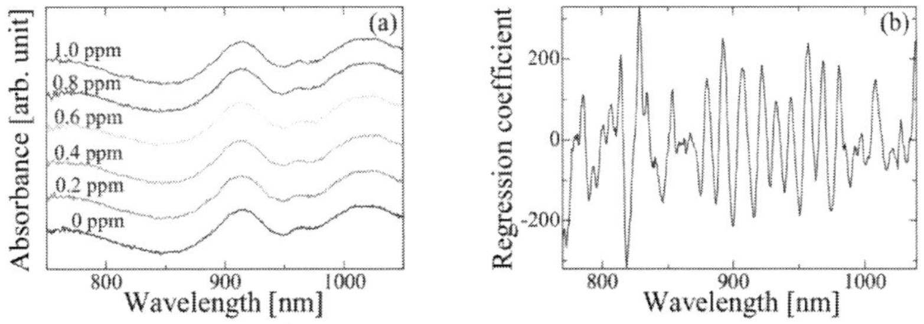

Figure 5. (a) Absorption spectra and (b) regression coefficient spectrum of Cu diluted solution.

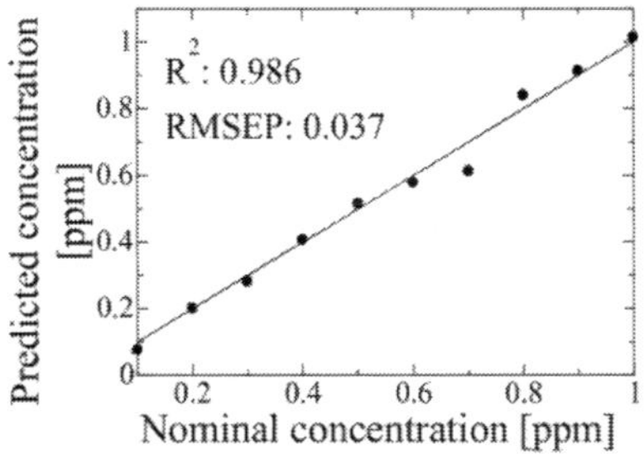

Figure 6. The relationship between nominal and predicted concentration of Cu.

5. Summary

In this study, the prototype of the NIR absorption spectrometry system by using our novel light source was developed and verified by using phosphoric acid and Cu diluted solutions. By using molybdenum blue method, phosphoric acid concentration of 0.01 ppm was detected. In the case of Cu solution, lower limit of Cu concentration was 0.1 ppm. It is concluded that the prototype of the NIR absorption spectrometry system has enough accuracy for the practical application.

Acknowledgement

This work was supported by Knowledge Hub Aichi.

References

[1] http://www.ffcr.or.jp/zaidan/FFCRHOME.nsf/pages/MRLs-p
[2] http://www.nihs.go.jp/hse/food-info/chemical/pest/mrl-link.html
[3] T. Okumura, Journal of Environmental Chemistry, **5** (1995) 597.
[4] K. Yamaguch, Jpn. J.Toxicl. Environ. Health, **42** (1996) 367.
[5] E. Ueno, *et. al.*, Food Hygiene and Safety Science, **49** (2008) 390.
[6] T. Tagami, *et. al.*,YAKUGAKU ZASSHI, **126** (2006) 991.
[7] K. Takeda, *et. al.*, Food Hygiene and Safety Science, **43** (2002) 280.
[8] K. Yoshii, *et. al.*, Food Hygiene and Safety Science, **40** (1999) 68.
[9] Y. Akiyama, *et. al.*, Food Hygiene and Safety Science, **37** (1996) 351.
[10] Y. Furukawa, *et.al*, *Infra-red and Raman Spectroscopy*, the Spectroscopical Society of Japan, Kodansha, 2009.
[11] J. Y. Hen *et al.*, Food Sci. Technol. Res **13** (2007) 291.
[12] H. W. Siesler, *et.al.*, *Near-Infrared Spectroscopy-Principles, Instruments, Applications*, WILEY-VCH, Weinheim, Germany, 2002.
[13] K. Oshima, *et al.*, Physica Status Solidi C, **9** (2012) 2340.
[14] S. Fuchi, *et al.*, Jpn. J. Appl.Phys., **47** (2008) 7932.
[15] S. Fuchi, *et al.*, Eur. J. Glass. Technol. A, **50** (2009) 319.
[16] S. Fuchi, *et al.*, Appl. Phys. Express, **2** (2009) 32102.
[17] S. Fuchi and Y. Takeda, Phys. Status Solidi C, **8** (2011) 2653.
[18] J. Copel and F. Regnaud, Anal. Chem. Acta, **27** (1962) 36.
[19] M. W. Pantoliano et al., J. Am. Chem. Soc., **104** (1982) 6310.

Acousto-Optic Tunable Filters (AOTFs) Optimised for Operation in the 2-4μm region

J. D. Ward[1], S. Valle[1,3], C. Pannell[2] and N. P. Johnson[3]

[1]Gooch and Housego (UK) Ltd, Dowlish Ford, Ilminster, Somerset, TA19 0PF, UK
[2]Gooch and Housego (Orlando), 4632 36th street, Orlando, Fl 32811, USA
[3]Optoelectronics Group, School of Engineering, University of Glasgow, Glasgow G12 8LT, UK

JWard@Goochandhousego.com

Abstract. Acousto-Optic Tunable Filters (AOTFs) are electronically-controlled bandpass optical filters. They are often preferred in applications in spectroscopy where their agility and rapid random-access tuning can be deployed to advantage. When used for spectral imaging a large aperture (typically 10mm or more) is desired in order to permit sufficient optical throughput. However, in the mid IR the λ^2 dependence on RF drive power combined with the large aperture can prove to be a hurdle, often making them impractical for many applications beyond about 2μm.We describe and compare a series of specialised free-space configurations of AOTF made from single crystal tellurium dioxide, that require relatively low RF drive power. We report on AOTFs specifically optimised for operation with a new generation of Supercontinuum source operating in the 2-4μm window and show how these may be used in a spectral imaging system. Finally, we describe an AOTF with an (acoustic) Fabry-Perot cavity operating at acoustic resonance rather than the conventional travelling-wave mode; the acoustic power requirement therefore being reduced. We present an analysis of the predicted performance. In addition, we address the practical issues in deploying such a scheme and outline the design of a prototype "resonant AOTF" operating in the 1-2μm region.

1. Introduction

The principles of acousto-optics have been known for some time, but it is only since the advent of the laser that the science has developed and practical devices found application. Bulk AO devices are used in many forms but principally Q-Switches, Mode-Lockers, Modulators, Deflectors, Frequency-Shifters [refs] and Tunable Filters, together with variations such as Cavity Dumpers and Pulse-Pickers [1].

An Acousto-Optic (AO) device is essentially an optical cell with an ultrasonic transducer (generally in the 10s/100s MHz region) bonded to one of its faces. The acoustic beam creates a sinusoidal perturbation of the refractive index of the medium due to the photoelastic effect which acts as a volume diffraction grating. An optical beam will interact with the induced grating when the angle of incidence and wavelength are matched to the grating periodicity, causing the optical beam to be diffracted at a given angle. The grating periodicity is determined by the acoustic velocity and frequency.

Most AO devices rely upon the "isotropic" AO interaction where the refractive indices of the incident and diffracted beams are the same; the acoustic beam propagates as a longitudinal mode. These devices often referred to as "Bragg Cells" form the basis of modulators and Q-switches etc [1]. However, there is another family of AO devices that utilise the "anisotropic" AO interaction [1]. As the term implies, this interaction requires an optically anisotropic (birefringent) medium, and the incident and diffracted beams have different (ordinary/extraordinary) refractive indices. In addition, the acoustic beam propagates in the slow-shear mode. As a result, the polarisation state of the optical beam is altered at the interaction. To a very good approximation one can say that a linearly polarised input results in a diffracted beam with the plane of polarisation rotated by 90°.

The intensity of the diffracted beam is a function of the power of the acoustic beam. It can be shown that [2]

$$DE = \frac{I_d}{I_0} \approx \sin^2 \sqrt{\frac{\pi^2 L M_2 P}{2H\lambda^2}} \qquad (1)$$

So that maximum diffraction efficiency (*DE*) occurs when

$$P_0 = \frac{H\lambda^2}{2M_2 L} \qquad (2)$$

Where H is the height of the acoustic field, L is the interaction-length and λ is the light vacuum wavelength. The parameter M_2 is known as the "Acousto-Optic figure of merit", it is material dependent and is a measure of the effectiveness or strength of the AO interaction. It is important to note that the required drive power increases with the square of the wavelength. This becomes very important when designing AOTFs to operate in the infra-red, particularly beyond about 2µm.

2. ACOUSTO OPTIC TUNABLE FILTER

The Acousto Optic Tunable Filter (AOTF) exploits the slow-shear anisotropic interaction to realise a solid-state agile random-access electronically addressable tunable filter, the wavelength being selected by the RF drive frequency. Most AOTFs are made from single crystal tellurium dioxide (TeO_2). TeO_2 is transparent over the range 400nm to 4·5µm and is available with good optical quality. Importantly it has a relatively high value of M_2. The acousto-optic interaction takes place when the phase matching condition between photon-phonon interaction as defined by the K-space diagram [2] is satisfied. The filtered wavelength depends on the RF frequency applied to the ultrasonic transducer.

2.1. *Non Collinear AOTF*

In theory an infinite number of configurations of incident and acoustic angle are possible; so long as the phase-matching condition in K-space is satisfied. However, in practice it is useful to introduce some constraint. Most AOTFs are designed to operate in the "parallel tangents" condition [3], which yields devices (figure 1-a) with the widest field of view and is generally preferred for this reason.

2.2. *Quasi Collinear AOTF*

It will be noted that the direction of energy propagation or group-velocity of the acoustic beam (ie walk-off) is in general different to the phase velocity. This is a consequence of TeO_2 being acoustically anisotropic. The effect is significant, so when designing a "parallel-tangents" AOTF care needs to be taken to manage the acoustic energy. An alternative constraint may be introduced such that the acoustic group-velocity direction is parallel with the input beam [4]. One way of achieving this is to reflect the acoustic beam at an optical interface (figure 1-b). In this case there will be excellent overlap between the acoustic and optical beams and the interaction-length may be made to be relatively long. This results in AOTFs with low drive power and narrow resolution [5]. By choosing the appropriate cut/orientation of TeO_2 the AOTF may be optimised to minimise drive power or to achieve narrower linewidth. A downside is that the field of view is restricted, so much so that in general the optical input needs to be at or very close to the diffraction-limit. Such AOTFs are known as Quasi-Collinear or sometimes Collinear Beam AOTFs.

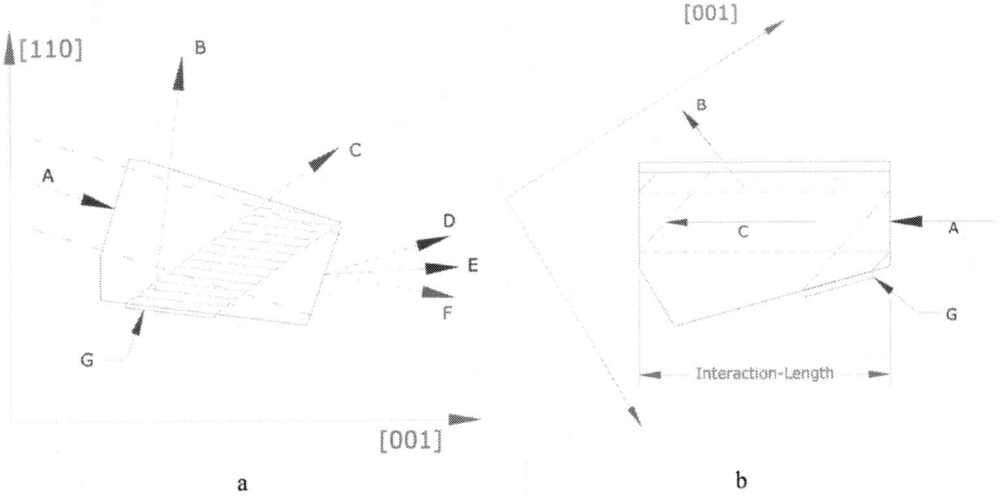

Figure 1. Configuration of (a) non-collinear AOTF & (b) Quasi-Collinear AOTF
A – unpolarised input optical beam, B - acoustic phase-velocity direction, C - acoustic group-velocity direction, D – vertically polarised diffracted order, E – 0-order, F - horizontally polarised diffracted order, G – acoustic transducer

3. OPERATION IN THE INFRA-RED

Tellurium dioxide is transparent to beyond 4µm. However, from equation 2 it is apparent that the drive power required increases with λ^2. This effect becomes noticeable at about 1µm and is compounded by the fact that the figure of merit (M_2) also decreases with increasing wavelength. The drive power becomes a major if not the overriding factor in the realisation of AOTFs at wavelengths beyond about 2µm.

3.1. Quasi-Collinear

For the reasons stated above, AOTF designs that inherently use lower power are of interest. In particular, the ongoing developments in ZBLAN based Supercontinuum sources operating beyond 2µm [6] have focussed attention on Quasi-Collinear AOTF designs since the Supercontinuum source has a single spatial mode output that aligns well with the angular aperture limitation of the Quasi-Collinear configuration. In addition, there was a desire to deliver an output with an unusually narrow spectral linewidth [7].

We have built a Quasi-Collinear AOTF that tunes from 2 to 4·5µm. The interaction-length at 40mm is longer than is normal resulting in a narrow passband; initial evaluation figures indicating $\delta\lambda$<1nm @ 2µm with modest drive power for an AOTF at this wavelength <0·25W @ 2µm giving excellent agreement with the design predictions. In addition to this AOTF, manufacture of a "cut-down" version with an interaction length of 20mm is well under way. This will have a passband approximately twice as broad and a slightly larger acceptance angle. Thus, it will be easier to align and allow a greater throughput making it more practical for a number of applications. When used in conjunction with the Supercontinuum this combination makes for an interesting illumination source for hyperspectral imaging. It is particularly appropriate for use on microscope based systems and dispenses with the need for a filter in front of the detector [7]. True hyperspectral operation is achievable since the AOTF is continuously tunable across the range; the system not being restricted to a fixed number of wavelength bands. In addition, the AOTF is truly random access and fast, characterized by a switching time of less than 100μs to tune between wavelengths. In actual fact, the long interaction length makes this AOTF relatively slow in comparison to other AOTFs.

3.2. Resonant Acoustic AOTF

An AOTF with a wider field of view and low drive power is also desirable. Returning to equation 2 it can be seen that the drive power requirement is inversely dependent upon the interaction length, thus one option would be to design a "long" AOTF. Eventually however practical limitations, not least the availability of good quality single crystal TeO_2 of sufficient dimensions restrict the ultimate size that is achievable. An alternative approach is to make better use of the acoustic energy, and rather than just "throw it away" at an acoustic dump to re-use it. To this end we have designed an acoustically resonant AOTF, where the opposite face to the transducer face is polished to be parallel. In this way we can create an acoustic resonant cavity and fill it with a standing acoustic wave, as near-perfect acoustic reflection will take place at the TeO_2/air boundary. Thus, when modelling the cavity the parameters of interest will principally be acoustic attenuation and divergence.

We have demonstrated the concept [8] by adapting a travelling-wave AOTF into a standing-wave AOTF, tuning across the 1-2μm region. Optical and pulse-echo acoustic measurements have indicated a gain of at least a factor of 4. This is in-line with our predictions based on acoustic attenuation and divergence. Resonance may be detected by monitoring the reflected RF signal from the transducer. The stability of the device was reasonable given that there is no active temperature control; experiments are in-hand to further investigate this point. Overall, the results were sufficiently encouraging for us to design a carefully optimised resonant AOTF to tune across the 2-4μm region. In addition, a specialised RF driver prototype to accompany the new device is also under design. This will continuously monitor the reflected RF signal and lock on to the resonance.

4. CONCLUSION

Despite the inherently high RF drive-power requirement at longer wavelengths, practical AOTFs may be built operating to the transmission window limit of 4·5μm. Further improvements are possible using carefully optimised resonant designs.

This research has been [partially] supported by the European Commission through the Framework Seven (FP7) project MINERVA (317803; www.minerva-project.eu).

[1] Goutzoulis, A. P. & Pape, D. R. *Design and Fabrication of Acousto-Optic Devices*. ISBN 0-8247-8930-X
[2] Xu, J. & Stroud, R. *Acousto-Optic Devices; Principles, Design & Applications*. ISBN 0-471-61638-9
[3] Chang, I.C., *Noncollinear acousto-optic filter with large angular aperture*. Appl. Phys. Lett., 1974. **25**(7): p. 370
[4] Voloshinov V, *Close to collinear Acousto-Optical interaction in paratellurite*, Opt Eng 31, 2089-2094 (1992)
[5] Kaplan D. Tournois P., *Theory and performance of the acousto optic programmable dispersive filter used for femtosecond laser pulse shaping* . Phys. IV France 12 (2002) Pr5-69
[6] Moselund P., Petersen C, Leick L., Seidelin Dam J, Tidemand-Lichtenberg P., Pedersen C., *Highly Stable, All-fiber, High Power ZBLAN Supercontinuum Source Reaching 4.75 μm used for Nanosecond mid-IR Spectroscopy*. Advanced Solid State Lasers, Paris, Oct/Nov 2013
[7] Farries M. et al, *Mid infra-red hyper-spectral imaging with bright super continuum source and fast acousto-optic tuneable filter for cytological applications*. ICOOPMA 2014
[8] Valle S. Ward J. Pannell C. Johnson N., *Acousto-Optic Tunable Filters (AOTFs) for imaging Applications in the 2-4μm range with Low RF Drive Power*. School on Acousto-Optics and Applications, Druskininkai, Lithuania 2014

Optical properties of the low-molecular amorphous azochromophores and their application in holography

A Gerbreders[*1], **J Aleksejeva**[1], **A Bulanovs** [2], **A Ogurcovs**[2], **E Zarins**[3], **A Tokmakovs**[1], **A Vembris**[1]
[1]Institute of Solid State Physics, University of Latvia, Latvia
[2]G.Liberts' Innovative Microscopy Centre, Daugavpils University, Latvia
[3]Institute of Applied Chemistry, Riga Technical University, Latvia

E-mail: andrejmah@gmail.com

Abstract. The films based on the low-molecular amorphous azochromophore 2-(3-(4-((4-(Ethyl(2-(trityloxy)ethyl)amino)phenyl)diazenyl)styryl)-5,5-dimethylcyclohex-2-enylidene) malononitrile (IWK-2M) were prepared. The optical properties of the material, such as transmittance and reflection spectra of the film, sensitivity to polarization holographic recording by two wavelengths (405 and 532 nm) were studied. The direct relief formation during the polarization holographic recording was explored, relief depth dependence on exposure and record beam intensity was investigated. The holographic matrix on this material base was produced without chemical etching process; the replication of holographic image was performed.

1. Introduction

Processes associated with the formation of surface relief during polarization holographic record (PHR) were studied for last several years [1-3]. According to some sources [4,5], this phenomenon is based on the change of the dipole moment of the photosensitive molecules and, as a consequence, the movement of particles of substance by an electric field. Similar effect was observed in the thin films of different compounds, but azochromophores in which under the action of the illumination occurs isomerization relative to N = N bond were recognized as the most effective [3].

One of the problems to be solved for the practical implementation of PHR is the formation of an optically homogeneous film of the azochromophore. In recent years, various options have been proposed to solve this problem. Different composite materials in which the diazo compound used as fillers have been tested [6, 7]. Also, various polymers containing N = N bond in the chain of the macromolecule were synthesized [8-10].However, despite the fact that some of these materials have high sensitivity and efficiency, their high cost puts into question the possibility of it for practical application.

In this paper, we considered one of the low-molecular diazocompounds forming amorphous film after solvent evaporation. Some authors have presented studies of these medium in holography [11, 12]. In our opinion, these materials are the most promising, due to the relative simplicity of synthesis and purification. Reference to a particular compound, we set out to show the possibility of recording

* Institute of Solid State Physics, University of Latvia,
8 Kengaraga str., Riga LV 1063, LATVIA

holographic images using polarization holography without etching operation and make a matrix for holographic image replication.

2. Experimental

As the low molecular azochromophore was chosen substance IWK-2M, having the chemical name 2 -(3 -(4-((4-(Ethyl(2-(trityloxy)ethyl)amino)phenyl)diazenyl)styryl)-5,5-dimethylcyclohex -2-enylidene) malononitrile, and the structure shown in Fig.1. The substance was synthesized as described in [13]. For film preparation, 9.66 mg of IWK-2M was dissolved in 580 mg of $CHCl_3$ (Chempur, CAS# 67-66-3, pure p.a.). The solution was applied to a glass substrate by spin-coating method. The thickness of dry films was 150-500 nm, determined by the Veeco Dektak 150 surface profiler.
Holographic record was made by polarization holography method by lasers at two wavelengths: 405 nm and 532 nm. P-p polarization of the recording beams was used for grating record by line 405 nm. In the case of wavelength 532 nm, experiments were performed with the polarization s-s, p-p and +45°/-45°. Record was done in the range of exposure from 0 to 800 kJ/cm^2.

3. Result and discussion

Figure 1 shows the structure of the compound. The presence of two double bonds, C = C and N = N does not allow to determine how spatial isomerization occurs during the material irradiation. Obviously, the possible existence of four spatial modifications and probably present in all four of the substance at each time instant in a ratio which varies by irradiation.

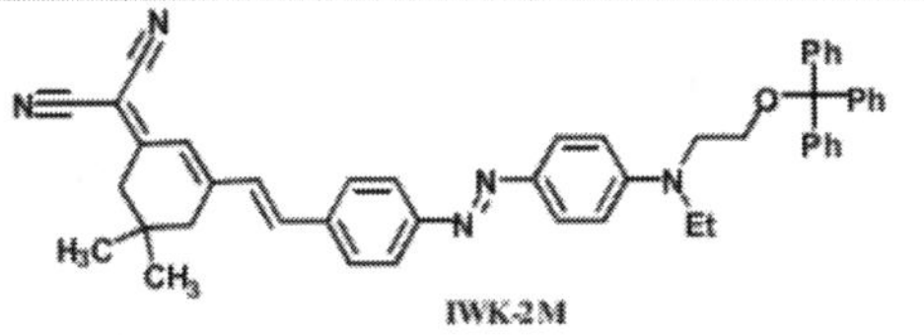

Figure 1. Structure of IWK-2M.

Figure 2 presents the transmission spectrum of the film. It shows that the wavelength selected for grating recording is completely absorbed by the film at a specimen thickness. Thus, the formation of a surface relief under the action of the electric field gradient of the recording beams occurs sequentially in depth from the surface of the film.

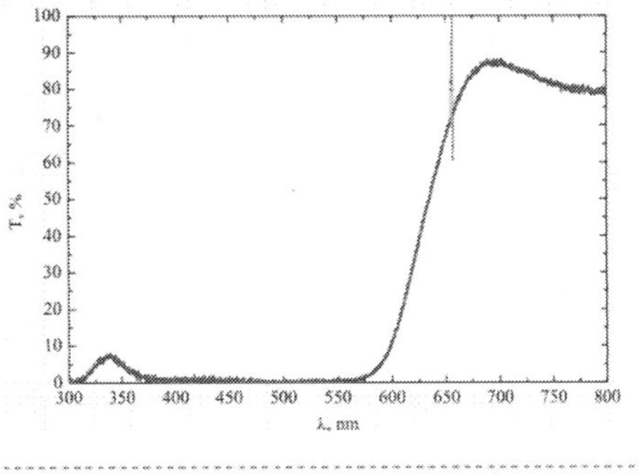

Figure 2. Transmission spectra of IWK-2M films in glass substrate. Film thickness - 300 nm.

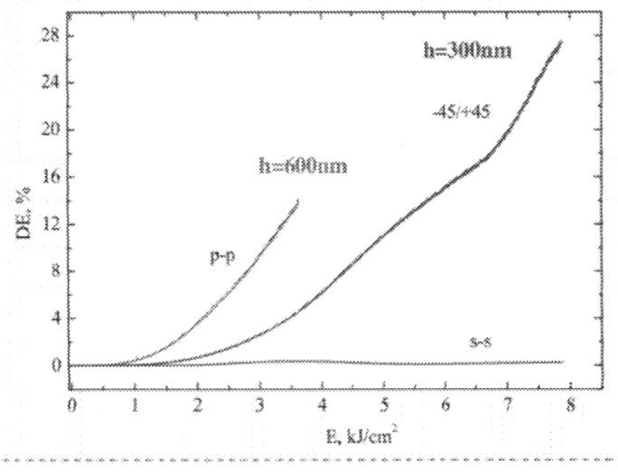

Figure 3. The transmission difraction efficiency dependence on recording exposure and beam polarization. Recording laser λ=532 nm, intensity of each beam is 0.25 W/cm^2, grating period is 1 μm. Reading laser λ=653 nm.

Figure 3 illustrates the increase in transmission diffraction efficiency dependence on the exposure at different polarizations of the recording beams. The most effective polarization direction for this matter is the electric field vector parallel to the plane of the recording (p-p). The relief depth of the each grating was determined after end of record, and is marked on the chart. The transmission diffraction efficiency after record with p-p polarization of the recording beams is 14% and corresponds to the relief depth 600 nm. But in the case of -45/+45 polarization of the beams, the transmission diffraction efficiency is 28% and comply with the relief depth 300 nm. So we can talk about that the transmission DE is not correlated with relief depth in this media.

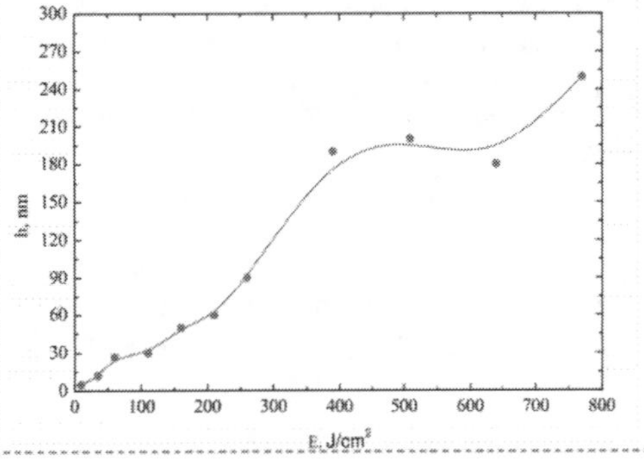

Figure 4. Dependence of grating relief depth on recording exposure. Recording laser λ=405 nm, p-p polarization, grating period is 1 μm.

Figure 5. Photo of the relief holographic image on IWK-2M film.

Figure 4 shows the dependence of the direct relief depth, which occurs after recording in the interval of exposure to 800 J/cm^2. Based on this graph, the recording conditions were chosen for producing of holographic images without the chemical etching on the film of IWK-2M. Image size 8x3 mm (Fig. 5) was recorded by the hologram recording setup, described in [14, 15]. The polarization of the recording beams of laser with a wavelength of 405 nm was set in p-p position.

The holographic matrix suitable for further replication of holograms has been successfully fabricated.

4. Conclusions

Azochromophore IWK-2M can be used as a photoresist for recording of relief holographic images using polarization holography method (without chemical etching operation) with the depth of the relief to 600 nm by exposures ranging from 400 to 1000 J/cm^2, for a laser with a wavelength of 405 nm. The most effective polarization of the recording beams for this material is p-p. Laser with a wavelength of 532 nm can also be used for recording of the relief holograms, but require higher exposure values - up to 3.5 kJ/cm^2.

Acknowledgment

This work has been supported by the European Social Fund within the project Nr. 2013/0045/1DP/1.1.1.2.0/13/APIA/VIAA/018.

5. References

[1] Matharu A at al 2007 *Chem. Soc. Rev.* 36, 1868
[2] Gertners U at al 2010 *Optical materials*, Vol.32, 8, 807-810
[3] Nikolava L, Ramanujam P 2009 "Polarization holography." Cambridge University Press, 239 p.
[4] Yager K at al. 2006 *J. of Photochemistry and Photobiology A: Chemistry* 182, 250
[5] Atassi Y at al 1998 *Pure Appl. Chem.* 70, 2157-2166
[6] Chan S at al 2006 *Chemical Physics Letters* 428, 371–375
[7] Gerbreders A at al 2010 *Lithuanian Journal of Physics* 50, 1, 47-53

[8] Cojocariu C at al 2004 J. *Mater. Chem.*14, 2909–2916
[9] Yang S at al 2001 *J. Macromol. Sci. - Pure Appl. Chem. A* 38(12), 1345–1354
[10] Gerbreders A at al 2011 *JOAM*, 13, No. 11-12, 1559-1562
[11] Ozols A at al 2012 *IOP Conf. Ser.: Mater. Sci. Eng.* 38, 012011
[12] Ozols A at al 2010 *Optical Materials* 32, 811-817
[13] Zarins E at al. 2013 *Proc. of SPIE* 8622 86221H-12
[14] Bulanovs A at al 2011 *Central European Journal of Physics* 9 (5) 1327 - 1333
[15] Bulanovs A at al 2011 *Latvian J. of Physics and Technical Sciences* 48 (5) 60 - 68

Controlling plasmonic properties of epitaxial thin films of indium tin oxide in the near-infrared region

R Kamakura, K Fujita, S Murai and K Tanaka

Department of Material Chemistry, Graduate School of Engineering, Kyoto University, Katsura, Nishikyo-ku, Kyoto 615-8510, Japan

Email: fujita@dipole7.kuic.kyoto-u.ac.jp, murai@dipole7.kuic.kyoto-u.ac.jp

Abstract. Epitaxial thin films of indium tin oxide (ITO) were grown on yttria-stabilized zirconia single-crystal substrates by using a pulsed laser deposition to examine their plasmonic properties. The dielectric function of ITO was characterized by spectroscopic ellipsometry. Through the concentration of SnO_2 in the target, the carrier concentration in the films was modified, which directly leads to the tuning of the dielectric function in the near-infrared region. Variable-angle reflectance spectroscopy in the Kretschmann geometry shows the dip in the reflection spectrum of p-polarized light corresponding to the excitation of surface plasmon polaritions (SPPs) in the near-infrared region. The excitation wavelength of the SPPs was shifted with changing the dielectric functions of ITO, which is reproduced by the calculation using transfer matrix method.

1. Introduction

Surface plasmon polaritons (SPPs) are the collective oscillations of conduction electrons coupled to the electromagnetic waves propagating at the interface between a metal and a dielectric. The electric field of SPPs is evanescently confined at the interface and accumulated in the scale below the diffraction limit of light [1]. These properties of SPPs have attracted attention in various fields [2, 3].

Noble metals, especially silver and gold, are almost exclusively used as conventional plasmonic materials. Their plasma frequency (ω_p) is suitable for SPPs applications in the visible region. In contrast, they do not work effectively as a plasmonic material in the infrared region [4]. Since the infrared region is technically important in the field such as telecommunication and sensing, complemental plasmonic materials are required that operate in this region [5]. Transparent conductive oxides, such as indium tin oxide (ITO) and gallium-doped zinc oxide (GZO), have been explored as alternative plasmonic materials because of their variable ω_p and their small optical losses in the infrared region [4-8].

ITO has been fabricated by various techniques such as sputtering and pulsed laser deposition (PLD) [6-9]. The lowest resistivity has been achieved for epitaxial thin films on yttria-stabilized zirconia (YSZ) substrates by using PLD technique [9]. Because the lattice constant of ITO matches well twice of the lattice constant of YSZ [7, 9], ITO grows epitaxially and shows high crystallinity with a low concentration of defects to minimize a scattering of electrons.

In the present study, we have fabricated epitaxial ITO thin films on YSZ substrates by PLD technique, and controlled the dielectric function by changing the concentration of SnO_2 in the target. The dielectric function in the near-infrared region was obtained from spectroscopic ellipsometry (SE). By increasing the SnO_2 concentration in the target, we have obtained thin film showing the highest ω_p among the ITO films for SPPs applications.

2. Experiment

Commercially available YSZ (111) single crystal substrate was annealed at 1350 °C in air to obtain atomically flat surface. In_2O_3 mixed with 10 and 20 wt% SnO_2 was pelletized and sintered at 1400 °C in air and used as a target. The deposition was done at the oxygen pressure of 1.2×10^{-3} Pa, and the substrate temperature of 600 °C. A KrF excimer laser ($\lambda = 248$ nm, repetition rate = 5 Hz, pulse energy = 4 $Jcm^{-2}pulse^{-1}$) was used to ablate the target. Two thin films with different SnO_2 concentrations have been prepared from the target with 10 and 20 wt% SnO_2, which are hereafter labelled as ITO–10 and ITO–20, respectively.

Crystallinity and orientation of ITO thin films were analyzed by x-ray diffraction (XRD) (ATX-G, Rigaku Co.). The thickness and the dielectric function of the thin films were measured by SE (M2000, J. A. Woollam Japan). The excitation of SPPs on the thin films was examined by variable-angle reflectance spectroscopy under the Kretschmann geometry [1].

3. Results and discussion

3.1. Structural characterization

Figures 1(a) and (b) show the out-of-plane XRD patterns of ITO–10 and ITO–20 grown on YSZ (111) substrates, respectively. The intense peaks of the ITO 222 and 444 are observed, along with the peaks of YSZ 111 and 222. In-plane XRD measurements (not shown here) confirm the intense peak of ITO 440 along with the peak of YSZ 220; other diffraction peaks were not observed. In-plane and out-of-plane XRD patterns demonstrate that the thin films were grown heteroepitaxially on YSZ substrates. The insets of Figures 1(a) and (b) show the rocking curve of the peak of ITO 222, and the full width at half maximum (FWHM) of ITO–10 and ITO–20 are 0.031° and 0.039°, respectively. These values are on the same order of the smallest value for atomically-flat ITO thin film (0.015°) [9]. The FWHM is larger for ITO–20, indicating the decrease of the crystallinity with increasing Sn concentration.

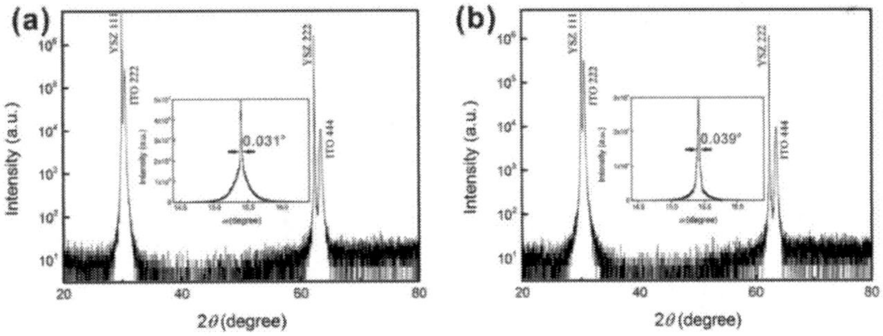

Figure 1. Out-of-plane XRD patterns of ITO–10 (a) and ITO–20 (b). The inset shows the out-of-plane rocking curve of the peak of ITO 222.

3.2. Optical properties

In analysis of the SE data, we employed a function with a Drude term and a Gaussian term to describe the dielectric function of the ITO. The Drude term and Gaussian term denote the contribution from the collective oscillation of free electrons and the interband transition of bound electrons, respectively [10]. The Drude term is given as follows:

$$\varepsilon_{\mathrm{Drude}} = \frac{\omega_p{}^2}{\omega^2+\gamma_p{}^2} - \frac{\gamma_p\omega_p{}^2}{\omega(\omega^2+\gamma_p{}^2)} i \qquad (1)$$

where ω is the frequency. Plasma frequency ω_p and damping factor γ_p are described as a function of carrier concentration n and mobility μ, respectively, as $\omega_p = [nq^2/(\varepsilon_0 m^*)]^{1/2}$ and $\gamma_p = q/(m^*\mu)$, with ε_0 the permittivity of free space, q and m^* the charge and the effective mass of electron.

Figure 2 shows the real (ε') and imaginary (ε'') parts of the dielectric functions of ITO–10 and ITO–20 calculated from the SE data. Also plotted are the dielectric functions of the ITO thin films reported in refs. 6 and 8. The crossover wavelength, where the ε' crosses zero, of ITO–10 and ITO–20 are 1290 and 1140 nm, respectively. The value for ITO–20 is the shortest among the ITO thin films prepared for plasmonic purpose to date. Figure 2(b) shows that the ε'' of ITO–10 is the smallest in the range of the plot.

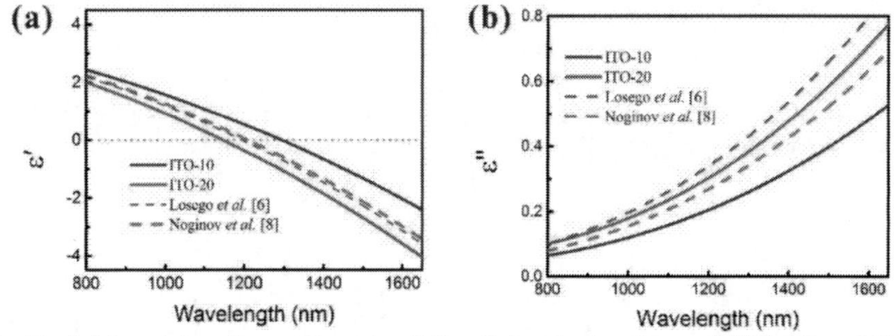

Figure 2. Real (a) and imaginary (b) parts of the dielectric functions of ITO–10 and ITO–20. Dielectric functions of ITO thin films reported in literatures are also plotted [6, 8].

Table 1 summarizes the n and μ for ITO–10 and ITO–20 evaluated from the Drude term of the fitting function, together with those in refs. 6 and 8. ITO–20 has the largest n, reflecting its shortest crossover wavelength, while ITO–10 shows the largest μ, which is in accordance with the lowest ε''. Also the high mobility for ITO–10 and ITO–20 agrees well with the crystallinity evaluated by XRD measurements.

Table 1. The carrier concentration (n) and mobility (μ) for ITO–10 and ITO–20 obtained from SE data. The values reported in the literatures [6, 8] are also tabulated.

	n (cm^{-3})	μ (cm^{2}V^{-1}s^{-1})
ITO–10	0.98×10^{21}	47.6
ITO–20	1.23×10^{21}	40.6
Losego *et al.* [6]	1.2×10^{21}	35
Noginov *et al.* [8]	1.17×10^{21}	43.2

3.3. Plasmonic properties

Figures 3(a) and (b) show the reflectance ratio as a function of incident angle and wavelength for ITO–10 and ITO–20, respectively, with the inset illustrating the schematic setup of the experiment. The p-polarized reflectance R_p is divided by s-polarized reflectance R_s because the SPPs are excited only by p-polarized waves. In Fig. 3(a), the dip in reflectance ratio, caused by the excitation of SPPs at the interface between the ITO–10 and air, is observed at 48° and around 1600 nm. In Fig. 3(b), that is, in the case of ITO–20, the dip is blue-shifted to 1400 nm.

The reflectance spectra of the samples were calculated by using transfer matrix method [11], as shown in Figs. 3(c) and (d). The inset illustrates the model for the calculation. The model consists of prism, substrate, ITO–10 or ITO–20, and air. The thickness (t) and the refractive index (n) of the layers were set as follows; prism (t = infinite, n = 1.78), YSZ substrate (t = 0.5 mm, n = 2.1), ITO–10 (t = 230 nm, n from Fig. 2), ITO–20 (t = 125 nm, n from Fig. 2), and air (t = infinite, n = 1). Comparing the experimental results with the calculations, it is found that the reduction of reflectance caused by the excitation of SPPs is observed at the similar wavelength region. The experimental data and the calculation agree well with each other.

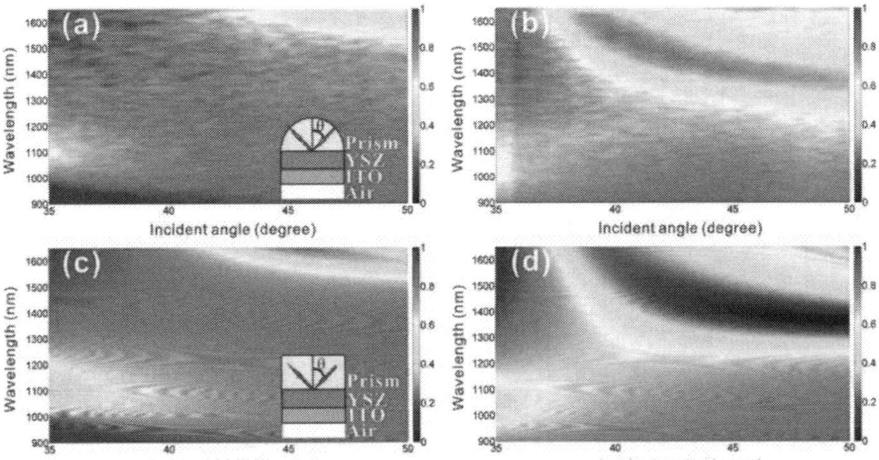

Figure 3. Reflectance spectra of ITO–10 (a) and ITO–20 (b) as a function of incident angle and wavelength. The inset shows the sample setup. Calculated reflectance spectra of ITO–10 (c) and ITO–20 (d) by using transfer matrix method. The inset shows the model for calculation.

4. Conclusion

We have experimentally demonstrated the tuning of SPPs wavelength in ITO thin film through n and μ. ITO thin films were epitaxially grown on YSZ substrates by PLD technique, and the dielectric functions in the range of 800-1700 nm were evaluated by SE measurement. The variable-angle reflectance spectroscopy shows the excitation of SPPs at the angle and the wavelength predicted from the dielectric functions of the thin films. The crossover wavelength for ITO–20 is shorter than those of ITO thin films reported so far.

Acknowledgments

The authors thank Y. Kuge at J. A. Woollam Japan for SE measurement. This study was supported by Grant-in-aid for Scientific Research (B, No 24350104 and C, No. 24560824) Grant-in-Aid for Challenging Exploratory Research (No. 24656385) from MEXT, and SCOPE (No. 132107004) from MIC, Japan. S. M. gratefully acknowledges the financial support from Yazaki Memorial Foundation for Science and Technology.

References

[1] Maier S A 2007 *Plasmonics: Fundamentals and Applications* (Springer, New York)
[2] Polman A 2008 *Science* **322**, 868
[3] Volker J. Sorger, Rupert F. Oulton, Ren-Min Ma and Xiang Zhang 2012 *MRS Bulletin* **37** 728
[4] West P R, Ishii S, Naik G V, Emani N K, Shalaev V M and Boltasseva A 2010 *Laser Photon. Rev.* **4** 795
[5] Boltasseva A and Atwater H A 2011 *Science* **331** 290
[6] Losego M D, Efremenko A Y, Rhodes C L, Cerruti M G, Franzen S, and Maria JP 2009 *J. Appl. Phys.* **106** 024903
[7] Yasuhara R, Murai S, Fujita K, and Tanaka K 2012 *Phys. Status Solidi C* **9** 2533
[8] Noginov M A, Gu L, Livenere J, Zhu G, Pradhan A K, Mundle R, Bahoura M, Barnakov Y A, and Podolskiy V A 2011 *Appl. Phys. Lett.* **99** 021101
[9] Ohta H, Orita M, Hirano M, and Hosono H 2002 *J. Appl. Phys.* **91** 3547
[10] Kim J, Naik G V, Gavrilenko A V, Dondapati K, Gavrilenko V I, Prokes S M, Glembocki O J, Shalaev V M, and Boltasseva A 2013 *Phys. Rev. X* **3** 041037
[11] Yeh P 1988 *Optical waves in Layerd Media* (Wiley-InterScience, College Station, Texas)

Journal of Physics: Conference Series **619** (2015) 012057 doi:10.1088/1742-6596/619/1/012057

Numerical study on rotational force in a bi-layered gammadion chiral metamaterial

Yohan Lee, Eui-Young Song, Seung-Yeol Lee, Joonsoo Kim, Jaebum Cho, and Byoungho Lee*

National Creative Research Center for Active Plasmonics Application Systems, Inter-University Semiconductor Research Center and School of Electrical Engineering, Seoul National University, Seoul 151-744, Korea

E-mail: byoungho@snu.ac.kr

Abstract. Recently it has been demonstrated that chiral metamaterials show giant optical activity and circular dichroism. However, it has been rarely known about rotational force generated by incident wave in chiral metamaterials. We calculate the rotational force in the bi-layered gammadion chiral metamaterial. In addition, we show that a bi-layered structure, which has greater optical activity than a single-layered structure, changes the rotational force tendency greatly. This result can be used as a possible optical manipulation in biological sciences.

1. Introduction

Chiral metamaterials have attracted much attention due to their interesting properties, such as negative refraction, giant optical activity, and circular dichroism. These properties have been used in many applications such as optical particle manipulation [1, 2], optical modulation [3, 4], display devices [5, 6], etc. Especially, the rotational force generated by incident light on the chiral metamaterials may have potential applications of optical micromanipulation which has a major impact on the biological and colloidal sciences [7-9]. A gammadion structure, one of the representative structures in chiral metamaterial, has been studied numerously due to its high potential for applications such as optically-driven motor and biological sensors [9, 10]. Moreover, it has been shown that a bi-layered metamaterial can be used for observing another kind of resonance which is different from the resonance observed in single-layered structure. Therefore, in this paper, we apply a bi-layered structure into gammadion-shaped chiral metamaterial and calculate absorption spectrum and rotational force. By observing the electric field and surface charge distribution near the gammadion structure, we classify the characteristics of resonance into electric and magnetic dipole types, and compare the aspect of rotational force on those resonance conditions.

2. Simulation conditions

Schematic diagram of proposed bi-layered chiral metamaterial is shown in figure 1(a). Planar metal structures of two identical gammadion-shaped gold resonators are separated by a SiO_2 spacer of which thickness is 10 μm. The thickness, width, and diameter of gold resonators are 200 nm, 8 μm, and 60 μm respectively. For reference structure, single-layered chiral metamaterial is also designed with the same dimension as proposed bi-layered chiral metamaterial as shown in figure 1(b). Finite difference time domain (FDTD) method is used to calculate the electromagnetic properties of these structures.

We calculate the torque generated by electromagnetic (EM) wave on single- and bi-layered structure by using Lorentz equation in harmonic incident wave. The time-averaged force per unit volume is

$$<\mathbf{f}(\mathbf{r},t)>_{avg}=\frac{1}{2}\mathrm{Re}[\rho^*(\mathbf{r})\mathbf{E}(\mathbf{r})+\rho(\mathbf{r})\mathbf{E}^*(\mathbf{r})]+\frac{1}{2}\mathrm{Re}[\mathbf{J}^*(\mathbf{r})\times\mathbf{B}(\mathbf{r})+\mathbf{J}(\mathbf{r})\times\mathbf{B}^*(\mathbf{r})] \quad (1)$$

We substitute electric and magnetic field distributions obtained by FDTD method into equation (1). Then, we also calculate surface charge density (ρ) and current density ($\vec{J}$) from the Maxwell equation reversely,

$$\rho(\mathbf{r})=\nabla\cdot\mathbf{D} \;, \quad (2)$$

$$\mathbf{J}(\mathbf{r})=\nabla\times\mathbf{H}-j\omega\mathbf{D} \;. \quad (3)$$

Finally, the torque per unit volume is calculated by cross product force with directional unit vector

$$\mathbf{N}=\iiint_V \tau dv=\iiint_V (\mathbf{r}\times\mathbf{f})dv \,. \quad (4)$$

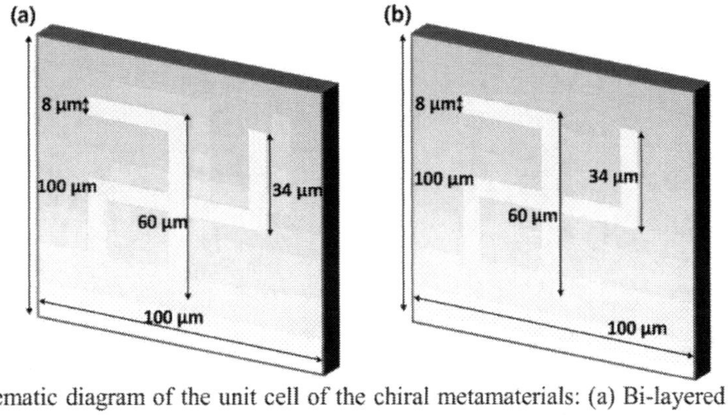

Figure 1. Schematic diagram of the unit cell of the chiral metamaterials: (a) Bi-layered gammadion-shaped gold resonators (yellow) separated by SiO_2 spacer (white blue) and (b) single-layered gammadion gold resonator laid on SiO_2 spacer.

3. Results and Discussions

We investigate absorption spectrum along frequency range from 0.4 THz to 1.3 THz. In the case of single-layered structure, the spectral response of resonance is almost independent of incident polarization. It is also shown that only a broad resonance exists on 1.05 THz for single-layered structure while there are two resonance peaks on 0.75 THz and 1.2 THz for bi-layered structure as shown in figure 2(b). The resonance peak on 0.75 THz in bi-layered structure is remarkably high and sharp compared with other resonances.

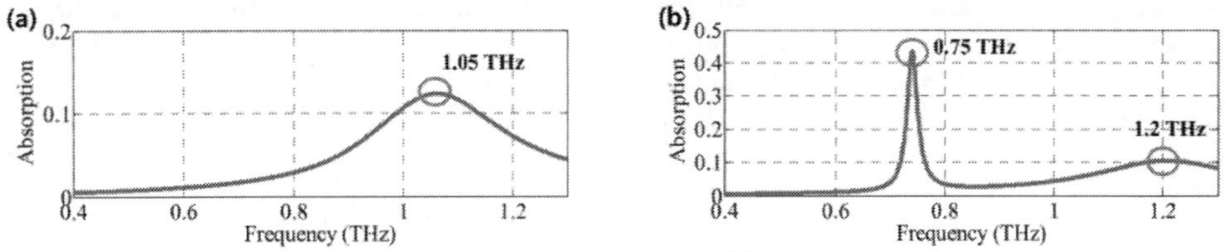

Figure 2. Absorption spectrum in (a) single-layered and (b) bi-layered gammadion chiral metamaterial.

In order to explain this tendency, distribution of E_x field, simplified surface charge, and surface current near the metal planes of bi-layered structure at two resonance conditions are shown in figure 3. As shown in figure 3(a), charge distribution of each resonator has opposite pattern at sharp resonance frequency (0.75 THz). In this condition, the electric field of the incident wave induces opposite direction of currents within each resonator and results in current loop in the central region between top and bottom resonators. Hence, a magnetic moments ***m*** is induced by this current loop. On the other hand, in figure 3(b), the directions of current on the two resonators are same at broad resonance condition (1.2 THz). In this case, charges oscillate and generate an electric moment ***p***. Since the charge distribution on the resonance condition of single layered structure cannot be separated into opposite directions like figure 3(a), only the broad resonance modelled by electric dipole moment can appear for single-layered structure, whereas sharp resonance caused by magnetic dipole resonance cannot be shown.

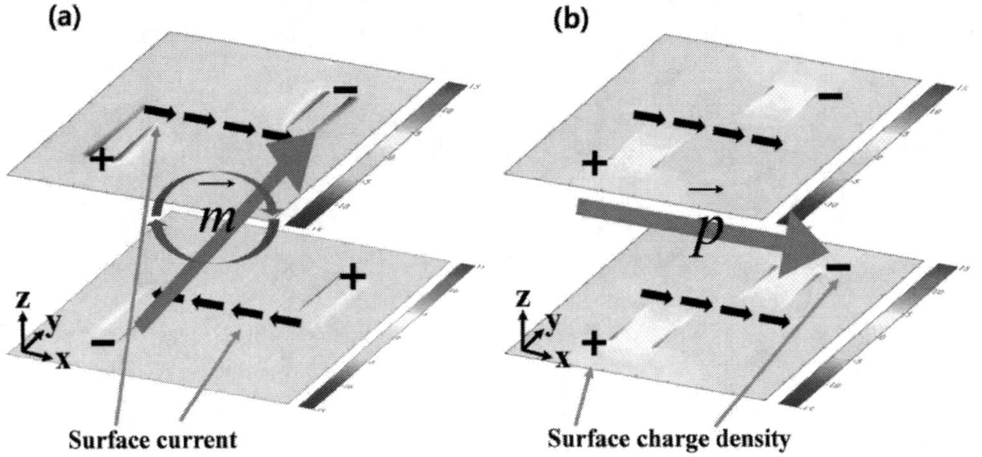

Figure 3. E_x field, surface charge and surface current at the resonance frequency of (a) 0.75 THz and (b) 1.2 THz in a unit-cell of bi-layered structure. The distance from each resonator to field plane is 500 nm. The phase is fixed at 0 degree.

According to this analysis, strong absorption peak shown in figure 2(b) comes from magnetic dipole moment and other peaks are based on electric dipole moment. Magnetic moment is formed by opposing electric dipoles. It is known that magnetic moment resonance generally has higher Q-factor as shown in figure 2(b) [11, 12].

Figure 4 shows the torque generated on the single- and bi-layered structure. Since intensity enhancement is the strongest at the resonance frequency, torque spectrum from numerical calculation shows similar tendency with absorption spectrum. Each peak value of torque by LCP or RCP in single- layered structure is almost same. On the other hand, the torque responses between LCP and RCP incidence are quite different. The black dotted line shown in figure 4(b) is the difference of torque between LCP and RCP incidences. It is shown that torque difference is remarkably increased for both resonances. Moreover, the sharpness of torque difference is also well-matched to that of resonance spectrum shown in figure 2. Since this kind of sharp torque differences are only achievable in bi-layered structure, we expect our structure can be applied to chiral metamaterial which has both highly frequency-selective and polarization sensitive characteristics.

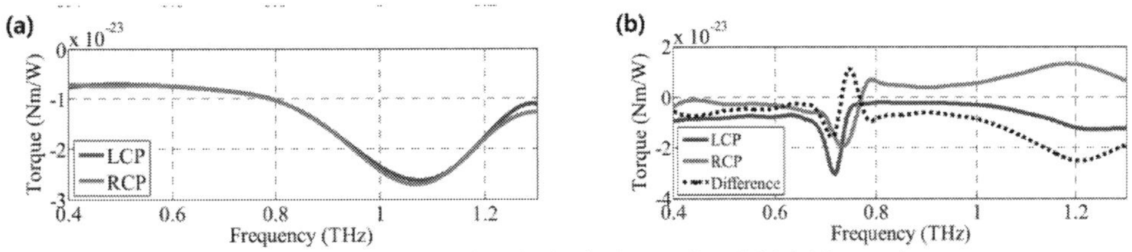

Figure 4. Torque plots along frequencies in (a) single-layered and (b) bi-layered structure.

4. Conclusion

We proposed a bi-layered gammadion chiral metamaterial which has sharp torque difference characteristics at around 0.7 THz. Our chiral metamaterial has strong rotational force on circular polarized EM wave incidence which is based on the magnetic moment resonance between bi-layered gammadion structures. In perspective of sharp peak, we can design microrotator which has the wave-length-sensitive property. Furthermore, this study on optical force can offer help in the field of optical micromanipulation , colloidal and biological sciences.

Acknowledgements
This work was supported by the National Research Foundation of Korea funded by Korean government (MSIP) through the Creative Research Initiatives Program (Active Plasmonics Application Systems).

References

[1] Baumgart K, Mazilu M and Dholakia K 2010 *Nat. Photonics* **2** 675-678
[2] Dholakia K, Reece P and Gu M 2008 *Chem. Soc. Rev.* **37** 42–55
[3] Yu N, Genevet P, Kats M A, Aieta F, Tetiennem J –P, Capasso F and Gaburro Z 2011 *Science* **334**, 333
[4] Luo J, Yu H, Song M and Zhang Z 2014 *Opt. Lett.* **39** 2229-2231
[5] Ni X, Kildishev A V and Shalaev V M 2013 *Nature commun.* **4** 2807
[6] Huang L, Chen X, Muhlenbernd H, Zhang H, Chen S, Bai B, Tan Q, Jin G, Cheah K -W, Qiu C -W, Li J, zentgraf T and Zhang S 2013 *Nature commun.* **4** 2808
[7] Fasman G D 1996 *Circular Dichroism and Conformational Analysis of Bio molecules* (New York: Plenum)
[8] Barron L D 2004 *Molecular Light Scattering and Optical Activity* (London: Cambridge Univ. Press)
[9] Hendry E, Carpy T, Johnston J, Popland M, Mikhaylovskiy R V, Lapthorn A J, Kelly S M, Barron L D, Gadegaard N and Kadodwala M 2010 *Nat. Nanotechnol.* **5** 783-787
[10] Liu M, Zentgraf T, Liu Y, Bartal G and Zhang X 2010 *Nat. Nanotechnol.* **5** 570-573
[11] Sersic I, Frimmer M, Verhagen E and Koenderink A F 2009 *Phys. Rev. Lett.* **103** 213902
[12] Evlyukhin A B, Novikov S M, Zywietz U, Eriksen R L, Reinhardt C, Bozhevolnyi S I and Chichkov B N 2012 *Nano Lett.* **12** 3749-3755

Microgrooved plasmonic bottle microresonator

M N Mohd Nasir, M Ding, G S Murugan, M N Zervas
Optoelectronics Research Centre, University of Southampton, Highfield, Southampton, SO17 1BJ, UK.

E-mail: mnmn1r11@orc.soton.ac.uk

Abstract. In this paper, we demonstrate an enhancement to SPW cavity through the incorporation of high-Q WGM bottle microresonator (BMR) with surface microgrooves. A standard BMR fabricated through the "soften-and-compress" technique with initial length of 280 μm, bottle diameter of 187 μm and stem diameter of 125 μm was utilized in the experiment for supporting WGMs. Thin gold film was deposited on top of the BMR for generating SPWs. 21 microgrooves was then inscribed on the metal surface of the BMR along the azimuthal direction with 10 μm length, 485 nm width, 6 μm depth and pitch of 1.5 μm. Due to surface curvature, the gold film only covered half of the BMR with a characteristic meniscus shape and maximum thickness of 30 nm. The meniscus provides appropriately tapered metal edges that facilitate the adiabatic transformation of BMR WGMs to SPWs and vice-versa. Lorentzian shape-line fit performed on the TM excited resonances show that plasmonic Q values in excess of 4000 could be achieved from such structure with ~ 25% coupling efficiency.

1. Introduction

The ability of surface plasmons in tailoring its interaction with light through the alteration of the metal's surface has opened up many great possibilities on the advancement of new types of photonics devices. However, in contrast to the high-Q factor of dielectric cavity resonators, surface plasmonic cavities performances has been reported to attain Q-factor values much lower than their theoretical values. The reported surface plasmonic cavity to date includes ring [1] and ring type v-groove [2], annular resonator [3], nanowire Fabry-Pérot (FP) [4] and periodic optical Bragg mirrors [5], [6]. The ability of whispering gallery (WG) structure in sustaining high-Q cavities has prompted extensive studies in the development of high-Q hybrid WGM-plasmonic microresonators. The demonstration of WG-plasmonic resonance in Au cylinder excited with tapered fibre has shown Q-factor value reaching the 300s [7]. A much more advanced WG dielectric microdisk resonator half-coated with silver has been demonstrated to attain Q-factor in excess of 1300 when excited from the dielectric side of the disk [8]. A study on WG microtoroid resonator coated with silver has also been anticipated to support plasmonic-WGMs with Q value reaching 1000s [9]. Recently we have proposed a high-Q hybrid plasmonic-BMR (PBMR) with a characteristic meniscus shape Au coating on top of the resonator [10]. TM-WGM excitation from the dielectric side of the plasmonic-BMR shows surface plasmon waves (SPWs) with composite Q-factor reaching 850 and individual Q-factor in excess of 2500 could be generated with coupling efficiency of ~ 75%.

In this paper, we demonstrate another type of hybrid plasmonic-BMR with the incorporation of microgroove gratings on the surface of the resonator. 21 microgrooves inscribed on the metal surface of the PBMR is demonstrated to excite WGMs which are very different form the original BMR. TM-excited WGMs is demonstrated to generate SPWs with Q-factor reaching 4000 and ~ 25% coupling efficiency.

2. Experimental setup

The BMR utilized in the experiment was fabricated using conventional silica fibre (SMF-28) through the "soften-and-compress" technique. As shown in figure 1 (a), the original BMR has initial bottle diameter D_b of 187 μm, length L_b of 280 μm and stem diameter D_s of 125 μm. Original WGMs excited at the centre of the BMR with tapered fibre of 2 μm waist diameter shows complex and dense spectra over a broad spectrum, as shown in figure 1 (b).

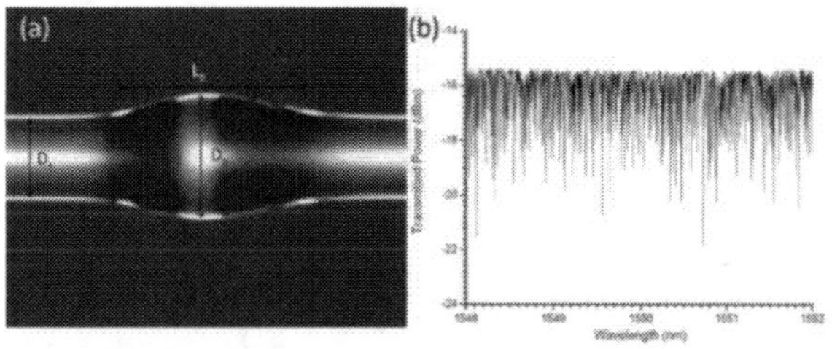

Figure 1 (a) BMR utilized before metal coating. (b) Initial BMR-WGMs excited at centre position.

The top surface of the BMR was then deposited with a thin gold layer. Due to the surface curvature of the BMR, the gold film only covered half of the resonator with a characteristic meniscus shape and maximum thickness of 30 nm. Adiabatic transformation of the BMR-WGMs into SPWs and vice-versa would be expedited by the tapered edges of the meniscus. Focused ion beam (FIB) milling with 50 nm precision were utilized in the experiment in order to inscribe 21 microgrooves on the azimuthal direction of the BMR metal surface. The microgrooves were inscribed with length of 10 μm, width of 485 nm, depth of 6 μm and pitch of 1.5 μm. Figure 2 (a) shows top side of the BMR with the scanning electron micrograph image of the microgrooves shown in figure 2 (b).

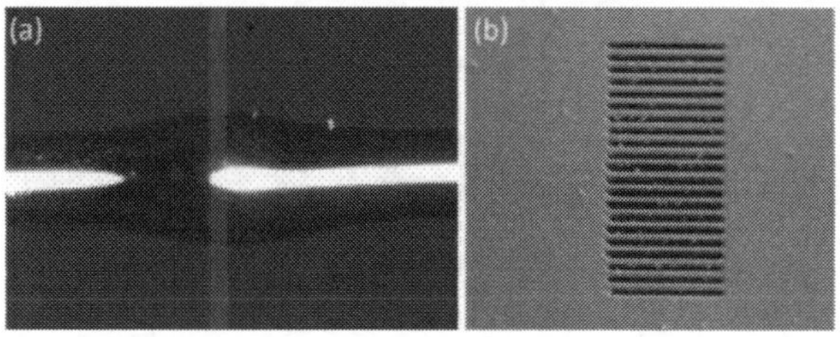

Figure 2 (a) Top side of the coated BMR with 30 nm thick gold. (b) SEM image of the inscribed microgrooves at the centre of BMR metal surface.

Figure 3 shows the schematic of the polarization-resolved measurement setup developed for characterizing the fabricated PBMR. A 2 μm tapered fibre was utilized in the experiment in order to couple light a from tuneable laser source (TLS) into the cavity of the PBMR. Since the decay length of light is high into the metal, the coupling was performed at the dielectric part of the resonator [8], [11]. To excite the PBMR with different light input polarizations, a polarization controller was placed in the system after the TLS with a high-extinction ratio polarizer (> 50 dB) placed after the coupling point. Micro positioning stages were utilized in the experiment for precise control of the PBMR coupling with the tapered fibre. The tapered fibre was cleaved at the full fibre diameter and brought in contact with the cavity by placing it on top of the resonator. A collimator was placed after the cleaved fibre in order to collimate the output light before it is aligned to pass through the polarizer. The polarizer was placed in the system in a manner where it could be freely rotated to 360°. The output light was then focused in to a fiber using a fibre launcher setup and measured using an InGaAs optical power meter for WGMs observation. The inset of figure 3 shows the scattering of light of the PBMR with an IR camera.

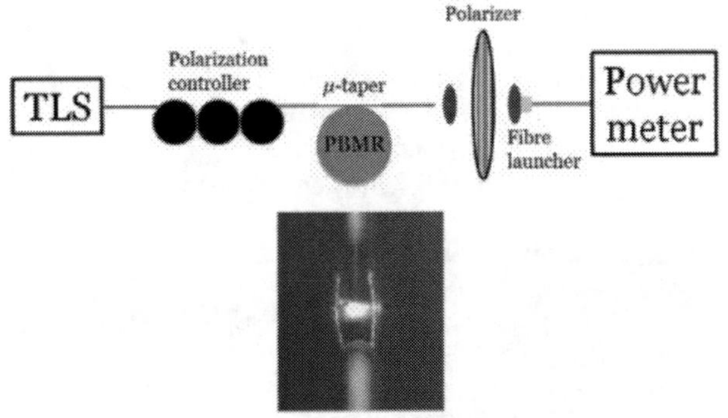

Figure 3. Schematic of the polarization-resolved measurement set-up for characterizing the fabricated PBMR with inset showing light scattering of the centrally excited PBMR.

3. Results and Discussion

The microtaper was coupled to the centre position of the dielectric part of the PBMR. The polarizer was then set to launch TE-polarized light into the cavity of the resonator. Figure 5 (a) shows the output resonance spectra of the TE-excited modes with maximum coupling efficiency of ~ 80% observed. The transmission spectra of the hybrid PBMR are dramatically different from the uncoated BMR as in figure 1 (b). Distinguishable family group-modes could also be observed from the output spectra with free spectral range (FSR) of ~ 2.9 nm; which matches the azimuthal resonances of the BMR. Fitting Lorentzian-shaped curves on the measured data revealed that the spectra actually comprise of overlapping resonances with higher Q value, as shown in Figure 5 (b). Maximum Q calculated is at resonance around 1550.2 nm with value of ~ 13000.

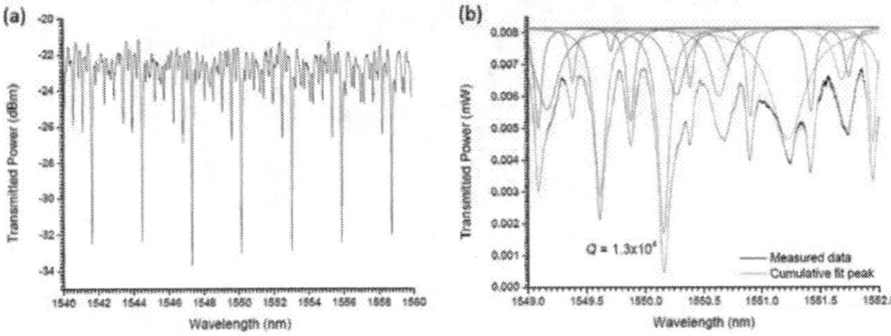

Figure 5 (a). Measured TE excited transmission spectra and (b) its Lorentzian-shaped curve fit.

Since SPWs are transverse magnetic in character, they require TM-polarized light as the source of excitation for electron density oscillations at the metal-dielectric interface. The polarization controller was then set to launch TM-polarized light into the cavity of the resonator. The output spectra observed shows maximum coupling efficiency is ~ 25%, as shown in Figure 6 (a). The low transmission depth of the resonances is mainly due to light loss into the metal as they interact. Distinguishable family group-modes of FSR ~ 2.9 nm were also observed from the spectra. Figure 6 (b) illustrates the Lorentzian-shaped curves fit of the TM-excited modes. From the fitting, the output spectra actually comprise of multiple overlapping resonances with narrower individual resonances. The strongest resonance dip which is at around 1556.1 nm has been calculated to attain Q in the range of 4700. However, the cumulative Q of the overlapping composite resonances at this wavelength is much lower, which is in the range of 700.

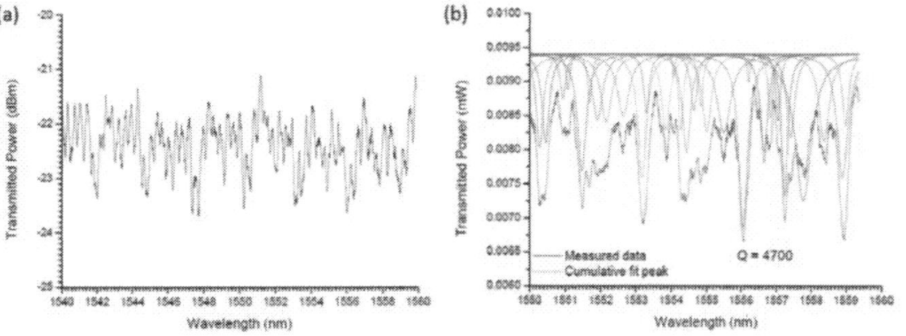

Figure 6 (a). Measured TM excited transmission spectra and (b) its Lorentzian-shaped curve fit.

4. Conclusion

We have studied experimentally the transmission properties of a microgrooved PBMR. An accurate polarization resolved measurement was developed in order to fully characterize the hybrid resonator. Different light polarization (TE and TM) coupled into the cavity of the resonator demonstrated that different modes are being excited by the cavity. Due to higher losses caused by SPWs, it was observed that TM-excited modes exhibited lower Q than TE-excited modes. Lorentzian-shaped curve fitting performed on the measured data reveals that TE/TM excitations are actually composite resonances comprising of multiple individual peaks with Qs in the range of 13000 and 4700, respectively.

References

[1] S. I. Bozhevolnyi, V. S. Volkov, E. Devaux, J.-Y. Laluet, and T. W. Ebbesen, "Channel plasmon subwavelength waveguide components including interferometers and ring resonators," Nature **440**, 508-511 (2006).

[2] A. Vardanyan, H. Haroyan, A. Babajanyan, K. Nerkararyan, and B. Friedman, "Ring-type V-groove surface plasmon microresonator: The modal structure and Q-factor," Journal of Applied Physics **111**, 053112 (2012).

[3] C. E. Hofmann, E. J. R. Vesseur, L. A. Sweatlock, H. J. Lezec, F. J. García de Abajo, A. Polman, and H. A. Atwater, "Plasmonic Modes of Annular Nanoresonators Imaged by Spectrally Resolved Cathodoluminescence," Nano Letters **7**, 3612-3617 (2007).

[4] H. Ditlbacher, A. Hohenau, D. Wagner, U. Kreibig, M. Rogers, F. Hofer, F. R. Aussenegg, and J. R. Krenn, "Silver Nanowires as Surface Plasmon Resonators," Physical Review Letters **95**, 257403 (2005).

[5] J.-C. Webber, Y. Lacroute, A. Dereux, E. Devaux, T. Ebbesen, C. Girard, M. U. González, and A.-L. Baudrion, "Near-field characterization of Bragg mirorrs engraved in surface plasmon waveguides," Physical Review B **70**, 235406 (2004).

[6] J.-C. Webber, A. Bouhelier, G. C. des Francs, L. Markey, and A. Dereux, "Submicrometer In-Plane Integrated Surface Plasmon Cavities," Nano Letters **7**, 1352-1359 (2007).

[7] X. Zhang, Z. Ma, H. Yu, X. Guo, Y. Ma, and L. Tong, "Plasmonic resonance of whispering gallery in an Au cylinder," Optics Express **19**, 3902-3907 (2011).

[8] B. Min, E. Ostby, V. Sorger, E. Ulin-Avila, L. Yang, X. Zhang, and K. Vahala, "High-Q surface-plasmon-polariton whispering-gallery microcavity," Nature **457**, 455-459 (2009).

[9] Y.-F. Xiao, C.-L. Zou, B.-B Li, Y. Li, C.-H. Dong, Z.-F. Han, and Q. Gong, "High-Q Exterior Whispering-Gallery Modes in a Metal-Coated Microresonator," Physical Review Letters **105**, 153902 (2010).

[10] M. N. Mohd Nasir, M. Ding, G. S. Murugan, and M. N. Zervas, "High-Q Plasmonic Bottle Microresonator," Proceedings of SPIE **8960**, 89600M (2014).

[11] W. L. Barnes, A. Dereux, and T. W. Ebbesen, "Surface plasmon subwavelength optics," Nature **424**, 824-830 (2003).

Induced correlation between molecules in different plasmonic fields

Yoshiki Osaka, Nobuhiko Yokoshi, Hajime Ishihara

Department of Physics and Electronics, Osaka Prefecture University, 1-1, Gakuen-cho, Naka-ku, Sakai, Osaka, 599-8531, Japan

E-mail: osaka@pe.osakafu-u.ac.jp

Abstract. We theoretically examine the optical responses of two molecules near gold nanoblocks which generate strongly localized near-fields due to localized surface plasmon resonance. The molecules located in the near-fields efficiently interact with light. As a result, we find that correlations between the molecules exist, even though the molecules are distantly positioned. Because these correlations are of importance to collective behaviors of the molecules such as superradiance and superfluorescence, this result implies the possibility of these collective behaviors being generated through the plasmonic fields. Besides, correlations between molecules must be taken into account with regards to device applications of metallic nanostructure arrays.

1. Introduction

Superfluorescence (SF) is the cooperative light emission from an ensemble of initially uncorrelated molecules (or atoms) [1, 2]. The molecule ensemble spontaneously creates a macroscopic dipole through the common emission field, and then gives rise to an SF pulse. The maximum intensity of the pulse is proportional to N^2 and the time duration is proportional to N^{-1}, where N represents the number of excited coherent molecules. Therefore, in general, SF occurs in high-density systems where the correlation between the molecules is large. On the other hand, metallic nanostructures, e.g., nanoparticles and nanoblocks, generate strongly localized near-fields due to localized surface plasmon resonance and the efficient interactions of molecules near the metal with light [3]. In other words, the metallic nanostructures play the role of an optical antenna and enlarge the effective scattering cross-section of the molecules for the incident light [4]. In particular, dimeric-arranged metal nanoblocks with a few nm-sized gap width provide enhanced near-fields near the gap with $\sim 10^4$ times greater intensities than that of the incident light [5]. An array of such the metallic nanostructure, which is actually produced experimentally, enables us to efficiently excite a number of molecules [6]. Then, we can expect that the molecules will exhibit collective behavior.

To date, we have theoretically examined optical responses of a single molecule located near the gap of such a single dimer-type metallic nanostructure and have found that a high-efficiency interaction between the incident light and the molecule is realized [7, 8]. In this work, we analyze the optical responses of two molecules near four gold nanoblocks arrayed in a dimer-type, and investigate the correlation between the efficiently excited molecules near the metallic antenna. As a result, we find that correlations between the molecules exist, even though the molecules are distantly positioned.

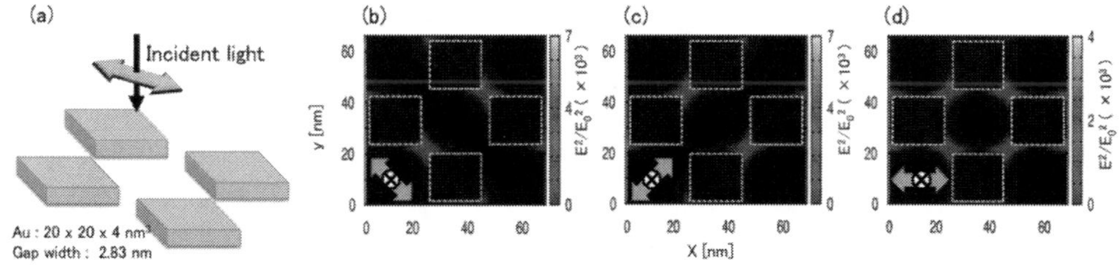

Figure 1. (a) Schematic figure of the assumed positions of the metallic nanostructures. We arrange four gold nanoblocks with a 2.83 nm gap. Each nanoblock is 20 nm × 20 nm × 4 nm. (b, c, d) Electric field intensity map of the metallic nanostructures. The orange arrow indicates the direction of polarization of the incident light.

2. Theory

We theoretically analyze the optical responses of four gold nanoblocks, arranged as shown in Fig. 1(a). For a clear demonstration, we assume the metal blocks to be thin so that the mode volume of localized light fields is not much larger than the molecular volume. The size of each block is $20 \times 20 \times 4$ nm^3, and the gap width is 2.83 nm. In order to obtain the optical response fields, we solve the discretized integral form of Maxwell's equation by applying the discrete dipole approximation (DDA) [9]. In this calculation, we divide the space containing the metal blocks into small cubic cells, where microscopic quantities such as the electric field and polarization in each cell are averaged by the volume of the cell. Additionally, the self-interaction can be evaluated analytically. The integral equation is given as

$$\mathbf{E}_i = \mathbf{E}_i^0 + \sum_{j=1}^{n} \mathbf{G}_{i,j}^f \mathbf{P}_j V_j, \tag{1}$$

$$\mathbf{P}_j = \chi_j \mathbf{E}_j, \tag{2}$$

where n represents the total number of cells, i and j $(= 1, ..., n)$ are the cell number. χ_j and V_j are the optical susceptibility and the volume of the j-th cell, respectively. $\mathbf{E}_i$ is the total response field and $\mathbf{E}_i^0$ is the incident field. The free space Green's function, $\mathbf{G}_{i,j}^f$, has both transverse and longitudinal electromagnetic components. The metal blocks are assumed to have a Drude-type dielectric function with the parameters of gold [10, 11]. As for the susceptibility of the molecule, we assume that it is of Lorentzian-type, i.e., $\chi_j^m = (|d|^2/V_j)/(\hbar\Omega^m - \hbar\omega - i\gamma)$, where d is the dipole moment, $\hbar\Omega^m$ is the eigenenergy, and γ is the nonradiative damping constant.

Figures 1(b, c, d) show the electric field intensity map of the metallic nanostructures when the light polarized along the direction of the orange arrows is irradiated. One can see the field enhancement at the gap. Therefore, when the molecules are near to the gap, the plasmonic fields excite the molecules efficiently.

In this work, we assume that two molecules, A and B, are positioned near the metal (Fig. 2(a)), and investigate the correlation between molecules A and B. We calculate the absorption cross section of molecule A according to the definition

$$\sigma_{abs} = 4\pi(\omega/c) \int_V d\mathbf{r}^3 |\mathbf{E}(\mathbf{r})|^2 \mathrm{Im}\{\chi(\mathbf{r})\}/|\mathbf{E}_0(\mathbf{r})|^2, \tag{3}$$

and estimate the correlation between molecules A and B from the spectra of the absorption cross section.

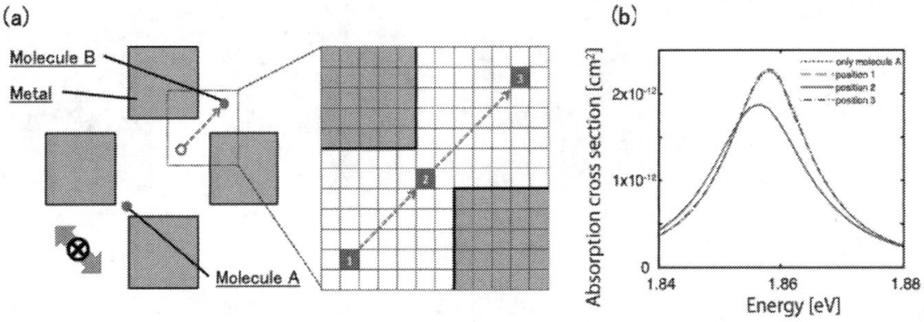

Figure 2. (a) Top view of the metallic nanostructures. The solid orange arrow indicates the direction of polarization of the incident light. The dashed blue arrow indicates the variation in the position of molecule B. Grids show the cells for the DDA calculation, which are 1 nm in length on each side. (b) Absorption spectra of molecule A when molecule B is located at positions 1, 2, and 3 as shown in Fig. 2(a), and when it is absent.

3. Results

Figure 2(a) gives the top view of the metallic nanostructures and demonstrates the variation of the position of molecule B. The position of molecule A is fixed near one of the gaps. In Fig. 2(b), we show the absorption spectra of molecule A when molecule B is placed at positions 1, 2, and 3 in Fig. 2(a). The black dotted line represents the absorption spectrum of molecule A when molecule B is absent. Only when molecule B is sited at position 2, corresponding to the neighborhood of the other gap, do we observe the red shift of absorption spectrum of molecule A. We can consider that this peak shift originates from the interaction energy with molecule B efficiently excited by enhanced field near the gap. Therefore, we can see that molecule A correlates with molecule B placed at position 2.

In Fig. 3, we discuss the relationship between the shift of the absorption spectrum of molecule A and the position of molecule B in more detail. Figures 3(b, c, d) show the absorption peak position of molecule A against the position of molecule B for three cases of polarization of incident light. The horizontal axis for the solid red (dashed blue) line represents the position of molecule B from the point P along the dashed red (blue) arrow in Fig. 3(a). As shown on the red line in Fig. 3(b), we find large shifts when molecule B is located near the gap. On the other hand, we find small shifts on the blue line in Fig. 3(b). Therefore, the condition dictating the appearance of the large shift corresponds to cases where both molecules are located in enhanced fields (see Fig. 1(b)). In Fig. 3(c), we cannot see the shifts, because molecule A is not located in the enhanced fields (see Fig. 1(c)). We find a large shift on both the red and blue lines in Fig. 3(d), because enhanced fields are generated in all of the gaps (Fig. 1(d)). In conclusion, these results state that the correlation between the molecules is strengthened when both molecule A and molecule B are located in enhanced fields near the gaps.

4. Conclusion

We have investigated the correlation between two molecules near four gold nanoblocks and found that these molecules are strongly correlated with each other when they are excited by the enhanced near-fields. Because the correlations between the molecules are of importance to their collective behavior, the results of this study suggest the possibility of generating cooperative phenomena through plasmonic fields. In addition, we believe that device applications of metallic nanostructure arrays must take correlation between molecules into account.

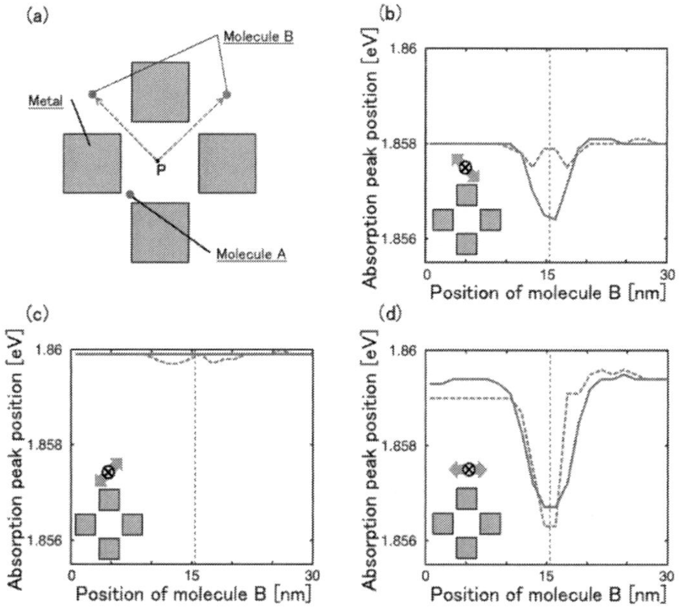

Figure 3. (a) Dashed red and blue arrows represent the variations in the position of molecule B. Figures 3(b, c, d) show the absorption peak position of molecule A against the position of molecule B for the incident light with different polarizations, represented by the orange arrows in the insets. The horizontal axis for the solid red (dashed blue) line represents the position of molecule B from the point P along the dashed red (blue) arrow in Fig. 3(a). The vertical dotted gray line denotes the gap position.

Acknowledgements
This work was partially supported by Grant-in-Aid for JSPS Fellows 25·9308 from MEXT, Japan.

References

[1] Dicke R H 1954 *Phys. Rev.* **93** 99
[2] Bonifacio R and Lugiato L A 1975 *Phys. Rev. A* **11** 1507
[3] Anger P, Bharadwaj P and Novotny L 2006 *Phys. Rev. Lett.* **96** 113002
[4] Bharadwaj P, Deutsch B and Novotny L 2009 *Adv. Opt. Photon.* **1** 438
[5] Tanaka Y, Ishiguro H, Fujiwara H, Yokota Y, Ueno K, Misawa H and Sasaki K 2011 *Opt. Express* **19** 7726
[6] Ueno K, Juodkazis S, Shibuya T, Yokota Y, Mizeikis V, Sasaki K and Misawa H 2008 *Journal of the American Chemical Society* **130** 6928
[7] Ishihara H, Nobuhiro A, Nakatani M and Mizumoto Y 2011 *Journal of Photochemistry and Photobiology A: Chemistry* **221** 148
[8] Osaka Y, Yokoshi N, Nakatani M and Ishihara H 2014 *Phys. Rev. Lett.* **112** 133601
[9] Purcell E M and Pennypacker C R 1973 *Astrophys. J.* **1186** 705
[10] Johnson P B and Christy R W 1972 *Phys. Rev. B* **6** 4370
[11] Antoine R, Brevet P F, Girault H H, Bethell D and Schiffrin D J 1997 *Chem. Commun.* 1901

Institute of Physics
Dirac House, Temple Back
Bristol BS1 6BE UK

CURRAN ASSOCIATES INC
proceedings.com